BARRON'S

ACCUPLACER®
MATH

Tyler S. Holzer and Todd C. Orelli

ACCUPLACER® is a registered trademark of the College Board, which was
not involved in the production of, and does not endorse, this product.

ABOUT THE AUTHORS

Tyler S. Holzer is the Adult Basic Education and High School Equivalency Program Coordinator at the Fifth Avenue Committee, a nonprofit agency serving central and south Brooklyn, NY. He is a founding member of the New York City Community of Adult Mathematics Instructors, an active member of the New York State Education Department's Teacher Leadership initiative, and the creator of *www.mathmemos.org*. He lives in Brooklyn with his wife, Rachel.

Todd C. Orelli is a mathematics teacher with the New York City Department of Education's Pathways to Graduation program. He supports students working toward a high school equivalency diploma and college and career readiness as well as fellow teachers as a staff developer, mentor teacher, and curriculum developer. He is a member of both the New York City Community of Adult Mathematics Instructors and the New York State Education Department's Teacher Leadership initiative, a contributor to *CollectEdNY.org*, and the founder of *p2gmath.com*. He lives on Long Island, New York with his wife Thayer and daughters Siena and Sadie.

ACKNOWLEDGMENTS

For Terri Martin, my high school math teacher. Very special thanks to Christina Curran, Brian Mendes, Carolyn Wright, Mickey Vizcarrondo, and Carla Jeanpierre at the Fifth Avenue Committee, whose support made it possible to write this book. Thank you to each and every one of the students I have had the privilege of working with and learning from over the years. Thanks also to the colleagues who help me become a better teacher every day: Mark Trushkowsky, Eric Appleton, Solange Farina, Parvoneh Shirgir, Jane Tarica, Denise Deagan, Kevin Winkler, Ramon Garcia, Nell Eckersley, Cynthia Bell, Usha Kotelawala, the New York City Community of Adult Math Instructors, and, of course, my good friend and coauthor, Todd. Huge thanks to our editor at Barron's, Samantha Karasik, for the attention, care, and love she put into turning our manuscript into the book you're reading today. My deepest gratitude to Gary and Roxanne Holzer for always pushing me to work hard and keep learning, and to Ryan Holzer for the years of laughs, music, and inspiration. Biggest thanks of all to my partner and best friend, Rachel. I'm so lucky.

—Tyler S. Holzer

Many thanks to Marie Polinsky, Roony Vizcaino, and Allan Baldassano at Pathways to Graduation for your support and trust and for allowing me the latitude to freely explore math in the classroom. Thank you to my hardworking students for your dedication and for believing in yourselves and me. You are an inspiration. Thank you to CUNY's Mark Trushkowsky and Eric Appleton, and all of our fellow New York State Ed. teacher leaders, for your insight, depth of knowledge, and commitment to sharing and promoting best practices in mathematics instruction. A special thank you to Tyler, my good friend and coauthor, and Samantha Karasik, our editor, for your diligence and attention, and, most of all, for being excellent teammates. Thank you to my parents, Kevin and Nancy Orelli, for giving me the tools to succeed and for teaching me that difficult problems are worth solving. And, last, but not least, thank you to my incredible wife, Thayer, for all that you do and all that you are. You make everything possible.

—Todd C. Orelli

All inquiries should be addressed to:
Barron's Educational Series, Inc.
250 Wireless Boulevard
Hauppauge, New York 11788
www.barronseduc.com

ISBN: 978-1-4380-0903-2

Library of Congress Control Number: 2017930892

PRINTED IN THE UNITED STATES OF AMERICA
9 8 7 6 5 4 3

10%
POST-CONSUMER
WASTE
Paper contains a minimum of 10% post-consumer waste (PCW). Paper used in this book was derived from certified, sustainable forestlands.

Contents

UNIT THREE: COLLEGE-LEVEL MATHEMATICS

PRACTICE TESTS

Introduction

Welcome! If you are reading this book, then you are on the path to furthering your education. Congratulations on taking these initial steps. Your journey will be a noble and exciting one, but it will also often be challenging and demanding.

Most colleges and universities require that their students demonstrate some level of proficiency in mathematics to graduate. For over 1,500 colleges today, the ACCUPLACER plays a key role in placing students in the correct math class. If you attend a school that uses the ACCUPLACER as a placement exam, you will have to score above a certain threshold to prove that you are ready for college-level math. If your score is not above this threshold, you may have to successfully complete a one- or two-semester developmental math course before you can enroll in a credit-bearing math course. This is the path for thousands of college-bound students every year, but it doesn't have to be your path. Studying hard and getting a high score on the ACCUPLACER math tests means that you can avoid spending your time, money, and effort on remedial math classes. We wrote this book to help you do just that.

THE THREE ACCUPLACER MATH TESTS

The ACCUPLACER math tests are actually three separate, untimed, computer-based, adaptive tests:

- Arithmetic
- Elementary Algebra
- College-Level Mathematics

The policies for which of these three tests you will need to take differ from school to school. Many, if not most, schools will require you to take all three tests back to back. Some institutions do not give the Arithmetic test, opting instead to start with Elementary Algebra. To make sure that you're studying the right material, contact your school to determine exactly which of the three exams you will be taking. Schools often make this information available on their websites, so you may be able to find the information there.

Here is a general overview of the content that each test will cover:

ARITHMETIC (17 MULTIPLE-CHOICE QUESTIONS)

- Operations with whole numbers and fractions
- Operations with decimals and percentages
- Word problems and problem solving questions that could include rate, percentages, measurement, simple geometry, and representing parts of a whole as a fraction

ELEMENTARY ALGEBRA (12 MULTIPLE-CHOICE QUESTIONS)

- Operations with integers (positive and negative numbers) and rational numbers (whole numbers, fractions, and decimals), absolute value, and ordering rational integers
- Evaluation of formulas and algebraic expressions that could include square roots and other radicals as well as exponents, fractions, and negative numbers. This section also includes performing operations on monomials and polynomials as well as factoring polynomials.
- Translation of written phrases into algebraic expressions and solving algebraic equations and inequalities. These problems include word problems, linear equations, and quadratic equations (that can be solved by factoring).

COLLEGE-LEVEL MATHEMATICS (20 MULTIPLE-CHOICE QUESTIONS)

- Simplifying rational expressions, multiplying and factoring polynomials, and working with roots and exponents
- Solving linear and quadratic equations and inequalities as well as systems of equations
- Coordinate geometry including points and graphs of functions on the coordinate plane that could include linear and quadratic functions and equations of circles
- Linear and quadratic functions as well as polynomial, exponential, and logarithmic functions
- Trigonometry and trigonometric functions
- Applications that could include complex numbers, series and sequences, determinants, permutations, combinations, factorials, and word problems

Whether you are required to take one, two, or all three of the tests, you will take the exam in one sitting. Taking all three tests takes about 1.5–2 hours on average, so you should be prepared to be there for at least that long. Note that this test is untimed, so you will have all the time you will need on each question. There is no need to rush. Take your time.

COMPUTER-ADAPTIVE TESTING

You will take the ACCUPLACER on a computer; that is, all of the questions will appear on a computer screen, and you will choose your answers by clicking on the correct answer. However, this does not mean that you will be doing your math on a computer. You will be allowed to use scrap paper, which you should use to perform computations, draw diagrams, and check and recheck your work. Although you can work out the problem on scrap paper, note that **you cannot go back to a question once you submit your answer**. Take all the time you need to be absolutely sure that the answer you choose is your final answer, since you will not be able to go back and revisit a question.

The ACCUPLACER is an adaptive test. This means that as you answer questions the computer will be using an algorithm to determine which question to give you next based on your previous answer. The first question given will be of medium difficulty. If you answer this question correctly, the next question you will be given will be as difficult or somewhat more difficult than the last question. However, if you answer the first question incorrectly, then the next question given to you will be less difficult. Because of this adaptive nature, no two tests are likely to be exactly alike. Each person who takes the test will have a slightly different experience. Similarly, if you take the test more than once, each successive test will be different than the one you took before. Regardless of difficulty, each test should have a balance of the math content covered. It's worth noting here that *difficulty* is a relative term. It's possible that you will get an ACCUPLACER question in the "difficult" category, but you will have no trouble with it all, just as it's possible for an "easy" question to seem impenetrable because you haven't studied the material. We encourage you to study as much of the material on the ACCUPLACER as possible and to aim for a deep understanding of the underlying math.

SCORING

Since the ACCUPLACER is a computer-based, multiple-choice test, your results will be available immediately after completing the exam. How and when your school reports those scores to you may differ from school to school.

Each institution that uses ACCUPLACER as a placement exam establishes its own metric for determining success or placement based on ACCUPLACER scores. Contact your school to determine how it measures success on ACCUPLACER. Some schools offer this information online. Official ACCUPLACER scores range from 20 to 120 for each of the three math tests. However, a school could opt to report scores on a percentage basis, or in any other way it deems appropriate.

CALCULATOR USE

ACCUPLACER recommends that institutions do not allow the use of personal calculators during the exam. The large majority of test centers follow this recommendation. Unless your school indicates otherwise, you should not bring your own calculator to the exam since you will not be allowed to use it.

On certain questions, an onscreen pop-up calculator will be available. However, this will only occur on questions where using the calculator will not allow you to correctly answer the question in and of itself. For example, on a College-Level Mathematics question focused on trigonometry, there may be a calculator that will allow you to add, subtract, multiply, and divide, but the calculator alone will not be enough for you to answer the question correctly without knowing key concepts of trigonometry. On the other hand, a calculator will not be available on an Arithmetic question that tests your ability to multiply decimal numbers, since providing you with the calculator would allow you to correctly answer the question without actually demonstrating that you have mastered this skill. In other words, you should be prepared to work without a calculator on almost every question.

HOW TO USE THIS BOOK

This book is divided into three units that address the content and skills needed to excel on all three of the ACCUPLACER Math exams: Arithmetic, Elementary Algebra, and College-Level Mathematics. Before you begin studying, contact your school and determine exactly which of the three exams you will be taking. If you are not required to take all three math tests, focus on the units of this book that correspond with the tests that you will be taking. However, remember that the Elementary Algebra test questions are written with the assumption that you have mastered everything in the Arithmetic section. If you will be taking the Elementary Algebra test but not the Arithmetic test, you still need to have deep knowledge of the material in the Arithmetic section. Therefore, don't overlook this part of the book if there are areas that you need to work on.

To determine the areas that you need to work on, we recommend that you first try the ACCUPLACER practice test on the College Board website. After you have attempted all the questions, look carefully at the ones you answered correctly and the ones you didn't. This will help you identify the content areas that you need to focus on the most.

As you then work through the chapters in the book, you will notice that each chapter contains some key features, in addition to examples and explanations.

- **ACCUPLACER TIPS:** These tips will specifically address crucial test-taking strategies that will help you improve your score on the ACCUPLACER.
- **SKILLS CHECK:** The Skills Check problems test your ability to perform key procedures needed to succeed on the test. That is, the Skills Check problems are quick drills that are designed to help you practice skills rather than concepts. They are not multiple-choice.
- **ACCUPLACER CHALLENGE:** These questions are designed to look like those that you will see on the actual tests. They are often more challenging and involved than Skills Check problems, as they ask you to synthesize your knowledge of math concepts while also demonstrating your proficiency with key skills. These problems are multiple-choice.
- **ANSWERS EXPLAINED:** This book contains full, stepped-out solutions for every single Skills Check and ACCUPLACER Challenge problem as well as all the questions on the practice tests.
- **SUMMARY OF FORMULAS:** Because geometry is not a major component of the ACCUPLACER, this book does not contain an entire chapter devoted to geometry. Instead, there are plenty of practice questions that involve geometry spread throughout the book. This summary includes key geometry formulas that will help you solve these problems. You should commit as many of these formulas to memory as possible! This summary also includes special factoring cases and other important formulas to remember.

BONUS ONLINE CONTENT

- **CHAPTER REVIEW:** Each Chapter Review consists of several questions that cover the entire range of content in a chapter. Like the ACCUPLACER Challenge problems, they are multiple-choice and require you to have an understanding of both math skills and concepts. Detailed answer explanations are provided for all questions.
- **UNIT REVIEW:** The Unit Reviews offer questions that cover the entire range of content in a unit. Like the ACCUPLACER Challenge problems and Chapter Review problems, they are multiple-choice and require you to have an understanding of both math skills and concepts. Detailed answer explanations are provided for all questions.

To access these features, go to: *barronsbooks.com/TP/accuplacer/math29qw/*

Also included in this book are three full-length practice tests, each of which contains sections for Arithmetic, Elementary Algebra, and College-Level Mathematics. Each section of each practice test has the same number of questions as each of the actual ACCUPLACER exams, and full solutions are provided for every question. We recommend that you do not attempt the practice tests until you are getting close to your ACCUPLACER test date. Take your time and work through the book first, taking care to spend extra time on topics that are challenging or new to you.

After you take the first full practice test, thoroughly evaluate your results. Which questions did you answer correctly? Which ones did you get wrong? Is there a pattern to the areas that you struggled with the most? Which specific examples from the book will help you get those kinds of problems right next time? After you have taken some time to address these questions, try the next practice test.

TEST-TAKING TIPS

Having a thorough understanding of the mathematics that you'll face on the ACCUPLACER is the best way to guarantee a good score. Being a savvy test taker, however, is almost as important. You'll find specific test-taking tips throughout this book, but we offer a few general strategies here:

- Take your time, and remember to double-check your work. Do not submit your answer until you are fully ready. You will not be able to go back.
- Use your scrap paper strategically and often. Write key things down, and perform calculations on paper. There are no extra points for doing math in your head.
- Questions sometimes ask for the one answer that is NOT correct or offer answers that are common misinterpretations. Before you submit your answer, reread the question and make sure that you are answering the question the test makers are asking.
- Do your best on every question, but don't lose confidence if you see a question you don't know how to answer. Remember that an incorrect answer will be followed by a somewhat easier question, so stay confident throughout.
- If you must guess, do so with purpose. Even if you can eliminate just one wrong answer, you greatly increase your chances of guessing correctly. Do not take "blind" guesses whenever possible.
- There are sometimes multiple pathways to a solution. If you are not sure you did a problem correctly, try another approach.
- Remember that multiple-choice tests often make it possible for you to work backwards from the answer choices. If you do not know how to do something, look at the answers and consider how they can fit into the problem.
- On algebra problems that ask for you to find equivalent expressions or equations, remember that you can sometimes substitute values for variables and evaluate in order to try and find a match.
- Solve an easier problem. Often the problems that are presented are quite complex, but you can try to create and solve a simpler problem to make sure that your mathematical thinking is correct. If your solution to the simpler problem works, then it is likely that the same plan will work on a more difficult problem.

HOW TO MAKE THE MOST OF YOUR TEST DAY

- Bring a photo ID. Some testing centers may require more than one form of identification.
- Get plenty of sleep and eat well on the days approaching the exam.
- Don't cram! Allow yourself some downtime before you take the test. Studying immediately before the exam may only serve to make it harder to concentrate.

- Make sure you have at least two writing utensils and scrap paper if your testing site does not provide them.
- Turn off your cell phone or leave it at home altogether.
- Expect to be at the test center for at least a few hours. Having important things to do immediately after an exam can sometimes cause you to rush at the end of an exam or lose focus. If possible, try to have a clear schedule after the exam so you are not focused on other commitments.
- Bring a light snack and something to drink.
- Some test centers might feel a little too cold for you to be comfortable. Bring a sweater just in case.

ADDITIONAL RESOURCES

The College Board, the maker of ACCUPLACER, offers online resources to help you study for the exam. At their website, *https://accuplacer.collegeboard.org/*, you will find practice tests as well as a link to a web-based ACCUPLACER study app. The study app includes even more practice tests with detailed answer feedback. Try taking them after you have worked through the material in this book.

UNIT ONE
Arithmetic

The Fundamentals of Arithmetic

1

On the ACCUPLACER, you will see all different kinds of math problems. Among other things, you'll be asked to perform difficult calculations on fractions and decimals, solve equations for an unknown variable, perform different operations on polynomials, and even demonstrate your knowledge of a little bit of trigonometry. Don't panic just yet! We'll get to all of these topics over the next several chapters. We only mention them here to illustrate an important point: All of these topics require strong computation skills, so to be successful on the ACCUPLACER, you will need to be able to add, subtract, multiply, and divide quickly and with precision.

Some ACCUPLACER problems will require you to call upon your knowledge of mathematical vocabulary and important math symbols, and throughout this book, we will also use a lot of math vocabulary as we introduce new concepts and ideas. At the end of this first chapter, we will give you an overview of key math words and symbols that you will see in future chapters and on the ACCUPLACER exam.

In this chapter, we will work to master

- adding and subtracting whole numbers,
- multiplying and dividing whole numbers,
- using estimation to solve multiple-choice problems,
- finding factors and multiples of numbers, and
- understanding the inequality symbols and the number line.

LESSON 1.1—ADDING WHOLE NUMBERS

We all add whole numbers every day, and depending on where and how you learned to add, you might have a different way of doing it. Here, we will review the standard algorithm for addition, which is what you most likely learned in school. When adding, we always use the + sign, and we refer to the answer we get when we add two or more numbers as the **sum**.

Example 1

Find 425 + 51.

Solution

First, set up the problem vertically. Make sure that the digits on the right line up, as shown below.

$$
\begin{array}{r}
425 \\
+\ 51 \\
\end{array}
$$

Then add, starting with the rightmost digits and working your way toward the left. Since we don't have anything to add to the 4, we just bring it down.

$$
\begin{array}{r}
425 \\
+\ 51 \\
\hline
476 \\
\end{array}
$$

Answer: The sum is 476.

 Example 2

Add 673 + 459.

Solution

Just like last time, line up the two numbers vertically.

$$
\begin{array}{r}
673 \\
+459 \\
\hline
\end{array}
$$

Notice that when we add the 3 and the 9 in the first column, we have a sum greater than 10, so we will need to carry the 1 into the next column to the left and then add it. We will need to carry the 1 again in the next column and then add it.

$$
\begin{array}{r}
11 \\
673 \\
+459 \\
\hline
1132 \\
\end{array}
$$

Answer: The sum is 1,132.

 Example 3

Add 175 + 84 + 97.

Solution

Even though there are three numbers, the process remains the same. Line up the rightmost digits, and then add in each column, carrying digits when you need to.

$$
\begin{array}{r}
2\,1 \\
175 \\
84 \\
+\,97 \\
\hline
356
\end{array}
$$

Notice that this time, when we added the digits in the second column, we got a sum of 25 and had to carry a 2.

Answer: The sum of these three numbers is 356.

LESSON 1.1—SKILLS CHECK

Directions: Find the sum for each of the addition problems below.

1. $45 + 52$
2. $67 + 25$
3. $811 + 19$
4. $909 + 36$

5. $365 + 365$
6. $784 + 216$
7. $2,576 + 1,443$
8. $7,224 + 8,155$

9. $9,350 + 5,651$
10. $4,369 + 3,972$
11. $1,024 + 512 + 256$
12. $79 + 2,506 + 683$

(Answers are on page 28.)

LESSON 1.2—SUBTRACTING WHOLE NUMBERS

There are also lots of ways to subtract whole numbers, but in this book, we are going to stick to the standard algorithm, as we did with addition. Using this method, subtracting whole numbers is mostly similar to adding. We still need to line up the numbers in the same way that we did before and then subtract by column, moving from right to left. When subtracting, we always use the – sign. The answer we get when we subtract two numbers is called the **difference**.

➡ Example 1

Subtract $895 - 361$.

Solution

First, line up the numbers vertically.

$$
\begin{array}{r}
895 \\
-361 \\
\hline
\end{array}
$$

Next, we subtract the numbers in each column, starting with the rightmost column and working our way toward the left.

$$
\begin{array}{r}
895 \\
-361 \\
\hline
534
\end{array}
$$

Answer: The difference of 895 and 361 is 534.

Subtraction gets significantly more challenging when borrowing is involved. For example, consider the problem 462 − 185. When we line up the two numbers vertically, we notice that we can't take 5 away from 2! This means that we'll need to borrow from the next column. Let's look at the steps for solving this one.

➡ Example 2

Find the difference of 462 and 185.

Solution

First, line up the numbers vertically.

$$
\begin{array}{r}
462 \\
-185
\end{array}
$$

Next, we need to subtract by column, but since we can't take 5 away from 2, we'll need to borrow from the 6 in the next column over. To do this, we cross out the 6 and decrease it by one, which leaves us with 5. Then we take the 1 that we borrowed and put it right in front of the 2 so that we now have 12.

$$
\begin{array}{r}
\overset{5\ 1}{4\cancel{6}2} \\
-185 \\
\hline
7
\end{array}
$$

Now move to the next column to the right. Notice that we can't subtract 8 from 5, so we will need to borrow again, this time from the 4. Finally, we subtract the leftmost column.

$$
\begin{array}{r}
\overset{3\ 15\ 1}{\cancel{4}\cancel{6}2} \\
-185 \\
\hline
277
\end{array}
$$

Answer: The difference is 277.

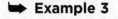 **Example 3**

Subtract 2,000 – 745.

Solution

We have to do a lot of borrowing here, since we can't take anything away from 0. Since we have zeros in the two middle columns, we can't borrow from them. We'll have to start by borrowing from the 2, as shown below.

$$
\begin{array}{r}
\overset{1}{\cancel{2}}\overset{1}{0}00 \\
-\ 745 \\
\end{array}
$$

See how we only carried the 1 over to the next column? We need to keep borrowing until we carry a 1 over to the rightmost column.

$$
\begin{array}{r}
\overset{1}{\ }\overset{9}{\ }\overset{9}{\ }\overset{1}{\ } \\
\cancel{2}\cancel{0}\cancel{0}0 \\
-\ 745 \\
\end{array}
$$

Now that all of the borrowing is complete, we can subtract by column, beginning at the right and working our way toward the left.

$$
\begin{array}{r}
\overset{1}{\ }\overset{9}{\ }\overset{9}{\ }\overset{1}{\ } \\
\cancel{2}\cancel{0}\cancel{0}0 \\
-\ 745 \\
\hline
1255 \\
\end{array}
$$

Answer: The difference of these numbers is 1,255.

LESSON 1.2—SKILLS CHECK

> **Directions**: Find the difference for each of the subtraction problems below.

1.	96 – 23	5.	220 – 158	9.	2,718 – 720
2.	54 – 30	6.	357 – 299	10.	6,552 – 2,133
3.	61 – 26	7.	4,215 – 319	11.	2,719 – 1,957
4.	94 – 58	8.	6,196 – 87	12.	9,281 – 8,888

(Answers are on page 28.)

LESSON 1.3—MULTIPLYING WHOLE NUMBERS

On the various parts of the ACCUPLACER, and in your future college classes, you can expect to see multiplication written a few different ways. For example, we could write "six times eight" as

$$6 \times 8 \quad \text{(the times sign)}$$

$$6 \cdot 8 \quad \text{(a raised dot)}$$

$$(6)(8) \quad \text{(parentheses touching)}$$

Later on in this book, we will begin to use parentheses more and more frequently, so you should start getting used to seeing multiplication written in all three ways. The answer to a multiplication problem is called the **product**. Let's review the mechanics of finding the product in a few multiplication problems.

➡ Example 1

Find the product of 65 and 9.

Solution

With multiplication problems, we always want to make sure that the number with the most digits goes on top. This might mean that you will sometimes need to change the order of the numbers, but it's okay—the order in which you multiply numbers doesn't matter. You will always get the same result.

First, line up the rightmost digits, with 65 on top.

To begin multiplying, you should start with the number 9. You will first multiply it by the 5 directly above it. Because $9 \times 5 = 45$, write a 5 below the line and carry the 4 above the 6.

Next, multiply the 9 by the next digit to the left: the 6. Remember to add the 4 to the product of 9 and 6. Since $9 \times 6 = 54 + 4 = 58$, write 58 in front of the 5.

Answer: The product of 65 and 9 is 585.

Now let's look at what happens if both numbers have two or more digits.

➡ Example 2

$27 \times 48 = ?$

Solution

Because both numbers have two digits, it doesn't matter which goes on top or bottom. First write out the problem vertically.

$$\begin{array}{r} 27 \\ \times\ 48 \\ \hline \end{array}$$

Again, we will start with the bottom-right number and multiply it by both numbers on top, starting with 8×7. We get $8 \times 7 = 56$, so we write a 6 below the line in the ones place and carry the 5. Then we multiply 8×2 and add the 5 that we carried.

$$\begin{array}{r} 5 \\ 27 \\ \times\ 48 \\ \hline 216 \end{array}$$

Notice that we just multiplied 27 by 8. This is the first step on our way to the solution, but we aren't finished yet! We have multiplied the 8 by both numbers, and now we need to do the same with the 4. To do this, we place a 0 in the ones place below the 6, and then multiply the 4 by both of the top digits, following the same process as we did before.

$$\begin{array}{r} 2 \\ 27 \\ \times\ 48 \\ \hline 216 \\ 1080 \\ \hline \end{array}$$

The last step is to add $216 + 1,080$. The result will be your product.

$$\begin{array}{r} 2 \\ 27 \\ \times\ 48 \\ \hline 216 \\ +1080 \\ \hline 1296 \end{array}$$

Answer: The product is 1,296.

Long multiplication can be tricky and time consuming. It's also very easy to make mistakes. Therefore, it's worth taking a minute to look at an alternate way of thinking about the previous example. Notice how we got started by multiplying 27 × 8 = 216. Then we essentially multiplied 27 × 40 = 1,080. In the end, we added the two. Here's why that works.

We could think of this problem as asking us to find the total number included in 48 groups of 27. Well, one group of 27 would be 27; two groups would be 54; three groups would be 81; and so on.

But what if we think of 48 as 40 plus 8 more? This way we can multiply 27 × 40 and 27 × 8 and then add the two results.

$$
\begin{array}{r}
\overset{2}{27} \\
\times\ \ 40 \\
\hline
00 \\
+\ 1080 \\
\hline
1080
\end{array}
\qquad \text{added to} \qquad
\begin{array}{r}
\overset{5}{27} \\
\times\ 8 \\
\hline
216
\end{array}
$$

1,080 + 216 = 1,296. Try this method on a few of the Skills Check problems in this section!

➥ Example 3

Multiply 32 × 651.

Solution

In this example, 651 has more digits than 32, so we need to put it on top and then multiply, following the steps that we did in the previous example.

$$
\begin{array}{r}
651 \\
\times\ \ \ 32 \\
\hline
1302 \\
+\ 19530 \\
\hline
20832
\end{array}
$$

Answer: The product is 20,832.

LESSON 1.3—SKILLS CHECK

Directions: Find the product for each of the multiplication problems below.

1. 98×7
2. 25×6
3. $9(45)$
4. $7(615)$

5. $26(99)$
6. $42 \cdot 48$
7. $90 \cdot 70$
8. $215 \cdot 12$

9. 585×30
10. $742(88)$
11. $611 \cdot 523$
12. $920(466)$

(Answers are on page 29.)

LESSON 1.4—DIVIDING WHOLE NUMBERS

Long division has given all of us a headache at some point. In the end, it just takes lots of practice. Once you get the steps down, though, it becomes pretty manageable. You should expect to see division problems on the Arithmetic section of the ACCUPLACER, and you might see them written a couple of ways. For example, 162 divided by 3 could be written either of the following ways:

$$162 \div 3 \qquad \text{(the division sign)}$$

$$\frac{162}{3} \qquad \text{(as a fraction)}$$

The answer to a division problem is called the **quotient**. Now let's practice some long division, starting with $162 \div 3$.

➡ Example 1

Find the quotient of $162 \div 3$.

Solution

When setting up a long division problem, the first number—or the number on top, if it is written as a fraction—always goes inside of the division bracket and the second number goes on the outside.

$$3\overline{)162}$$

We need to see how many 3s there are in 162. To start, we check to see if 3 can be divided evenly into the first digit in 162. In this case, 3 can't divide evenly into 1. Next, we check to see if 3 can divide evenly into the first two digits: 16. Three goes into 16 five times, so we write a 5 above the 6.

$$3\overline{)\overset{5}{162}}$$

Three times 5 is 15, so we write a 15 below the 16, and then subtract.

$$
\begin{array}{r}
5 \\
3\overline{)\,162} \\
-15 \\
\hline
1
\end{array}
$$

Next, we bring down the 2 so that it falls right after the 1.

$$
\begin{array}{r}
5 \\
3\overline{)\,162} \\
-15\downarrow \\
\hline
12
\end{array}
$$

Now repeat the whole process again. How many times can 3 go into 12? Three goes into 12 four times, so we write a 4 next to the 5 above the division bracket. Because $3 \times 4 = 12$, we write another 12 below the existing 12, and then subtract.

$$
\begin{array}{r}
54 \\
3\overline{)\,162} \\
-15\downarrow \\
\hline
12 \\
-12 \\
\hline
0
\end{array}
$$

The remainder is 0, which tells us that we are finished.

<u>Answer:</u> Therefore, $162 \div 3 = 54$.

➡ Example 2

What is $\dfrac{1728}{8}$?

Solution

First, write the division bracket, and then follow the same steps as we did before.

$$
8\overline{)\,1728}
$$

Notice that 8 cannot divide evenly into 1, so we first have to check how many times 8 goes into 17.

$$
\begin{array}{r}
2 \\
8\overline{)\,1728} \\
-16 \\
\hline
1
\end{array}
$$

Bring down the 2, and then repeat the process.

$$
\begin{array}{r}
216 \\
8\overline{)1728} \\
-16\downarrow \\
\hline
12 \\
-8\downarrow \\
\hline
48 \\
-48 \\
\hline
0
\end{array}
$$

<u>Answer:</u> The quotient is 216.

Now let's look at one more example—one in which we have to divide by a number with more than two digits.

➡ Example 3

What is $432 \div 12$?

Solution

Even though we're dividing by a bigger number, the main idea stays the same.

$$
\begin{array}{r}
3 \\
12\overline{)432} \\
-36\downarrow \\
\hline
72
\end{array}
$$

Now we need to repeat the process.

$$
\begin{array}{r}
36 \\
12\overline{)432} \\
-36\downarrow \\
\hline
72 \\
-72 \\
\hline
0
\end{array}
$$

<u>Answer:</u> The quotient is 36.

You've probably noticed that all of these problems work out nice and neat and have no remainder. In Chapters 2 and 3, we'll look at how to deal with remainders.

ACCUPLACER TIPS

There's a great shortcut for multiplying by numbers that end in zeros, and it can help you to move more quickly through some ACCUPLACER questions.

Let's say you need to multiply 215×200. The shortcut is to multiply 215×2 and then add the remaining two zeros to the end of your answer.

$$215 \times 2 = 430, \text{ so } 215 \times 200 = 43,000$$

Or, if you needed to multiply $63 \times 1,000$, you could just multiply $63 \times 1 = 63$, and then add three zeros at the end. So, $63 \times 1,000 = 63,000$.

There's also a very helpful shortcut for dividing by numbers that end in zeros. Let's say we wanted to solve $644,000 \div 200$. We would first write the problem as a fraction.

$$\frac{644,000}{200}$$

Notice that there are three zeros at the end of the first number and two zeros at the end of the second number. We can cross off one zero on top for every zero on the bottom.

$$\frac{644,0\cancel{00}}{2\cancel{00}}$$

Now we're left with $6,440 \div 2$, which is $3,220$.

LESSON 1.4—SKILLS CHECK

Directions: Find the quotient for each of the division problems below.

1. $189 \div 9$	5. $1,518 \div 6$	9. $352 \div 11$
2. $385 \div 7$	6. $6,368 \div 8$	10. $630 \div 21$
3. $\dfrac{332}{4}$	7. $\dfrac{1426}{2}$	11. $2,688 \div 24$
4. $\dfrac{950}{5}$	8. $1,250 \div 5$	12. $15,775 \div 25$

(Answers are on page 30.)

LESSON 1.5—USING ESTIMATION

All of the math questions on the ACCUPLACER are multiple-choice. This means that in some cases, you can predict what the answer should be close to and eliminate one, two, or even three of the choices! We can do this by **estimating**. Estimation is the process of rounding numbers up or down, just a little bit, to make the numbers easier to compute.

➡ Example 1

Find $2,601 \div 9$.

(A) 189
(B) 209
(C) 289
(D) 389

Solution

We could do the long division here, or we could try estimating. Notice how 2,601 is very close to 2,600, and 9 is very close to 10. Let's try dividing $2,600 \div 10$. A quick shortcut for dividing by 10 is to set the problem up as a fraction, then cross off one zero on top and one zero on bottom.

$$\frac{260\cancel{0}}{1\cancel{0}} = \frac{260}{1} = 260$$

Answer: The closest answer to 260 is choice C, which is 289.

Estimating won't always make the correct answer immediately clear like it did in this case, but it can definitely help! Try to choose nice numbers that are easy to calculate quickly, but don't stray too far from the original number. If you do, you risk getting an estimate that's just too far off to be effective.

➡ Example 2

What is the difference of 746 and 297?

(A) 449
(B) 481
(C) 551
(D) 569

Solution

If we set this up and subtract it, we will have to do a lot of borrowing, which can produce errors if you're not careful. It's a good idea to estimate first and see if that can get us close. 746 is very close to 750, and 297 is very close to 300. Let's instead find the difference of 750 and 300.

$$750 - 300 = 450$$

Answer: Choice A, 449, is very close to our estimation, so it must be the right answer.

LESSON 1.5—ACCUPLACER CHALLENGE

Directions: Using estimation, choose the best answer from the four choices given.

1. $789 + 47 =$

   (A) 806
   (B) 836
   (C) 876
   (D) 1,259

2. $396 \times 19 =$

   (A) 3,564
   (B) 3,960
   (C) 7,364
   (D) 7,524

3. $703 \times 688 =$

   (A) 403,664
   (B) 423,664
   (C) 483,664
   (D) 563,664

4. $20,045 - 1,979 =$

   (A) 17,724
   (B) 18,066
   (C) 18,526
   (D) 19,006

5. $1,029 \div 21 =$

   (A) 39
   (B) 43
   (C) 49
   (D) 53

6. Which of the following choices is the best estimate for the division problem $11,386 \div 195$?

   (A) 57
   (B) 55
   (C) 53
   (D) 51

7. Which is the best estimate of 253×49?

   (A) 1,250
   (B) 12,500
   (C) 125,000
   (D) 1,250,000

8. The population of Newbridge is 929,456, and the population of Osage is 403,920. About how many more people live in Newbridge than Osage?

   (A) 530,000
   (B) 540,000
   (C) 550,000
   (D) 570,000

(Answers are on page 31.)

LESSON 1.6—FACTORS, PRIME NUMBERS, AND MULTIPLES

A factor is a whole number that divides evenly into another whole number. When we say that a number *divides evenly* into a number, we mean that there will be no remainder after dividing. All positive whole numbers—with the exception of 0 and 1—have at least two factors: 1 and the number itself. To find the complete set of factors for a particular number, we need to find every number that divides evenly into it.

➡ Example 1

Find all the factors of 16.

Solution

We know that 16 has to have at least two factors: 1 and 16. A great way to organize lists of factors is in a factor tree, like the one shown below.

What other numbers divide evenly into 16? Since 16 is an even number, it must be divisible by 2. Sixteen is also divisible by 4 and by 8.

Answer: The complete set of factors of 16 is 1, 2, 4, 8, 16.

➡ Example 2

Find all the factors of 27.

Solution

The number 27 has to have 1 and 27 as factors. We can think about our times tables to find the rest. The only other numbers that divide evenly into 27 are 3 and 9.

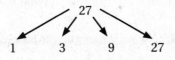

Answer: Therefore, the complete set of factors of 27 is 1, 3, 9, 27.

Because 16 and 27 have more than two factors, we refer to them as composite numbers. A **composite number** is any number with more than two factors. A number that has exactly two factors is called a **prime number**. A prime number has only 1 and itself as factors.

➡ Example 3

Is the number 11 a prime number or a composite number?

Solution

For 11 to be a composite number, it would need to have more than two factors. However, the only numbers that divide evenly into 11 are 1 and 11.

Answer: Therefore, 11 is a prime number.

Now let's talk about multiples. We have established that a factor is a whole number that divides evenly into another whole number. A **multiple** is the result you get when you start with a number and then multiply it by any whole number. For example, the multiples of 3 and the multiples of 5 would be

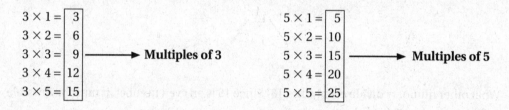

$$3 \times 1 = \boxed{3}$$
$$3 \times 2 = \boxed{6}$$
$$3 \times 3 = \boxed{9} \longrightarrow \textbf{Multiples of 3}$$
$$3 \times 4 = \boxed{12}$$
$$3 \times 5 = \boxed{15}$$

$$5 \times 1 = \boxed{5}$$
$$5 \times 2 = \boxed{10}$$
$$5 \times 3 = \boxed{15} \longrightarrow \textbf{Multiples of 5}$$
$$5 \times 4 = \boxed{20}$$
$$5 \times 5 = \boxed{25}$$

We could extend these lists forever because there are infinitely many multiples of 3 and infinitely many multiples of 5.

LESSON 1.6—SKILLS CHECK

> **Directions**: Create factor trees for the problems below.

1. Find all the factors of 15.

2. List all the factors of 28.

3. List all the factors of 36.

4. What are all the factors of 48?

5. Find all the factors of 100.

6. List all the prime numbers less than 40.

7. Which factors of 30 are prime numbers?

8. Which number has more factors: 24 or 32?

9. What are the first 5 multiples of 15?

10. Is 86 a multiple of 12?

(Answers are on page 31.)

LESSON 1.7—THE INEQUALITY SYMBOLS

On the ACCUPLACER tests, you can expect to see a few questions involving inequalities. The word **inequality** refers to the mathematical words *greater than, less than, greater than or equal to,* and *less than or equal to.* Inequalities are important mathematical ideas that we actually use all the time when thinking about things like money, especially when budgeting. When shopping for something like a new television, you might say, "I would prefer to spend less than $300," which means that you would be willing to pay $100, $225, or even $299—as long as it is below $300. Or you might say, "If I'm going to accept this job, I need to make more than $10 per hour." If the employer offered you $7.75 per hour, you wouldn't take the job. If the employer offered you $11.50 or $12 per hour, it would be *greater than* the $10 per hour that you were hoping for.

The inequality symbols are:

Less than: <
Greater than: >
Less than or equal to: ≤
Greater than or equal to: ≥
Not equal: ≠

We can use these symbols to compare numbers. For example, we know that 1 < 3, or *one is less than three.* We can also plot them on a number line.

➦ Example 1

Let's revisit the example of accepting a job. Your reasoning is that you need to make *greater than* $10 per hour. This means that if a job offered you exactly $8 per hour, you wouldn't accept it because you want to make more than $10 per hour. Show this on a number line.

Solution

To plot this on the number line, you would put an open circle at 10. The open circle means that exactly $10 per hour is *not* included in the set of wages you would be willing to accept. You would be willing to take anything that is greater than 10. On the number line below, you would draw an arrow from your open circle that points to the right of 10.

Answer:

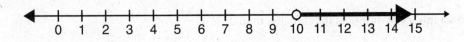

Let's try another one.

➡ **Example 2**

This time, plot the set of numbers that are *less than or equal to* 5.

Solution

Because the inequality says *less than or equal to*, we will place a filled-in circle at 5. This indicates that the number 5 is included in our set. Since we're looking at the numbers less than or equal to 5, we will draw an arrow to the left, which indicates that every number to the left of 5 on our number line is included in the set.

Answer:

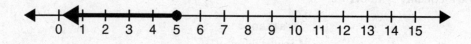

LESSON 1.7—SKILLS CHECK

Directions: Insert the correct "greater than" or "less than" symbol.

1. 5 _____ 9

2. 11 _____ 2

3. 1,001 _____ 10,001

4. 9×7 _____ $248 \div 4$

5. On the number line, plot the set of numbers that are greater than or equal to –9.

6. Plot the set of numbers that are less than 2.

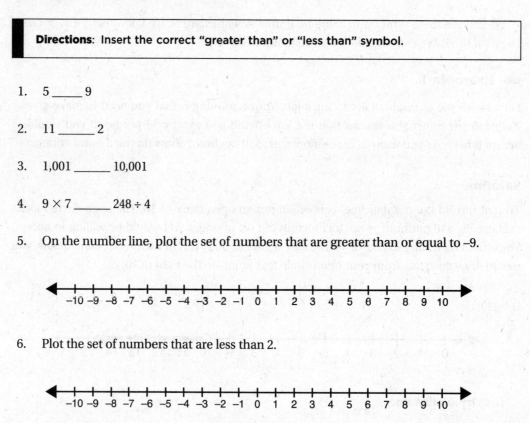

7. Plot the set of numbers that are greater than –2 and less than 4.

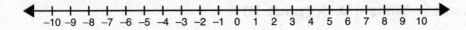

8. Plot the set of numbers that are less than or equal to 10 and greater than 0.

9. Plot the set of numbers that are between 6 and 9, inclusive.

10. Plot the set of numbers that are less than or equal to –7 or greater than or equal to –2.

(Answers are on page 32.)

To complete a set of practice questions that reviews all of Chapter 1, go to:
barronsbooks.com/TP/accuplacer/math29qw/

Lesson 1.1—Skills Check (page 11)

1. **97**

$$\begin{array}{r} 45 \\ + 52 \\ \hline 97 \end{array}$$

2. **92**

$$\begin{array}{r} {}^{1} \\ 67 \\ + 25 \\ \hline 92 \end{array}$$

3. **830**

$$\begin{array}{r} {}^{1} \\ 811 \\ + 19 \\ \hline 830 \end{array}$$

4. **945**

$$\begin{array}{r} {}^{1} \\ 909 \\ + 36 \\ \hline 945 \end{array}$$

5. **730**

$$\begin{array}{r} {}^{11} \\ 365 \\ + 365 \\ \hline 730 \end{array}$$

6. **1,000**

$$\begin{array}{r} {}^{11} \\ 784 \\ + 216 \\ \hline 1000 \end{array}$$

7. **4,019**

$$\begin{array}{r} {}^{11} \\ 2576 \\ + 1443 \\ \hline 4019 \end{array}$$

8. **15,379**

$$\begin{array}{r} 7224 \\ + 8155 \\ \hline 15379 \end{array}$$

9. **15,001**

$$\begin{array}{r} {}^{11} \\ 9350 \\ + 5651 \\ \hline 15001 \end{array}$$

10. **8,341**

$$\begin{array}{r} {}^{111} \\ 4369 \\ + 3972 \\ \hline 8341 \end{array}$$

11. **1,792**

$$\begin{array}{r} {}^{1} \\ 1024 \\ 512 \\ + 256 \\ \hline 1792 \end{array}$$

12. **3,268**

$$\begin{array}{r} {}^{111} \\ 2506 \\ 683 \\ + 79 \\ \hline 3268 \end{array}$$

Lesson 1.2—Skills Check (page 13)

1. **73**

$$\begin{array}{r} 96 \\ - 23 \\ \hline 73 \end{array}$$

2. **24**

$$\begin{array}{r} 54 \\ - 30 \\ \hline 24 \end{array}$$

3. **35**

$$\begin{array}{r} {}^{5}{}^{1} \\ \cancel{6}1 \\ - 26 \\ \hline 35 \end{array}$$

4. **36**

$$\begin{array}{r} {}^{8}{}^{1} \\ \cancel{9}4 \\ - 58 \\ \hline 36 \end{array}$$

5. **62**

$$\begin{array}{r} {}^{1}{}^{11}{}^{1} \\ \cancel{2}\cancel{2}0 \\ - 158 \\ \hline 62 \end{array}$$

6. **58**

$$\begin{array}{r} {}^{2}{}^{14}{}^{1} \\ \cancel{3}\cancel{5}7 \\ - 299 \\ \hline 58 \end{array}$$

7. **3,896**

$$
\begin{array}{r}
{}^{3}\cancel{4}\,{}^{11}\cancel{2}\,{}^{10}\cancel{1}5 \\
-\;\;3\,1\,9 \\
\hline
3\,8\,9\,6
\end{array}
$$

8. **6,109**

$$
\begin{array}{r}
{}^{8}\cancel{6}\,1\,\cancel{9}6 \\
-\;\;\;\;8\,7 \\
\hline
6\,1\,0\,9
\end{array}
$$

9. **1,998**

$$
\begin{array}{r}
{}^{1}\,{}^{16}\cancel{2}\cancel{7}18 \\
-\;\;7\,2\,0 \\
\hline
1\,9\,9\,8
\end{array}
$$

10. **4,419**

$$
\begin{array}{r}
{}^{4}\,6\cancel{5}\cancel{5}\,{}^{1}2 \\
-\;2\,1\,3\,3 \\
\hline
4\,4\,1\,9
\end{array}
$$

11. **762**

$$
\begin{array}{r}
{}^{1}\,{}^{16}\cancel{2}\cancel{7}\,{}^{1}9 \\
-\;1\,9\,5\,7 \\
\hline
7\,6\,2
\end{array}
$$

12. **393**

$$
\begin{array}{r}
{}^{8}\,{}^{11}\,{}^{17}\cancel{9}\cancel{2}\cancel{8}1 \\
-\;8\,8\,8\,8 \\
\hline
3\,9\,3
\end{array}
$$

Lesson 1.3—Skills Check (page 17)

1. **686**

$$
\begin{array}{r}
{}^{5}\;\; \\
98 \\
\times\;\;7 \\
\hline
686
\end{array}
$$

2. **150**

$$
\begin{array}{r}
{}^{3}\;\; \\
25 \\
\times\;\;6 \\
\hline
150
\end{array}
$$

3. **405**

$$
\begin{array}{r}
{}^{4}\;\; \\
45 \\
\times\;\;9 \\
\hline
405
\end{array}
$$

4. **4,305**

$$
\begin{array}{r}
{}^{1\,3}\;\; \\
615 \\
\times\;\;\;7 \\
\hline
4305
\end{array}
$$

5. **2,574**

$$
\begin{array}{r}
{}^{5}\;\; \\
26 \\
\times\;\;99 \\
\hline
234 \\
+\,2340 \\
\hline
2574
\end{array}
$$

6. **2,016**

$$
\begin{array}{r}
{}^{1}\;\; \\
42 \\
\times\;\;48 \\
\hline
336 \\
+\,1680 \\
\hline
2016
\end{array}
$$

7. **6,300**

$$
\begin{array}{r}
90 \\
\times\;\;70 \\
\hline
00 \\
+\,6300 \\
\hline
6300
\end{array}
$$

8. **2,580**

$$
\begin{array}{r}
{}^{1}\;\; \\
215 \\
\times\;\;12 \\
\hline
430 \\
+\,2150 \\
\hline
2580
\end{array}
$$

9. **17,550**

$$
\begin{array}{r}
{}^{2\,1}\;\; \\
585 \\
\times\;\;30 \\
\hline
000 \\
+\,17550 \\
\hline
17550
\end{array}
$$

10. **65,296**

$$
\begin{array}{r}
742 \\
\times\;\;88 \\
\hline
5936 \\
+\,59360 \\
\hline
65296
\end{array}
$$

11. **319,553**

$$
\begin{array}{r}
611 \\
\times\;\;523 \\
\hline
1833 \\
12220 \\
+\,305500 \\
\hline
319553
\end{array}
$$

12. **428,720**

$$
\begin{array}{r}
920 \\
\times\;\;466 \\
\hline
5520 \\
55200 \\
+\,368000 \\
\hline
428720
\end{array}
$$

Lesson 1.4—Skills Check (page 20)

1. 21

```
        21
    9) 189
       -18↓
        09
       -09
         0
```

2. 55

```
        55
    7) 385
       -35↓
        35
       -35
         0
```

3. 83

```
        83
    4) 332
       -32↓
        12
       -12
         0
```

4. 190

```
        1 9 0
    5) 9 5 0
       -5↓
        45
       -45↓
        00
        00
         0
```

5. 253

```
        253
    6) 1518
       -12↓↓
        31
       -30↓
        18
       -18
         0
```

6. 796

```
        796
    8) 6368
       -56↓
        76
       -72↓
        48
       -48
         0
```

7. 713

```
        713
    2) 1426
       -14↓
        02
       -02↓
        06
       -06
         0
```

8. 250

```
        250
    5) 1250
       -10↓
        25
       -25↓
        00
       -00
         0
```

9. 32

```
        32
    11) 352
        -33↓
         22
        -22
          0
```

10. 30

```
        30
    21) 630
       -63↓
        00
       -00
         0
```

11. 112

```
         112
    24) 2688
       -24↓
        28
       -24↓
        48
       -48
         0
```

12. 631

```
          631
    25) 15775
       -150↓
        77
       -75↓
        25
       -25
         0
```

Lesson 1.5—ACCUPLACER Challenge (page 22)

1. **(B)** Round 789 up to 800, and round 47 up to 50. 800 + 50 = 850. Choice B is the closest answer to this estimate.

2. **(D)** Round 396 up to 400, and round 19 up to 20. 400 $\times$ 20 = 8,000. Choice D is closest to this estimate.

3. **(C)** Round 703 down to 700, and round 688 up to 700. 700 $\times$ 700 = 490,000. Choice C is closest to this estimate.

4. **(B)** Round 20,045 down to 20,000, and round 1,979 up to 2,000. 20,000 – 2,000 = 18,000. Choice B is closest to this estimate.

5. **(C)** Round 1,029 down to 1,000, and round 21 down to 20. 1,000 $\div$ 20 = 50. Choice C is closest to this estimate.

6. **(A)** Round 11,386 up to 11,400. Then round 195 up to 200. 11,400 $\div$ 200 = 57. Note that the actual answer to 11,386 $\div$ 195 will not result in a whole number. Of the choices given, 57 is the closest.

7. **(B)** Round 253 down to 250, and round 49 up to 50. 250 $\times$ 50 = 12,500. Note that the actual product of 253 $\times$ 49 = 12,397. Choice B is the closest approximation of this answer.

8. **(A)** First, round 929,456 up to 930,000. Then round 403,920 down to 400,000. The difference in these two estimates is 530,000.

Lesson 1.6—Skills Check (page 24)

1. **1, 3, 5, 15**

$$1 \times 15 = 15$$
$$3 \times 5 = 15$$

2. **1, 2, 4, 7, 14, 28**

$$1 \times 28 = 28$$
$$2 \times 14 = 28$$
$$4 \times 7 = 28$$

3. **1, 2, 3, 4, 6, 9, 12, 18, 36**

$$1 \times 36 = 36$$
$$2 \times 18 = 36$$
$$3 \times 12 = 36$$
$$4 \times 9 = 36$$
$$6 \times 6 = 36$$

4. **1, 2, 3, 4, 6, 8, 12, 16, 24, 48**

$$1 \times 48 = 48$$
$$2 \times 24 = 48$$
$$3 \times 16 = 48$$
$$4 \times 12 = 48$$
$$6 \times 8 = 48$$

5. **1, 2, 4, 5, 10, 20, 25, 50, 100**

$$1 \times 100 = 100$$
$$2 \times 50 = 100$$
$$4 \times 25 = 100$$
$$5 \times 20 = 100$$
$$10 \times 10 = 100$$

6. **2, 3, 5, 7, 11, 13, 17, 19, 23, 29, 31, 37**

Note that the number 1 is not considered prime since it only has one factor.

7. **2, 3, 5**

The factors of 30 are 1, 2, 3, 5, 6, 10, 15, 30. Of those, only 2, 3, and 5 are prime numbers.

8. **24**

24 has 8 factors: 1, 2, 3, 4, 6, 8, 12, 24. 32 has 6 factors: 1, 2, 4, 8, 16, 32.

9. **15, 30, 45, 60, 75**

$$15 \times 1 = 15$$
$$15 \times 2 = 30$$
$$15 \times 3 = 45$$
$$15 \times 4 = 60$$
$$15 \times 5 = 75$$

10. **No**

The multiples of 12 are 12, 24, 36, 48, 60, 72, 84, 96 . . . .

Lesson 1.7—Skills Check (pages 26-27)

1. **5 < 9**

5 is less than 9.

2. **11 > 2**

11 is greater than 2.

3. **1,001 < 10,001**

1,001 is less than 10,001.

4. **$9 \times 7 > 248 \div 4$**

$9 \times 7 = 63$ and $248 \div 4 = 62$. Therefore, $63 > 62$.

5.

Draw a closed circle at –9, and then draw an arrow extending to the right.

6.

Draw an open circle at 2, and draw an arrow extending to the left.

7.

Draw an open circle at –2 and another at 4. The set of numbers greater than –2 and less than 4 fall in between those open circles.

8.

Draw an open circle at 0 and a closed circle at 10. The solution set falls in between these two points.

9.

Draw two closed circles: one at 6 and another at 9. The set of numbers greater than or equal to 6, and less than or equal to 9, fall in between these circles.

10.

Draw a closed circle at –7. The set of numbers less than or equal to –7 would be all the numbers to the left. Draw another closed circle at –2. The set of numbers greater than or equal to –2 would be all the numbers to the right of this circle.

Fractions

2

When we first learned to count, it was all pretty straightforward. There's one apple, two apples, three apples, and so forth. But what if we have two people who want to share a single apple? How much do they each get? We realize that we have a need to count things that are greater than nothing, but also less than one whole. We need numbers between 0 and 1, and even between 1 and 2, and between all of the whole numbers really. Because not all quantities can be considered whole, we have a need for fractions.

Something to keep in mind: fractions are part of a fundamental area of mathematics critical to the workings of proportional reasoning, data analysis, coordinate geometry, and higher mathematics. Fractions also give many of us serious headaches and difficulties. Because of these two facts, you can be sure that questions involving fractions will be well represented on the Arithmetic section of the ACCUPLACER exam. In other words, you will not likely score well on this section without a strong working knowledge of fractions, so you should place high emphasis on this chapter.

In this chapter, we will work to master

- raising and lowering fractions to equivalents,
- comparing fractions,
- ordering fractions,
- converting mixed numbers and improper fractions, and
- adding, subtracting, multiplying, and dividing fractions and mixed numbers.

LESSON 2.1—UNDERSTANDING FRACTIONS

One of the most common uses for fractions is to compare a part of something to its whole. The part becomes the numerator, and the whole becomes the denominator. The fraction bar, which separates the numerator and the denominator, is another symbol for division. The fraction $\frac{5}{24}$ can also be written as $5 \div 24$.

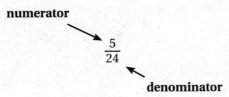

A whole day, for example, has 24 hours. If we spend part of one day working, say 5 hours, then we could say that we spent $\frac{5}{24}$ of the day at work. We would read this as "five twenty-fourths of the day."

Another example we can examine is a pizza that looks like this:

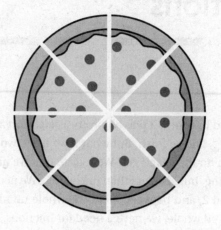

The pizza is divided into eight equal parts, or eighths. If I ate one slice, or part, then I ate $\frac{1}{8}$ of the entire pizza. If I ate 3 parts, then I ate $\frac{3}{8}$ of the whole pizza. If I ate the whole pizza, then I ate $\frac{8}{8}$ of it. Since the fraction bar means division, $\frac{8}{8}$ is equal to 1 whole pizza.

Equivalent Fractions

Some fractions can look different but represent the same numerical relationship. We call these fractions **equivalent**. For example, let's say two friends, Jill and Roman, go out for pizza. Jill orders $\frac{1}{2}$ of a pizza, and Roman orders 4 slices out of an 8-piece pizza pie. Here's what each of their orders would look like.

<u>**Jill's Pizza**</u>   <u>**Roman's Pizza**</u>

Slices Ordered

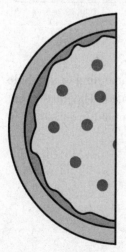

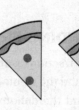

Slices Not Ordered

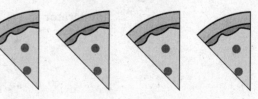

We can see from this example that $\frac{1}{2} = \frac{4}{8}$! The fractions may look different, but we can see that Jill and Roman are getting the same amount of pizza.

You may notice that both the numerator and the denominator in $\frac{1}{2}$ became 4 times as big to produce $\frac{4}{8}$. That is to say $\frac{1}{2} \times \frac{4}{4} = \frac{4}{8}$. But, isn't $\frac{4}{4} = 1$? Remember that when we multiply by 1, we get the same number that we started with. Multiplying by $\frac{4}{4}$ is the same as multiplying by 1. When you do this, you are not changing the value of the fraction; you are only changing the *appearance* of the fraction. When we do this, we say that we are **raising a fraction to an equivalent**.

Here are some other fractional equivalents:

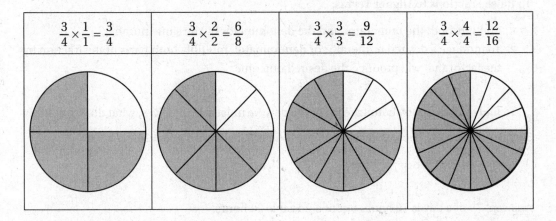

$\frac{3}{4} \times \frac{1}{1} = \frac{3}{4}$	$\frac{3}{4} \times \frac{2}{2} = \frac{6}{8}$	$\frac{3}{4} \times \frac{3}{3} = \frac{9}{12}$	$\frac{3}{4} \times \frac{4}{4} = \frac{12}{16}$

This is to say that you can have three out of four, six out of eight, nine out of twelve, or twelve out of sixteen pieces of a whole pizza, and you would have exactly the same amount of pizza.

ACCUPLACER TIPS

While you're studying, you can use a calculator to check to see that fractions are equivalent by dividing the numerator by the denominator. Equivalent fractions will result in the same decimal quotient when you divide. Remember, though, you won't have a calculator to do this on the ACCUPLACER!

$$\frac{3}{4} = 3 \div 4 = 0.75 \qquad \frac{6}{8} = 6 \div 8 = 0.75$$

$$\frac{9}{12} = 9 \div 12 = 0.75 \qquad \frac{12}{16} = 12 \div 16 = 0.75$$

Most of the time, you will need to raise a fraction to an equivalent with a desired numerator or denominator.

➡ **Example 1**

Knowing that there are 24 hours in a day, how many hours are in $\frac{2}{3}$ of a day?

Solution

We know that there are 24 hours in a day, so if we want to know how many hours are in $\frac{2}{3}$ of a day, we essentially need to raise $\frac{2}{3}$ to an equivalent fraction with 24 as the denominator. The question is $\frac{2}{3} = \frac{?}{24}$.

To Raise Fractions to Higher Terms:

1. Multiply both the numerator and the denominator by the same number.
2. To produce a desired numerator or denominator, multiply both parts of the fraction by the factor that will produce the desired outcome.

To find out how many hours are in $\frac{2}{3}$ of a day, we must ask ourselves, what do we multiply 3 by to produce 24? The answer to that question is 8, because $3 \times 8 = 24$.

Therefore, $\frac{2}{3} \times \frac{8}{8} = \frac{2 \times 8}{3 \times 8} = \frac{16}{24}$.

Answer: $\frac{2}{3}$ of a day is 16 hours out of a possible 24 hours.

Reducing (Simplifying) Fractions

Using $\frac{4}{8}$ of a pizza and $\frac{1}{2}$ of a pizza, we previously saw that equivalent fractions can be used to represent the same mathematical idea. To keep the numbers we are working with smaller and to improve communication, we generally reduce fractions to their simplest form. A fraction is in its simplest form, or **lowest terms**, when the numerator and denominator share no common factor other than 1.

It's important to know that ACCUPLACER answer choices involving fractions will almost always be in lowest terms. You could do a problem and find that your correct answer of $\frac{8}{10}$ is not available as a choice. In that case, you would choose $\frac{4}{5}$, the simplest form of $\frac{8}{10}$.

To Reduce a Fraction to Lowest Terms:

1. List the common factors of both the numerator and the denominator.
2. Divide both the numerator and the denominator by the greatest common factor (GCF) found in both lists.

➡ Example 2

Express $\frac{12}{18}$ in lowest terms.

Solution

First, separately list the factors of 12 and 18.

Factors of 12:   1, 2, 3, 4, ⑥, 12

Factors of 18:   1, 2, 3, ⑥, 9, 18

The greatest common factor is 6, so we divide the numerator and the denominator by 6.

$$\frac{12}{18} \div \frac{6}{6} = \frac{12 \div 6}{18 \div 6} = \frac{2}{3}$$

<u>Answer:</u> The fraction $\frac{12}{18}$ is equal to $\frac{2}{3}$.

➡ Example 3

What fraction of the day do you spend at work if you work for 8 hours each day? Write your answer in lowest terms.

Solution

Begin by setting up the fraction as a part over a whole. In this case, 8 hours is the part of the day you work. Twenty-four is the whole because there are 24 hours in a whole day. Therefore, our fraction is $\frac{8}{24}$.

Next, list the factors of 8 and 24.

Factors of 8:   1, 2, 4, ⑧

Factors of 24:   1, 2, 3, 4, 6, ⑧, 12, 24

Then you need to divide the numerator and denominator by 8, the GCF.

$$\frac{8}{24} \div \frac{8}{8} = \frac{1}{3}$$

<u>Answer:</u> Expressed as a fraction, if you work for 8 hours, then you work $\frac{1}{3}$ of a day.

Note that some fractions, such as $\frac{21}{23}$, have a numerator and a denominator with no common factor other than 1. These fractions are already in simplest form.

LESSON 2.1—SKILLS CHECK

Directions: Write the numerator or denominator that makes the two fractions equivalent.

1. $\dfrac{1}{5} = \dfrac{}{20}$

4. $\dfrac{4}{5} = \dfrac{}{20}$

7. $\dfrac{3}{5} = \dfrac{}{10}$

10. $\dfrac{5}{6} = \dfrac{}{18}$

2. $\dfrac{2}{5} = \dfrac{}{20}$

5. $\dfrac{5}{5} = \dfrac{}{20}$

8. $\dfrac{2}{5} = \dfrac{12}{}$

3. $\dfrac{3}{5} = \dfrac{}{20}$

6. $\dfrac{1}{2} = \dfrac{}{12}$

9. $\dfrac{2}{3} = \dfrac{14}{}$

Directions: Reduce each fraction to lowest terms.

11. $\dfrac{3}{12}$

13. $\dfrac{12}{18}$

15. $\dfrac{175}{225}$

12. $\dfrac{5}{15}$

14. $\dfrac{24}{36}$

16. $\dfrac{56}{140}$

(Answers are on page 68.)

LESSON 2.1—ACCUPLACER CHALLENGE

Directions: For each of the questions below, choose the best answer from the four choices given.

1. The Tigers won 8 of their first 12 hockey games. What fraction of their games did they win so far?

    (A) $\dfrac{4}{12}$

    (B) $\dfrac{1}{2}$

    (C) $\dfrac{2}{3}$

    (D) $\dfrac{3}{4}$

2. Tyleek ate $\dfrac{3}{4}$ of a box of candies. How many candies did Tyleek eat if there were 32 candies in the box?

    (A) 3

    (B) 8

    (C) 24

    (D) 29

3. Theophilus has to drive 60 miles to pick up a microwave. After 45 miles, he stops for gasoline. What fraction of the total distance has he driven so far?

    (A) $\frac{1}{4}$

    (B) $\frac{15}{45}$

    (C) $\frac{1}{2}$

    (D) $\frac{3}{4}$

4. The Blue Devils lost 7 of their first 22 baseball games. About what fraction of their games did they lose so far?

    (A) $\frac{1}{22}$

    (B) $\frac{1}{5}$

    (C) $\frac{1}{4}$

    (D) $\frac{1}{3}$

(Answers are on page 68.)

LESSON 2.2—COMPARING FRACTIONS

The two diagrams below represent two equal-sized cakes. If you could choose a piece from either cake, which would give you a bigger piece?

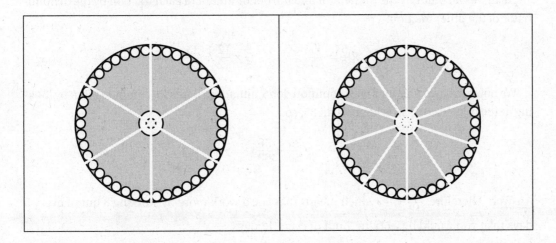

Representing the pieces as a fraction, the cake on the left has pieces that are $\frac{1}{6}$ of the cake, and the cake on the right has pieces that are $\frac{1}{10}$ of the cake. You may have noticed that the cake split into 6 pieces has bigger pieces than the cake split into 10 pieces. From our cakes, we can conclude that $\frac{1}{6}$ is greater than $\frac{1}{10}$. Mathematically, that is written as $\frac{1}{6} > \frac{1}{10}$. Wait, isn't 6 smaller than 10? Hmm, let's think about that.

The denominator represents how many pieces something is split into or how many pieces we need to make the whole. The more pieces we cut something up into, the smaller the pieces get. When comparing fractions with the same numerator, the larger the number in the denominator, the smaller the fraction becomes. That means the following order would be going from greatest to least moving left to right:

<u>Greatest</u>　　　　　　　<u>Least</u>

$$\frac{1}{2}, \frac{1}{3}, \frac{1}{4}, \frac{1}{5}, \frac{1}{6}, ..., \frac{1}{\infty}$$

But how do we compare fractions with different denominators?

To Compare Fractions with Different Denominators:

1. First, multiply the numerator and the denominator of each fraction by the denominator of the other fraction.
2. Once the denominators are the same for both fractions, the fraction with the larger numerator is the larger of the two fractions.

➡ Example 1

You have the option to work 7 out of every 12 days or 3 out of every 5 days. Which option has you working more?

Solution

First, we will write each relationship as a fraction. Seven out of every 12 becomes $\frac{7}{12}$, and 3 out of every 5 days becomes $\frac{3}{5}$.

Next, we will multiply the numerator and the denominator of each fraction by the denominator of the other fraction.

$$\frac{7}{12} \times \frac{5}{5} = \frac{35}{60} \qquad\qquad \frac{3}{5} \times \frac{12}{12} = \frac{36}{60}$$

We now have two fractions with common denominators, so we can compare them by looking at the numerators. 36 is larger than 35, so

$$\frac{35}{60} < \frac{36}{60}$$

<u>Answer:</u> Therefore, $\frac{7}{12} < \frac{3}{5}$, which means that you'd work more by working 3 out of every 5 days than you would by working 7 out of every 12 days.

To Order Three or More Fractions:

1. Raise each fraction to an equivalent with a common denominator.
2. Put the fractions with the common denominators in order according to their numerators.
3. Establish the order according to the original fractions by writing them under their equivalent form in your ordered list.

➡ **Example 2**

Which of the following choices lists the fractions in order from least to greatest?

(A) $\dfrac{3}{8}, \dfrac{1}{4}, \dfrac{5}{16}, \dfrac{1}{2}$

(B) $\dfrac{1}{2}, \dfrac{1}{4}, \dfrac{3}{8}, \dfrac{5}{16}$

(C) $\dfrac{5}{16}, \dfrac{3}{8}, \dfrac{1}{4}, \dfrac{1}{2}$

(D) $\dfrac{1}{4}, \dfrac{5}{16}, \dfrac{3}{8}, \dfrac{1}{2}$

Solution

To start, we can use the list from choice A, $\dfrac{3}{8}, \dfrac{1}{4}, \dfrac{5}{16}, \dfrac{1}{2}$, to look for a common denominator. Let's begin by listing the multiples of each denominator to find one they all share in common.

Multiples of 8:	8, (16), 24, 32, 40, 48,…
Multiples of 4:	4, 8, 12, (16), 20, 24,…
Multiples of 16:	(16), 32, 48, 64,…
Multiples of 2:	2, 4, 6, 8, 10, 12, 14, (16), 18,…

The least multiple that they share in common is 16, so we'll go ahead and change each fraction to an equivalent with 16 as the common denominator.

$$\dfrac{3}{8}, \quad \dfrac{1}{4}, \quad \dfrac{5}{16}, \quad \dfrac{1}{2}$$
$$\downarrow \quad\quad \downarrow \quad\quad \downarrow \quad\quad \downarrow$$
$$\dfrac{6}{16}, \quad \dfrac{4}{16}, \quad \dfrac{5}{16}, \quad \dfrac{8}{16}$$

Now that we have common denominators, we can list the fractions in order from least to greatest according to their numerators. After we do that, we can write the original fractions underneath so that we can have our list from least to greatest using the fractions we started with.

$$\dfrac{4}{16}, \quad \dfrac{5}{16}, \quad \dfrac{6}{16}, \quad \dfrac{8}{16}$$
$$\downarrow \quad\quad \downarrow \quad\quad \downarrow \quad\quad \downarrow$$
$$\dfrac{1}{4}, \quad \dfrac{5}{16}, \quad \dfrac{3}{8}, \quad \dfrac{1}{2}$$

<u>Answer:</u> The correct answer is choice D, which has the correct order, $\dfrac{1}{4}, \dfrac{5}{16}, \dfrac{3}{8}, \dfrac{1}{2}$.

LESSON 2.2—SKILLS CHECK

> **Directions:** Compare each set of fractions by writing either the <, >, or = sign in between them.

1. $\frac{1}{3} \square \frac{1}{4}$

2. $\frac{3}{4} \square \frac{7}{8}$

3. $\frac{3}{9} \square \frac{1}{3}$

4. $\frac{2}{3} \square \frac{1}{2}$

5. $\frac{5}{6} \square \frac{15}{18}$

6. $\frac{9}{12} \square \frac{3}{4}$

7. $\frac{7}{10} \square \frac{2}{3}$

8. $\frac{7}{15} \square \frac{2}{5}$

9. $\frac{9}{10} \square \frac{3}{4}$

> **Directions:** Put the following fractions in order from least to greatest.

10. $\frac{2}{3}, \frac{1}{2}, \frac{5}{6}, \frac{1}{3}, \frac{1}{6}$

11. $\frac{1}{3}, \frac{1}{12}, \frac{1}{2}, \frac{1}{6}, \frac{1}{4}$

12. $\frac{1}{2}, \frac{2}{3}, \frac{5}{8}, \frac{5}{6}, \frac{3}{4}$

(Answers are on page 69.)

LESSON 2.2—ACCUPLACER CHALLENGE

> **Directions:** For each of the questions below, choose the best answer from the four choices given.

1. Hal is looking for a socket wrench to use to fix his car. He tried his $\frac{1}{2}$-inch socket wrench, but it was too large. He tried his $\frac{3}{8}$-inch socket wrench, but it was too small. Which of the following could be the correct sized socket wrench that Hal is looking for?

   (A) $\frac{5}{16}$ inch

   (B) $\frac{11}{32}$ inch

   (C) $\frac{15}{32}$ inch

   (D) $\frac{5}{8}$ inch

2. Which of the following choices lists the fractions in order from least to greatest?

(A) $\dfrac{1}{2}, \dfrac{7}{12}, \dfrac{2}{3}, \dfrac{3}{4}, \dfrac{5}{6}$

(B) $\dfrac{7}{12}, \dfrac{5}{6}, \dfrac{3}{4}, \dfrac{2}{3}, \dfrac{1}{2}$

(C) $\dfrac{5}{6}, \dfrac{7}{12}, \dfrac{1}{2}, \dfrac{3}{4}, \dfrac{2}{3}$

(D) $\dfrac{1}{2}, \dfrac{2}{3}, \dfrac{3}{4}, \dfrac{5}{6}, \dfrac{7}{12}$

3. Which of the following lists the fractions in order from least to greatest?

(A) $\dfrac{2}{7}, \dfrac{5}{8}, \dfrac{1}{3}, \dfrac{3}{5}$

(B) $\dfrac{2}{7}, \dfrac{1}{3}, \dfrac{3}{5}, \dfrac{5}{8}$

(C) $\dfrac{5}{8}, \dfrac{3}{5}, \dfrac{1}{3}, \dfrac{2}{7}$

(D) $\dfrac{1}{3}, \dfrac{2}{7}, \dfrac{3}{5}, \dfrac{5}{8}$

(Answers are on page 70.)

LESSON 2.3—IMPROPER FRACTIONS AND MIXED NUMBERS

A **proper fraction** has a numerator that is less than its denominator, as in the fractions $\dfrac{1}{2}, \dfrac{3}{4}, \dfrac{1}{100}$, and $\dfrac{999}{1000}$. All proper fractions have a value that is greater than 0 but less than 1.

An **improper fraction**, on the other hand, has a numerator that is greater than or equal to its denominator. For example, $\dfrac{3}{2}, \dfrac{7}{4}, \dfrac{100}{100}$, and $\dfrac{900,000}{1000}$ are all examples of improper fractions. These fractions have values that are greater than or equal to 1.

Mixed numbers are numbers that consist of a whole number and a proper fraction, such as $10\dfrac{1}{2}, 5\dfrac{1}{4}$, and $250\dfrac{3}{4}$. Mixed numbers have a value greater than 1 and can also be written as improper fractions.

When we have at least one whole, we can use either an improper fraction or a mixed number to represent the same amount. We experience this with money.

Using dollar bills and change, there are many ways we can make $1.25. Two ways are shown below:

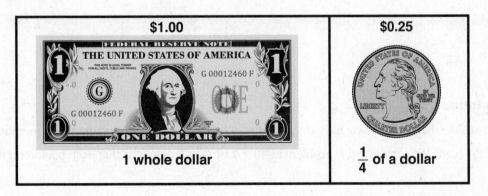

$1.00	$0.25
1 whole dollar	$\dfrac{1}{4}$ of a dollar

It just so happens that we call the $0.25 coin a quarter for a very good reason: it is worth $\frac{1}{4}$ of a dollar. Considering that, $1.25 can also be called $1\frac{1}{4}$ of a dollar.

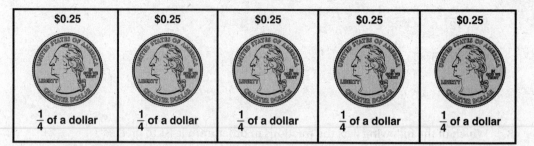

$0.25	$0.25	$0.25	$0.25	$0.25
$\frac{1}{4}$ of a dollar	$\frac{1}{4}$ of a dollar	$\frac{1}{4}$ of a dollar	$\frac{1}{4}$ of a dollar	$\frac{1}{4}$ of a dollar

Thinking about quarters as $\frac{1}{4}$ of a dollar, we can see from this example that five quarters, which is also worth $1.25, can be thought of as $\frac{5}{4}$ of a dollar.

From these examples, we can see that $1\frac{1}{4} = \frac{5}{4}$. Here the same amount can be represented using a mixed number and using an equivalent improper fraction. This may seem confusing at first, but this flexibility, to use either an improper fraction or a mixed number to represent the same quantity, makes other computations a lot easier, especially subtraction with mixed numbers. It's also convenient that we can make the same amount of money several different ways when we're trying to make purchases.

To Change an Improper Fraction to a Mixed Number:

1. Divide the denominator into the numerator.
2. Make the whole quotient the whole number in front of the fraction sign.
3. The remainder becomes the new numerator.
4. Keep the same denominator.
5. Reduce fractions to simplest form whenever possible.

➡ Example 1

A group of seven friends has decided to run a 26-mile marathon as a relay team. How many miles does each person have to run individually to complete a total of 26 miles as a team?

(A) $2\frac{6}{7}$

(B) $3\frac{5}{7}$

(C) $3\frac{5}{26}$

(D) $7\frac{5}{26}$

Solution

To answer this problem, we have to break the 26 miles up evenly among the seven members of the team. We can write this problem as $26 \div 7$ or $\frac{26}{7}$. This means that each person on the relay team must run $\frac{26}{7}$ miles, but that's not an answer choice. Since the answer choices are written as mixed numbers, we will need to change $\frac{26}{7}$ into a mixed number.

We need to divide 7 into 26 to see how many miles each person must run individually, written as a mixed number.

Using long division:

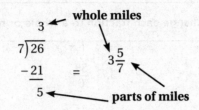

Therefore, $\frac{26}{7} = 3\frac{5}{7}$. Notice how the denominator of the fraction is the same whether it's written as an improper fraction or as a mixed number.

<u>Answer:</u> The correct answer is choice B.

To Change a Mixed Number to an Improper Fraction:

1. Multiply the denominator by the whole number, and then add the numerator. Make this your new numerator.
2. Keep the same denominator.

➡ Example 2

To follow a recipe, Shalinda needs to put $2\frac{3}{4}$ cups of flour into a bowl using a $\frac{1}{4}$-cup measuring cup, which is the only one she has. How many full scoops of flour should she put into the bowl using the $\frac{1}{4}$-cup measuring cup?

(A) 5
(B) 9
(C) 11
(D) 23

Solution

Essentially, the question is asking us how many fourths are in $2\frac{3}{4}$. We could set this up as a division problem, but we can also change $2\frac{3}{4}$ into an improper fraction to determine how many $\frac{1}{4}$ scoops she would need to use.

To change $2\frac{3}{4}$ to an improper fraction we perform: $\frac{(4 \times 2) + 3}{4} = \frac{11}{4}$.

Alternatively, we can symbolize the steps like this:

$$2 \overset{+}{\underset{\times}{\rightleftarrows}} \frac{3}{4} = \frac{11}{4}$$

Answer: Having changed the mixed number to $\frac{11}{4}$, we can see that the correct answer is choice C, since there are $\frac{11}{4}$ in $2\frac{3}{4}$.

LESSON 2.3—SKILLS CHECK

Directions: Change each fraction to a whole or mixed number.

1. $\frac{14}{5}$

2. $\frac{14}{8}$

3. $\frac{33}{6}$

4. $\frac{12}{3}$

5. $\frac{30}{9}$

6. $\frac{44}{16}$

Directions: Change each mixed number to an improper fraction.

7. $2\frac{3}{4}$

8. $1\frac{4}{7}$

9. $5\frac{1}{3}$

10. $6\frac{2}{7}$

11. $9\frac{1}{2}$

12. $8\frac{3}{4}$

(Answers are on page 70.)

LESSON 2.3—ACCUPLACER CHALLENGE

Directions: For each of the questions below, choose the best answer from the four choices given.

1. In woodworking, every time you cut a board, you lose some wood equal to the entire width of the saw blade. This is where sawdust comes from, and it is referred to as *kerf*. You need to make twelve cuts into a 60-inch board with a saw blade that is $\frac{1}{8}$-inch thick. How many inches of wood will you lose to the kerf?

(A) $1\frac{1}{2}$

(B) $4\frac{1}{2}$

(C) $7\frac{1}{2}$

(D) $58\frac{1}{2}$

2. All of the following are equal to $\frac{27}{5}$ EXCEPT

    (A) $1\frac{20}{7}$

    (B) $3\frac{12}{5}$

    (C) $4\frac{7}{5}$

    (D) $5\frac{2}{5}$

3. Lucy walks $\frac{1}{3}$ of a mile from home to work each day and another $\frac{1}{3}$ of a mile back from work each day. How many total miles does Lucy walk traveling from home to work and back during a five-day, Monday–Friday workweek?

    (A) $\frac{2}{3}$

    (B) $1\frac{2}{3}$

    (C) $3\frac{1}{3}$

    (D) $\frac{1}{3}$

4. Which of the following lists the numbers in order from least to greatest?

    (A) $3\frac{1}{5}, \frac{10}{3}, \frac{7}{2}, 3\frac{2}{3}$

    (B) $\frac{7}{2}, \frac{10}{3}, 3\frac{2}{3}, 3\frac{1}{5}$

    (C) $\frac{7}{2}, \frac{10}{3}, 3\frac{1}{5}, 3\frac{2}{3}$

    (D) $3\frac{1}{5}, 3\frac{2}{3}, \frac{7}{2}, \frac{10}{3}$

(Answers are on page 71.)

LESSON 2.4—ADDING AND SUBTRACTING FRACTIONS

When adding or subtracting fractions, it is important to remember the roles of the numerator and the denominator. The denominator tells us how many parts are needed to make one whole, and the numerator tells us how many parts we have. The fraction $\frac{3}{10}$ indicates that 10 parts are needed to make a whole, and we have 3 of those parts.

numerator $\longrightarrow$ $\frac{3}{10}$ $\longleftarrow$ number of parts we have

denominator $\longrightarrow$ $\frac{3}{10}$ $\longleftarrow$ number of parts needed to make a whole

To Add or Subtract Fractions with Common Denominators:

1. Add or subtract the numerators.
2. Keep the common denominator.
3. Reduce whenever possible.

➡ Example 1

Find the sum of $\dfrac{3}{10} + \dfrac{2}{10}$.

Solution

By adding $\dfrac{3}{10}$ to $\dfrac{2}{10}$, we are changing the total number of parts that we have, but not the number of parts we need to make the whole. In other words, we have 3 of 10 parts and we're adding on 2 of the 10 parts, so we now have 5 of the 10 parts. Mathematically, it looks like this:

$$\frac{3}{10} + \frac{2}{10} = \frac{3+2}{10} = \frac{5}{10}$$

Since $\dfrac{5}{10}$ can be reduced, we write $\dfrac{5}{10} = \dfrac{1}{2}$.

Answer: $\dfrac{3}{10} + \dfrac{2}{10} = \dfrac{1}{2}$

➡ Example 2

Find the difference of $\dfrac{3}{10} - \dfrac{2}{10}$.

Solution

Subtracting these fractions looks like this:

$$\frac{3}{10} - \frac{2}{10} = \frac{3-2}{10} = \frac{1}{10}$$

Answer: $\dfrac{3}{10} - \dfrac{2}{10} = \dfrac{1}{10}$

In Examples 1 and 2, the fractions had common denominators. But what happens in a situation when the denominators are different, like $\dfrac{2}{5} + \dfrac{1}{3}$? We can't say that this would be equal to $\dfrac{3}{8}$, because that would be the result of $\dfrac{2}{8} + \dfrac{1}{8}$, and that's not the problem we have. Likewise, it can't be $\dfrac{3}{5}$ or $\dfrac{3}{3}$, because those would be the results of $\dfrac{2}{5} + \dfrac{1}{5}$, and $\dfrac{2}{3} + \dfrac{1}{3}$, respectively. We don't have those problems either. There must be another way.

To Add or Subtract Fractions with Unlike Denominators:

1. Change each fraction to an equivalent fraction with a common denominator.
2. Add or subtract the numerators.
3. Keep the common denominator.
4. Reduce whenever possible.

➥ Example 3

Each month, Jessa spends $\frac{2}{5}$ of her total income on rent. She also spends $\frac{1}{3}$ of her total income on bills and other expenses. After paying her rent, bills, and expenses, what fraction of her income does she have left?

(A) $\frac{1}{15}$

(B) $\frac{4}{15}$

(C) $\frac{3}{8}$

(D) $\frac{11}{15}$

Solution

First, we need to find the total fraction of her income that she spends on rent, bills, and other expenses combined. To do that, we have to add both fractions together. However, $\frac{2}{5}$ and $\frac{1}{3}$ have different denominators.

First, find a common denominator. To find a common denominator for $\frac{2}{5} + \frac{1}{3}$, we list the multiples of 5 and 3.

$$\text{Multiples of 5:} \quad 5, 10, \textcircled{15}, 20, 25, 30,\dots$$
$$\text{Multiples of 3:} \quad 3, 6, 9, 12, \textcircled{15}, 18, 21, 24, 27, 30,\dots$$

Fifteen is the first multiple we find in both lists, so we will use 15 as the common denominator. You may notice that 30 is also a common multiple, so we could also use 30. However, 15 is what we call the **lowest common denominator** or **LCD**. Using the LCD is not mandatory, but it will usually save us from having to reduce at the end.

To make finding equivalents and adding the two fractions easier, you can write the operation vertically.

$$
\begin{array}{l}
\frac{2}{5} \implies \frac{2}{5} \times \frac{3}{3} \implies \frac{6}{15} \\[2mm]
+\frac{1}{3} \implies \frac{1}{3} \times \frac{5}{5} \implies +\frac{5}{15} \\[1mm]
\hline
\phantom{+\frac{1}{3} \implies \frac{1}{3} \times \frac{5}{5} \implies }\frac{11}{15}
\end{array}
$$

Now that we have established that Jessa spends $\frac{11}{15}$ of her income on rent, bills, and expenses, we have to figure out how much she has left over. To do this we have to consider that her entire income is considered 1 whole income. To find what's left over, we essentially have to subtract what she spends, the part, from 1. This leads us to the expression $1 - \frac{11}{15}$. We must again have common denominators to complete this subtraction problem. Since we need to write 1 as a fraction with a denominator of 15, we can change 1 to $\frac{1}{1} = \frac{15}{15}$.

Line the problem up vertically and subtract.

$$1 \Rightarrow \frac{15}{15}$$
$$-\frac{11}{15} \Rightarrow -\frac{11}{15}$$
$$\frac{4}{15}$$

<u>Answer:</u> The correct answer is choice B.

ACCUPLACER TIPS

To find a common denominator, we can always find the product of two denominators. For example, if you needed to add $\frac{1}{6} + \frac{1}{4}$, you could multiply the denominators, 6 and 4, to choose a common denominator. Since $6 \times 4 = 24$, the number 24 could serve as a common denominator here. The LCD would be 12, not 24, but 24 will still work.

$$\frac{1}{6} \Rightarrow \frac{4}{24}$$
$$+\frac{1}{4} \Rightarrow +\frac{6}{24}$$
$$\frac{10}{24} = \frac{5}{12}$$

Not using the LCD means that we had to reduce at the end, but that's okay. This technique will also work for finding the common denominator for three or more fractions. For example, consider the expression $\frac{1}{2} + \frac{1}{3} - \frac{1}{5}$. Since $2 \times 3 \times 5 = 30$, the number 30 could serve as a common denominator for this expression.

LESSON 2.4—SKILLS CHECK

Directions: Add or subtract each fraction as directed. Express each sum as a proper fraction or mixed number in simplest form.

1. $\frac{5}{9} + \frac{2}{9}$

2. $\frac{5}{8} - \frac{1}{8}$

3. $\frac{7}{12} + \frac{5}{6}$

4. $\frac{2}{3} + \frac{3}{4}$

5. $\frac{5}{6} - \frac{2}{5}$

6. $\frac{3}{4} + \frac{2}{7}$

7. $\frac{5}{6} - \frac{4}{9}$

8. $\frac{1}{2} + \frac{2}{5} + \frac{3}{4}$

9. $\frac{2}{3} + \frac{5}{8} + \frac{3}{4}$

10. $\frac{3}{8} - \frac{1}{4} + \frac{1}{2}$

(Answers are on page 71.)

LESSON 2.4—ACCUPLACER CHALLENGE

> **Directions:** For each of the questions below, choose the best answer from the four choices given.

1. Theophilus has to drive 90 miles to drop off a canoe. After 60 miles, he stops for gasoline. What fraction of the total distance does he have left to drive?

   (A) $\frac{1}{3}$

   (B) $\frac{2}{3}$

   (C) $\frac{60}{90}$

   (D) 30

2. Samantha had $\frac{3}{4}$ pounds of butter. She used $\frac{1}{2}$ pound to bake a batch of muffins. How much butter was left in pounds?

   (A) $\frac{1}{8}$

   (B) $\frac{1}{4}$

   (C) $\frac{3}{4}$

   (D) $\frac{2}{2}$

3. A bag contains red, green, and yellow candies. If half of the candies are red and one-third of the candies are green, what fraction of the candies is yellow?

   (A) $\frac{1}{6}$

   (B) $\frac{2}{5}$

   (C) $\frac{3}{5}$

   (D) $\frac{5}{6}$

4. In a recent survey of Americans, $\frac{1}{3}$ said that they like to watch soccer more than baseball, and $\frac{3}{5}$ said that they like to watch baseball more than soccer. The rest were undecided. What fraction of the Americans surveyed were undecided?

(A) $\frac{1}{15}$

(B) $\frac{1}{5}$

(C) $\frac{1}{2}$

(D) $\frac{14}{15}$

(Answers are on page 72.)

LESSON 2.5—ADDING AND SUBTRACTING MIXED NUMBERS

When adding and subtracting mixed numbers, especially on the ACCUPLACER, you will find that your work sometimes results in mixed numbers that contain improper fractions, such as $10\frac{9}{8}$. Answer choices will not look like this, so we need to be prepared to deal with situations that end up this way as well as other situations that require working with mixed numbers that contain an improper fraction to find a solution. When we change a number like $10\frac{9}{8}$ to $11\frac{1}{8}$, we call it **regrouping**. We will now look at how to add and subtract mixed numbers and how to deal with regrouping.

To Add or Subtract Mixed Numbers:

1. Change each fraction to an equivalent fraction with a common denominator.
2. Add the whole numbers to the whole numbers and the fractions to the fractions for addition problems.
3. Subtract the whole numbers from the whole numbers and the fractions from the fractions for subtraction problems.
4. Whenever appropriate, regroup improper fractions, and reduce fractions to simplest form.

➥ Example 1

A carpenter is increasing the length of a $10\frac{7}{8}$-inch long piece of wood by adding a $1\frac{1}{4}$-inch long piece of wood to it. What is the total length, in inches, of the piece of wood after the carpenter has joined the two pieces together?

(A) $11\frac{9}{16}$

(B) $11\frac{2}{3}$

(C) 11.98

(D) $12\frac{1}{8}$

Solution

The total length of the two pieces of wood joined together would be found using the expression: $10\frac{7}{8}$ inches + $1\frac{1}{4}$ inches. To do this, we need to first change $1\frac{1}{4}$ to $1\frac{2}{8}$, so that we can have common denominators to make our addition possible. Once we do that, we can then add our mixed numbers together: whole numbers + whole numbers, and fractions + fractions.

Written vertically, those steps look like this:

$$10\frac{7}{8} \Rightarrow 10\frac{7}{8}$$
$$+ \; 1\frac{1}{4} \Rightarrow + \; 1\frac{2}{8}$$
$$\overline{\hspace{2cm} 11\frac{9}{8}}$$

The answer is $11\frac{9}{8}$ inches, but something is wrong. We can't leave the answer as $11\frac{9}{8}$ because the mixed number contains an improper fraction. We still need to regroup the $11\frac{9}{8}$ to a mixed number that does not contain an improper fraction. We know that $\frac{9}{8} = 1\frac{1}{8}$. Since we have $11\frac{9}{8}$, we must combine the 11 we already have with the $1\frac{1}{8}$ that comes from the $\frac{9}{8}$.

$$11\frac{9}{8} = 11 + 1\frac{1}{8} = 12\frac{1}{8}$$

<u>Answer:</u> The correct answer is choice D.

Mixed number subtraction problems, on the other hand, often involve the regrouping of mixed numbers into a mixed number with an improper fraction. When subtracting mixed numbers, we sometimes encounter situations in which the fraction in the top mixed number is less than the fraction in the bottom mixed number. In these cases, we need to borrow from the whole number to make subtraction possible. For example, we may need to regroup $20\frac{1}{2}$ to become $19\frac{3}{2}$. We do this whenever we need to borrow from the whole number.

To Subtract Mixed Numbers Using Borrowing:

1. Write the subtraction problem vertically.
2. Change each fraction to an equivalent with a common denominator.
3. Subtract 1 from the whole number in the mixed number on top and give it to the fraction by adding the value of the denominator to the numerator. Keep the denominator the same.
4. Subtract the whole numbers from the whole numbers and the fractions from the fractions.
5. Reduce whenever possible.

➡ Example 2

Dan bought $20\frac{1}{2}$ square yards of carpet to cover his living room floor. In the end, he used only $18\frac{3}{4}$ square yards. In square yards, how much carpet does he have left over?

Solution

In this problem, we're given the total, the $20\frac{1}{2}$ square yards of carpet he starts with, and a part, the $18\frac{3}{4}$ square yards he uses, and we're asked to find how much is left over.

Before we subtract, we must first change $20\frac{1}{2}$ to $20\frac{2}{4}$ so that we have a common denominator with $18\frac{3}{4}$. Now that we have a plan, we write the problem vertically and then subtract.

$$20\frac{1}{2} \Rightarrow 20\frac{2}{4}$$
$$-18\frac{3}{4} \Rightarrow -18\frac{3}{4}$$

We can't subtract in the usual way since the mixed number on top, $20\frac{2}{4}$, has a smaller numerator than the mixed number on the bottom, $18\frac{3}{4}$. Just like any other time this happens in subtraction, we have to borrow from the larger place to the left.

To complete the subtraction problem that we started, we first borrow 1 from the 20, and make it 19. Next, we add the 1 that we borrowed to the $\frac{2}{4}$ in the form of $\frac{4}{4}$. Finally, we subtract using our new equivalent fractions:

$$20\frac{2}{4} \Rightarrow (20-1)+\frac{4+2}{4} \Rightarrow 19\frac{6}{4}$$
$$-18\frac{3}{4} \xRightarrow{\hspace{2cm}} -18\frac{3}{4}$$
$$\overline{\hspace{4cm} 1\frac{3}{4}}$$

Answer: Therefore, the answer is $1\frac{3}{4}$ yards of fabric are left over.

LESSON 2.5—SKILLS CHECK

> **Directions:** Add or subtract each fraction as directed. Express each sum as a proper fraction or a mixed number in simplest form.

1. $2\frac{1}{5}+3\frac{2}{5}$ 4. $1\frac{1}{2}+5\frac{3}{8}$ 7. $8\frac{5}{6}-2\frac{1}{4}$

2. $2\frac{1}{5}+6\frac{4}{5}$ 5. $3\frac{3}{4}+4\frac{1}{3}$ 8. $2\frac{3}{10}+9\frac{4}{5}$

3. $8\frac{6}{7}-2\frac{2}{7}$ 6. $20\frac{1}{3}-8\frac{2}{3}$ 9. $6\frac{1}{2}-3\frac{1}{3}$

10. $11\frac{1}{4} - 3\frac{2}{5}$ 12. $25\frac{1}{3} - 17\frac{4}{7}$ 14. $6\frac{2}{3} - 3\frac{3}{4}$

11. $5\frac{2}{3} + 3\frac{3}{4}$ 13. $40\frac{3}{4} - 15\frac{7}{8}$

(Answers are on page 72.)

LESSON 2.5—ACCUPLACER CHALLENGE

Directions: For each of the questions below, choose the best answer from the four choices given.

1. A recipe calls for $2\frac{3}{4}$ cups of milk, $\frac{1}{2}$ cup of cooking oil, and $3\frac{1}{2}$ cups of water. In cups, what is the total amount of liquid used in the recipe?

   (A) $5\frac{3}{4}$

   (B) $5\frac{5}{8}$

   (C) $6\frac{1}{4}$

   (D) $6\frac{3}{4}$

2. From a 16-foot pipe, a plumber cut $6\frac{3}{4}$ feet. How many feet long is the remaining piece of pipe?

   (A) $9\frac{1}{4}$

   (B) $9\frac{3}{4}$

   (C) $10\frac{3}{4}$

   (D) $11\frac{1}{4}$

3. Melissa weighed $158\frac{1}{4}$ pounds in October. By December 1, she weighed $5\frac{3}{4}$ pounds less. In pounds, what was her weight on December 1?

   (A) $152\frac{1}{2}$

   (B) $153\frac{1}{4}$

   (C) $153\frac{1}{2}$

   (D) 164

(Answers are on page 74.)

LESSON 2.6—MULTIPLYING FRACTIONS

If you were to get paid $200 for one day of work, then how much should you get paid for a half day of work? Most people would say $100, which seems fair given that $100 is half of $200. There are a couple of ways to write this mathematically. One way would be to write $200 ÷ 2 = $100. To take "half of" something is the same as dividing that something by 2.

Another way to translate "half of" something is to multiply that something by $\frac{1}{2}$. In other words, we're saying that $200 × $\frac{1}{2}$ = $100. We know it must be true, so let's look at how to get there.

To Multiply a Whole Number by a Fraction:

1. Write the whole number as a fraction by placing the number over 1.
2. Multiply the numerator by the numerator, and multiply the denominator by the denominator.
3. Reduce whenever possible.

To find $\frac{1}{2}$ of $200, we translate the "of" to mean multiply, and write $200 over 1.

Then multiply the numerator by the numerator and the denominator by the denominator.

$$\frac{1}{2} \times \frac{\$200}{1} = \frac{\$200}{2}$$

Finally, reduce.

$$\frac{\$200}{2} = \frac{\$100}{1} = \$100$$

$\frac{1}{2}$ of $200 is $100.

What about $\frac{1}{2}$ of $\frac{1}{3}$? Again, "$\frac{1}{2}$ of" translates to "$\frac{1}{2}$ ×." Therefore, we're essentially trying to find $\frac{1}{2} \times \frac{1}{3}$. To help illustrate this, we can look at a square that measures one foot on each side. If we divide the square in half horizontally, and into thirds vertically, we end up with six smaller rectangles. It appears that the original square has been split into six equal pieces, or sixths. Here's where it gets exciting. To find the area of one of the smaller rectangles, we multiply length × width. Here, that would be $\frac{1}{2} \times \frac{1}{3}$. We know we split the larger square into sixths and it works out! $\frac{1}{2} \times \frac{1}{3} = \frac{1}{6}$.

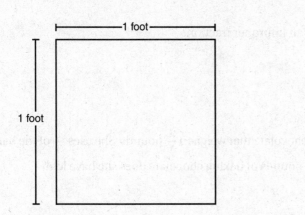

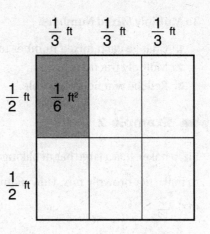

We can now write our rule for multiplying fractions.

To Multiply Fractions:

1. Multiply the numerator by the numerator.
2. Multiply the denominator by the denominator.
3. Reduce when possible.

➡ Example 1

$\frac{3}{5}$ of $\frac{1}{4}$ is ?

(A) $\frac{3}{20}$

(B) $\frac{4}{9}$

(C) $1\frac{5}{12}$

(D) 3

Solution

As we saw earlier, to find a fraction of another quantity, we just need to multiply the two. We need to multiply the numerators, multiply the denominators, and then reduce.

$$\frac{3}{5} \times \frac{1}{4} = \frac{3}{20}$$

In this case, we can't reduce the fraction $\frac{3}{20}$, so we're finished.

Answer: The correct answer is choice A.

To Multiply Mixed Numbers:

1. Change each mixed number to an improper fraction.
2. Multiply fractions across.
3. Reduce whenever possible.

➡ Example 2

Liz, a baker, has a large bar of baking chocolate that weighs $1\frac{3}{4}$ pounds. She uses $\frac{1}{3}$ of the bar to make her brownie mix. How many pounds of baking chocolate does she have left?

(A) $\frac{7}{12}$

(B) $\frac{2}{3}$

(C) $1\frac{1}{6}$

(D) $1\frac{7}{12}$

Solution

In this problem, you are asked to find how much chocolate Liz has left after using $\frac{1}{3}$ of her baking chocolate bar. If you immediately think that 1 whole chocolate bar $-$ $\frac{1}{3}$ chocolate bar $= \frac{2}{3}$ chocolate bar, then you are thinking wisely. You do need to subtract to find out how much she has left, and $\frac{2}{3}$ of the bar remains. Be careful, though, and remember that you are asked to find how many *pounds* she has left.

Having figured out that Liz has $\frac{2}{3}$ of the chocolate bar left, we need to now figure out what $\frac{2}{3}$ of the chocolate bar weighs in pounds. Considering that the whole chocolate bar started out as $1\frac{3}{4}$ pounds, we need to figure out what $\frac{2}{3}$ of $1\frac{3}{4}$ pounds is.

To find $\frac{2}{3}$ of $1\frac{3}{4}$ pounds:

First, translate "of" to "multiply" and write an expression:

$$\frac{2}{3} \times 1\frac{3}{4}$$

Next, change $1\frac{3}{4}$ to a mixed number:

$$1\frac{3}{4} = \frac{(4 \times 1) + 3}{4} = \frac{7}{4}$$

Rewrite the expression:

$$\frac{2}{3} \times \frac{7}{4}$$

Finally, multiply across and then reduce:

$$\frac{2}{3} \times \frac{7}{4} = \frac{14}{12} = \frac{7}{6} = 1\frac{1}{6}$$

<u>Answer:</u> Therefore, the correct answer is choice C.

LESSON 2.6—SKILLS CHECK

Directions: Express the product as a fraction reduced to its simplest form. Change improper fractions to mixed numbers or whole numbers where appropriate.

1. $\frac{2}{3} \times \frac{4}{5}$

2. $\frac{3}{4} \times \frac{2}{7}$

3. $\frac{7}{5} \times \frac{3}{4}$

4. $\frac{8}{15} \times \frac{5}{4}$

5. $\frac{5}{2} \times \frac{2}{5}$

6. $3\frac{2}{5} \times 1\frac{3}{4}$

7. $2\frac{2}{5} \times 2\frac{1}{2}$

8. $2\frac{3}{4} \times 6$

9. $3\frac{1}{9} \times 2\frac{1}{4}$

10. $\frac{4}{7}$ of $1\frac{2}{5} = ?$

(Answers are on page 74.)

LESSON 2.6—ACCUPLACER CHALLENGE

Directions: For each of the questions below, choose the best answer from the four choices given.

1. Five hundred questionnaires were sent out asking people their reaction to a plan to build a new cement plant in their community. In one month, about $\frac{1}{6}$ of the questionnaires were returned. Approximately how many of the questionnaires were returned in one month?

   (A) 40
   (B) 80
   (C) 120
   (D) 150

2. Last year, approximately $\frac{1}{5}$ of Javier's income was withheld for federal taxes, and approximately $\frac{1}{8}$ of his income was withheld for state taxes. Last year, approximately how much money was withheld from Javier's income in total if he made $80,000?

(A) $10,000

(B) $12,000

(C) $16,000

(D) $26,000

3. Erin bought $5\frac{1}{2}$ yards of fabric. How much fabric does she have left after using $\frac{2}{3}$ of it to make curtains?

(A) $\frac{1}{3}$ yd

(B) $1\frac{5}{6}$ yd

(C) $3\frac{2}{3}$ yd

(D) $4\frac{5}{6}$ yd

(Answers are on page 74.)

LESSON 2.7—DIVIDING FRACTIONS

It's important for the discussion on how to divide fractions that we introduce the idea of a **reciprocal**. Two numbers are reciprocals of one another when their product is 1. Some examples are:

$$\frac{1}{2} \times \frac{2}{1} = \frac{2}{2} = 1$$

$$\frac{3}{4} \times \frac{4}{3} = \frac{12}{12} = 1$$

$$\frac{9}{5} \times \frac{5}{9} = \frac{45}{45} = 1$$

These numbers are also called **multiplicative inverses**. The less mathematical definition would be the **flip** of a fraction. The reciprocal of $\frac{2}{3}$ is $\frac{3}{2}$. To write the reciprocal of a whole number, such as 8, we must write 8 as a fraction by placing it over 1, and then we *flip* that fraction to find the reciprocal. In other words, the reciprocal of $\frac{8}{1}$ is $\frac{1}{8}$.

This brings us back to our pizzas. If there are 8 slices in a pizza pie, then there are 16 slices in two pies.

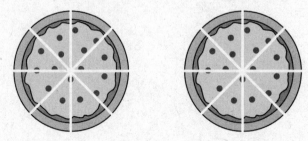

Earlier, we called these slices $\frac{1}{8}$ of a pizza. Here, we could also represent the question of how many slices are in 2 pizza pies as how many eighths are in 2. This would be written as $2 \div \frac{1}{8}$. Well, we already know that there are 8 slices in 1 of our pizzas, and 16 slices in 2 of our pizzas, so we already know that the answer is 16. Therefore, we have just established that $2 \div \frac{1}{8} = 16$ and $2 \times 8 = 16$.

If both expressions equal 16, then both expressions must equal one another. In other words, $2 \div \frac{1}{8} = 2 \times 8$.

Making our whole numbers into fractions by placing them over 1, we can more easily see the relationship of the two sides of the equation.

$$\frac{2}{1} \div \frac{1}{8} = \frac{2}{1} \times \frac{8}{1}$$

$$16 = 16$$

If we compare the left side of the equation to the right, we can see that the $\frac{2}{1}$ stayed the same, the $\div$ became $\times$, and $\frac{1}{8}$ flipped to its reciprocal $\frac{8}{1}$.

Now we have our next rule.

To Divide Fractions:

1. Keep the first fraction.
2. Change the $\div$ to $\times$.
3. Flip the second fraction to its reciprocal.
4. Multiply across.
5. Reduce whenever possible.

This is also known as **Keep**, **Change**, and **Flip** or **KCF**.

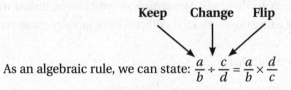

As an algebraic rule, we can state: $\dfrac{a}{b} \div \dfrac{c}{d} = \dfrac{a}{b} \times \dfrac{d}{c}$

➡ Example 1

If $\frac{9}{5} \div \frac{1}{2} = p$, then which of the following inequalities is true of p?

(A) $3 < p < 4$
(B) $4 < p < 5$
(C) $5 < p < 6$
(D) $6 < p < 7$

Solution

First, we rewrite the division problem as a multiplication problem using **Keep**, **Change**, and **Flip**.

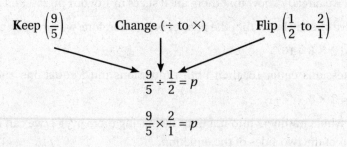

Then, we multiply across.

$$p = \frac{9}{5} \times \frac{2}{1} = \frac{18}{5}$$

Finally, we change $\frac{18}{5}$ to a mixed number.

$$p = \frac{18}{5} = 3\frac{3}{5}$$

We know that $3\frac{3}{5}$ is larger than 3, but smaller than 4. Remember that when using the < (less than) and > (greater than) symbols, the open side of the symbol always faces the greater of the two quantities.

<u>Answer:</u> Therefore, the correct answer is choice A. We can read this as p is between 3 and 4, or p is greater than 3, but less than 4.

Dividing Mixed Numbers

Using mixed numbers to divide is messy unless we change mixed numbers to improper fractions and divide using the same KCF skills we have already mastered.

To Divide with Mixed Numbers:

1. Change mixed numbers to improper fractions.
2. Change whole numbers to fractions by placing them over 1.
3. Follow the rules for dividing fractions.

➡ Example 2

Charlene needs $3\frac{1}{4}$ yards of fabric to make a coat. How many coats can she make from $15\frac{1}{2}$ yards of material?

(A) 2

(B) 3

(C) 4

(D) 5

Solution

Since Charlene is literally cutting up the whole $15\frac{1}{2}$ yards of fabric into equal parts, the $3\frac{1}{4}$ yards of fabric needed to make a coat, we are talking about division. Remember, to divide means to cut, or split, something into equal pieces.

First, write an expression.

$$15\frac{1}{2} \div 3\frac{1}{4}$$

Next, change each mixed number to an improper fraction and rewrite the expression.

$$\frac{31}{2} \div \frac{13}{4}$$

Now, change the division problem to a multiplication problem using **Keep**, **Change**, and **Flip**.

$$\frac{31}{2} \times \frac{4}{13}$$

Multiply across.

$$\frac{31}{2} \times \frac{4}{13} = \frac{31 \times 4}{2 \times 13} = \frac{124}{26}$$

Change the resulting improper fraction to a mixed number.

$$4\frac{20}{26}$$

(Note that we can reduce $4\frac{20}{26}$ to $4\frac{10}{13}$, but it won't change the answer here.)

Then, having determined our quotient, we need to now consider our answer choices. All of the choices are whole numbers, but this makes sense. The question asked how many coats Charlene could make. She's closest to making 5 coats, but a part of a coat is no good to her. Therefore, she can only make 4 whole coats.

<u>Answer:</u> The correct answer is choice C.

LESSON 2.7—SKILLS CHECK

> **Directions:** Find the product or quotient as directed.

1. $\dfrac{2}{3} \div \dfrac{2}{3}$

2. $\dfrac{1}{2} \div \dfrac{2}{3}$

3. $\dfrac{5}{7} \div \dfrac{3}{14}$

4. $\dfrac{7}{8} \div \dfrac{1}{8}$

5. $\dfrac{14}{48} \div \dfrac{7}{8}$

6. $2\dfrac{1}{3} \div 3\dfrac{4}{9}$

7. $3\dfrac{3}{4} \div 1\dfrac{2}{3}$

8. $16 \div \dfrac{4}{5}$

9. $4\dfrac{1}{2} \div \dfrac{1}{8}$

10. Find $\dfrac{2}{3} \div 1\dfrac{1}{2} \times 5\dfrac{2}{3}$.

(Answers are on page 75.)

LESSON 2.7—ACCUPLACER CHALLENGE

> **Directions:** For each of the questions below, choose the best answer from the four choices given.

1. $\dfrac{8}{9} \div \dfrac{4}{3}$ is equal to

   (A) $\dfrac{8 \div 4}{9 \div 3}$

   (B) $\dfrac{8 \times 4}{9 \times 3}$

   (C) $\dfrac{8 \times 3}{9 \times 4}$

   (D) $\dfrac{4}{3} : \dfrac{8}{9}$

2. Takehiro needs to build a 6-foot-high wall out of bricks that are each $\dfrac{2}{3}$ of a foot in height. How many bricks will Takehiro need?

   (A) 6

   (B) 7

   (C) 8

   (D) 9

3. A math tutor in Alaska needs $1\frac{2}{3}$ hours to tutor a student and then drive to the next student. What is the maximum number of students this tutor can schedule in a night if he has $5\frac{1}{2}$ hours available for tutoring?

(A) 3
(B) 4
(C) 5
(D) 6

4. Anna spent $41.40 for $4\frac{1}{2}$ yards of material. What was the cost of one yard of the material?

(A) $2.30
(B) $4.60
(C) $8.60
(D) $9.20

(Answers are on page 75.)

To complete a set of practice questions that reviews all of Chapter 2, go to:
barronsbooks.com/TP/accuplacer/math29qw/

Lesson 2.1—Skills Check (page 40)

1. $\frac{1}{5} = \frac{4}{20}$ $5 \times 4 = 20$. Multiply $\frac{1}{5} \times \frac{4}{4} = \frac{1 \times 4}{5 \times 4} = \frac{4}{20}$.

2. $\frac{2}{5} = \frac{8}{20}$ $5 \times 4 = 20$. Multiply $\frac{2}{5} \times \frac{4}{4} = \frac{2 \times 4}{5 \times 4} = \frac{8}{20}$.

3. $\frac{3}{5} = \frac{12}{20}$ $5 \times 4 = 20$. Multiply $\frac{3}{5} \times \frac{4}{4} = \frac{3 \times 4}{5 \times 4} = \frac{12}{20}$.

4. $\frac{4}{5} = \frac{16}{20}$ $5 \times 4 = 20$. Multiply $\frac{4}{5} \times \frac{4}{4} = \frac{4 \times 4}{5 \times 4} = \frac{16}{20}$.

5. $\frac{5}{5} = \frac{20}{20}$ $5 \times 4 = 20$. Multiply $\frac{5}{5} \times \frac{4}{4} = \frac{5 \times 4}{5 \times 4} = \frac{20}{20}$.

6. $\frac{1}{2} = \frac{6}{12}$ $2 \times 6 = 12$. Multiply $\frac{1}{2} \times \frac{6}{6} = \frac{1 \times 6}{2 \times 6} = \frac{6}{12}$.

7. $\frac{3}{5} = \frac{6}{10}$ $5 \times 2 = 10$. Multiply $\frac{3}{5} \times \frac{2}{2} = \frac{3 \times 2}{5 \times 2} = \frac{6}{10}$.

8. $\frac{2}{5} = \frac{12}{30}$ $2 \times 6 = 12$. Multiply $\frac{2}{5} \times \frac{6}{6} = \frac{2 \times 6}{5 \times 6} = \frac{12}{30}$.

9. $\frac{2}{3} = \frac{14}{21}$ $2 \times 7 = 14$. Multiply $\frac{2}{3} \times \frac{7}{7} = \frac{2 \times 7}{3 \times 7} = \frac{14}{21}$.

10. $\frac{5}{6} = \frac{15}{18}$ $6 \times 3 = 18$. Multiply $\frac{5}{6} \times \frac{3}{3} = \frac{5 \times 3}{6 \times 3} = \frac{15}{18}$.

11. $\frac{3}{12} = \frac{1}{4}$ The GCF is 3. Divide $\frac{3}{12} \div \frac{3}{3} = \frac{3 \div 3}{12 \div 3} = \frac{1}{4}$.

12. $\frac{5}{15} = \frac{1}{3}$ The GCF is 5. Divide $\frac{5}{15} \div \frac{5}{5} = \frac{5 \div 5}{15 \div 5} = \frac{1}{3}$.

13. $\frac{12}{18} = \frac{2}{3}$ The GCF is 6. Divide $\frac{12}{18} \div \frac{6}{6} = \frac{12 \div 6}{18 \div 6} = \frac{2}{3}$.

14. $\frac{24}{36} = \frac{2}{3}$ The GCF is 12. Divide $\frac{24}{36} \div \frac{12}{12} = \frac{24 \div 12}{36 \div 12} = \frac{2}{3}$.

15. $\frac{175}{225} = \frac{7}{9}$ The GCF is 25. Divide $\frac{175}{225} \div \frac{25}{25} = \frac{175 \div 25}{225 \div 25} = \frac{7}{9}$.

16. $\frac{56}{140} = \frac{2}{5}$ The GCF is 28. Divide $\frac{56}{140} \div \frac{28}{28} = \frac{56 \div 28}{140 \div 28} = \frac{2}{5}$.

Lesson 2.1—ACCUPLACER Challenge (pages 40-41)

1. **(C)** The question asks, "What fraction of their games did they win?" The number of games becomes the denominator (whole), and the wins become the numerator (part). We write $\frac{8}{12}$ and then reduce. $\frac{8}{12} \div \frac{4}{4} = \frac{2}{3}$.

2. **(C)** The whole box contains 32 candies, so 32 becomes the denominator, and the number of candies he ate, which is what we need to find, becomes the numerator. We know that he ate $\frac{3}{4}$ of the box, so we need to raise $\frac{3}{4}$ to an equivalent fraction with 32 as the denominator. $\frac{3}{4} \times \frac{8}{8} = \frac{3 \times 8}{4 \times 8} = \frac{24}{32}$.

3. **(D)** The whole trip is 60 miles, and 45 miles is the part of the trip driven so far. The fraction of the total distance driven is $\frac{45}{60}$. Since this is not an answer choice, we must reduce to find an equivalent answer. $\frac{45}{60} \div \frac{15}{15} = \frac{45 \div 15}{60 \div 15} = \frac{3}{4}$.

4. **(D)** The question asks, "About what fraction <u>of their games</u> did they lose?" The number of games becomes the denominator (whole), and the losses become the numerator (part). We write $\frac{7}{22}$. This is not an answer choice, and, more importantly, the answer asks, "About what fraction." Therefore, we need to find the answer choice that is closest to being equivalent to $\frac{7}{22}$. The fraction with the closest equivalent is $\frac{1}{3} \times \frac{7}{7} = \frac{7}{21}$.

Lesson 2.2—Skills Check (page 44)

1. $\frac{1}{3} > \frac{1}{4}$ Use equivalent fractions with a common denominator. $\frac{4}{12} > \frac{3}{12}$; therefore, $\frac{1}{3} > \frac{1}{4}$.

2. $\frac{3}{4} < \frac{7}{8}$ Use equivalent fractions with a common denominator. $\frac{24}{32} < \frac{28}{32}$; therefore, $\frac{3}{4} < \frac{7}{8}$.

3. $\frac{3}{9} = \frac{1}{3}$ Reducing $\frac{3}{9}$ yields $\frac{1}{3}$; therefore, $\frac{3}{9} = \frac{1}{3}$.

4. $\frac{2}{3} > \frac{1}{2}$ Use equivalent fractions with a common denominator. $\frac{4}{6} > \frac{3}{6}$; therefore, $\frac{2}{3} > \frac{1}{2}$.

5. $\frac{5}{6} = \frac{15}{18}$ Reducing $\frac{15}{18}$ yields $\frac{5}{6}$; therefore, $\frac{5}{6} = \frac{15}{18}$.

6. $\frac{9}{12} = \frac{3}{4}$ Reducing $\frac{9}{12}$ yields $\frac{3}{4}$; therefore, $\frac{9}{12} = \frac{3}{4}$.

7. $\frac{7}{10} > \frac{2}{3}$ Use equivalent fractions with a common denominator. $\frac{21}{30} > \frac{20}{30}$; therefore, $\frac{7}{10} > \frac{2}{3}$.

8. $\frac{7}{15} > \frac{2}{5}$ Use equivalent fractions with a common denominator. $\frac{35}{75} > \frac{30}{75}$; therefore, $\frac{7}{15} > \frac{2}{5}$.

9. $\frac{9}{10} > \frac{3}{4}$ Use equivalent fractions with a common denominator. $\frac{36}{40} > \frac{30}{40}$; therefore, $\frac{9}{10} > \frac{3}{4}$.

10. $\frac{1}{6}, \frac{1}{3}, \frac{1}{2}, \frac{2}{3}, \frac{5}{6}$ Change all the fractions to equivalents with a common denominator. Then order them according to their numerators from least to greatest:

$$\frac{1}{6}, \frac{1}{3} = \frac{2}{6}, \frac{1}{2} = \frac{3}{6}, \frac{2}{3} = \frac{4}{6}, \frac{5}{6}$$

11. $\frac{1}{12}, \frac{1}{6}, \frac{1}{4}, \frac{1}{3}, \frac{1}{2}$ Change all the fractions to equivalents with a common denominator. Then order them according to their numerators from least to greatest:

$$\frac{1}{12}, \frac{1}{6} = \frac{2}{12}, \frac{1}{4} = \frac{3}{12}, \frac{1}{3} = \frac{4}{12}, \frac{1}{2} = \frac{6}{12}$$

12. $\frac{1}{2}, \frac{5}{8}, \frac{2}{3}, \frac{3}{4}, \frac{5}{6}$ Change all the fractions to equivalents with a common denominator. Then order according to their numerators from least to greatest:

$$\frac{1}{2} = \frac{12}{24}, \frac{5}{8} = \frac{15}{24}, \frac{2}{3} = \frac{16}{24}, \frac{3}{4} = \frac{18}{24}, \frac{5}{6} = \frac{20}{24}$$

Lesson 2.2—ACCUPLACER Challenge (pages 44–45)

1. **(C)** The $\frac{3}{8}$-inch wrench is too small, and the $\frac{1}{2}$-inch wrench is too large. We need a wrench that is between $\frac{3}{8}$ and $\frac{1}{2}$. To compare the wrenches, we can change all our fractions to equivalents with an LCD of 32. $\frac{3}{8} = \frac{12}{32}$ and $\frac{1}{2} = \frac{16}{32}$. Only $\frac{15}{32}$ is greater than $\frac{3}{8} = \frac{12}{32}$ and less than $\frac{1}{2} = \frac{16}{32}$.

2. **(A)** To order each fraction, we change each to an equivalent fraction with a common denominator of 12. Then we order the fractions from least to greatest:

$$\frac{1}{2} = \frac{6}{12}, \frac{7}{12}, \frac{2}{3} = \frac{8}{12}, \frac{3}{4} = \frac{9}{12}, \frac{5}{6} = \frac{10}{12}$$

3. **(B)** To order each fraction, we change each to an equivalent fraction with a common denominator of $7 \times 8 \times 3 \times 5 = 840$. Then we order the fractions from least to greatest:

$$\frac{2}{7} = \frac{240}{840}, \frac{1}{3} = \frac{280}{840}, \frac{3}{5} = \frac{504}{840}, \frac{5}{8} = \frac{525}{840}$$

Lesson 2.3—Skills Check (page 48)

1. $2\frac{4}{5}$ 5 goes into 14 two times with a remainder of 4. $\frac{14}{5} = 2\frac{4}{5}$.

2. $1\frac{3}{4}$ 8 goes into 14 one time with a remainder of 6; then reduce. $\frac{14}{8} = 1\frac{6}{8} = 1\frac{3}{4}$.

3. $5\frac{1}{2}$ 6 goes into 33 five times with a remainder of 3; then reduce. $\frac{33}{6} = 5\frac{3}{6} = 5\frac{1}{2}$.

4. 4 3 goes into 12 four times with no remainder. $\frac{12}{3} = 4$.

5. $3\frac{1}{3}$ 9 goes into 30 three times with a remainder of 3; then reduce. $\frac{30}{9} = 3\frac{3}{9} = 3\frac{1}{3}$.

6. $2\frac{3}{4}$ 16 goes into 44 two times with a remainder of 12; then reduce. $\frac{44}{16} = 2\frac{12}{16} = 2\frac{3}{4}$.

7. $\frac{11}{4}$ $\frac{(4 \times 2) + 3}{4} = \frac{8 + 3}{4} = \frac{11}{4}$

8. $\frac{11}{7}$ $\frac{(7 \times 1) + 4}{7} = \frac{7 + 4}{7} = \frac{11}{7}$

9. $\frac{16}{3}$ $\frac{(3 \times 5) + 1}{3} = \frac{15 + 1}{3} = \frac{16}{3}$

10. $\frac{44}{7}$ $\frac{(7 \times 6) + 2}{7} = \frac{42 + 2}{7} = \frac{44}{7}$

11. $\frac{19}{2}$ $\frac{(2 \times 9) + 1}{2} = \frac{18 + 1}{2} = \frac{19}{2}$

12. $\frac{35}{4}$ $\frac{(4 \times 8) + 3}{4} = \frac{32 + 3}{4} = \frac{35}{4}$

Lesson 2.3—ACCUPLACER Challenge (pages 48–49)

1. **(A)** With twelve cuts, you will lose $12 \times \frac{1}{8} = \frac{12}{1} \times \frac{1}{8} = \frac{12}{8} = 1\frac{4}{8} = 1\frac{1}{2}$ inches.

2. **(A)** Changing each of the answer choices from mixed numbers to improper fractions, we see that they are all equal to $\frac{27}{5}$ except for $1\frac{20}{7}$, which equals $\frac{27}{7}$.

3. **(C)** Each day of the workweek, Lucy travels $\frac{1}{3} + \frac{1}{3} = \frac{2}{3}$ miles from home to work and back. She does this for five days, so she will travel $5 \times \frac{2}{3} = \frac{5}{1} \times \frac{2}{3} = \frac{10}{3} = 3\frac{1}{3}$ miles in total.

4. **(A)** To compare, we need to put the numbers in order from least to greatest by first changing them all to equivalent improper fractions with a common denominator. The LCD is 30, so we have: $3\frac{1}{5} = \frac{96}{30}, \frac{10}{3} = \frac{100}{30}, \frac{7}{2} = \frac{105}{30}, 3\frac{2}{3} = \frac{110}{30}$.

Lesson 2.4—Skills Check (page 52)

1. $\frac{7}{9}$ $\frac{5}{9} + \frac{2}{9} = \frac{5 + 2}{9} = \frac{7}{9}$

2. $\frac{1}{2}$ $\frac{5}{8} - \frac{1}{8} = \frac{5 - 1}{8} = \frac{4}{8} = \frac{1}{2}$

3. $\frac{17}{12}$ or $1\frac{5}{12}$ $\frac{7}{12} + \frac{5}{6} = \frac{7}{12} + \frac{10}{12} = \frac{17}{12} = 1\frac{5}{12}$

4. $\frac{17}{12}$ or $1\frac{5}{12}$ $\frac{2}{3} + \frac{3}{4} = \frac{8}{12} + \frac{9}{12} = \frac{17}{12} = 1\frac{5}{12}$

5. $\frac{13}{30}$ $\frac{5}{6} - \frac{2}{5} = \frac{25}{30} - \frac{12}{30} = \frac{13}{30}$

6. $\frac{29}{28}$ or $1\frac{1}{28}$ $\frac{3}{4} + \frac{2}{7} = \frac{21}{28} + \frac{8}{28} = \frac{29}{28} = 1\frac{1}{28}$

7. $\frac{7}{18}$ $\frac{5}{6} - \frac{4}{9} = \frac{15}{18} - \frac{8}{18} = \frac{7}{18}$

8. $\dfrac{33}{20}$ or $1\dfrac{13}{20}$ $\dfrac{1}{2}+\dfrac{2}{5}+\dfrac{3}{4}=\dfrac{10}{20}+\dfrac{8}{20}+\dfrac{15}{20}=\dfrac{33}{20}=1\dfrac{13}{20}$

9. $\dfrac{49}{24}$ or $2\dfrac{1}{24}$ $\dfrac{2}{3}+\dfrac{5}{8}+\dfrac{3}{4}=\dfrac{16}{24}+\dfrac{15}{24}+\dfrac{18}{24}=\dfrac{49}{24}=2\dfrac{1}{24}$

10. $\dfrac{5}{8}$ $\dfrac{3}{8}-\dfrac{1}{4}+\dfrac{1}{2}=\dfrac{3}{8}-\dfrac{2}{8}+\dfrac{4}{8}=\dfrac{5}{8}$

Lesson 2.4—ACCUPLACER Challenge (pages 53-54)

1. **(A)** Theophilus drove 60 miles out of 90 miles, so he has $90-60=30$ miles left to drive. As a fraction, he has $\dfrac{30}{90}=\dfrac{1}{3}$ of the total distance left to drive.

2. **(B)** To find how much was left, we subtract $\dfrac{3}{4}-\dfrac{1}{2}$. To do so, we change each fraction to equivalent fractions with common denominators.

$$\dfrac{3}{4}-\dfrac{1}{2}=\dfrac{3}{4}-\dfrac{2}{4}=\dfrac{1}{4}$$

3. **(A)** To determine the fraction of candies that are yellow, we need to first determine the fraction of the bag that is not yellow, which is the fraction of candies that are red plus the fraction of candies that are green. $\dfrac{1}{2}+\dfrac{1}{3}=\dfrac{3}{6}+\dfrac{2}{6}=\dfrac{5}{6}$. The rest of the candies are yellow, which can be found by subtracting from 1, which represents the whole bag.

$$1-\dfrac{5}{6}=\dfrac{6}{6}-\dfrac{5}{6}=\dfrac{1}{6}$$

4. **(A)** To determine the fraction of Americans who were undecided, we need to first determine the fraction of Americans who were decided, which is the fraction of Americans who said they prefer soccer plus the fraction of Americans who said they prefer baseball. $\dfrac{1}{3}+\dfrac{3}{5}=\dfrac{5}{15}+\dfrac{9}{15}=\dfrac{14}{15}$. The rest of the Americans who were surveyed were undecided, which can be found by subtracting from 1, which represents the whole survey.

$$1-\dfrac{14}{15}=\dfrac{15}{15}-\dfrac{14}{15}=\dfrac{1}{15}$$

Lesson 2.5—Skills Check (pages 56-57)

1. $5\dfrac{3}{5}$

2. **9**

3. $6\frac{4}{7}$

$$8\frac{6}{7}$$
$$- \ 2\frac{2}{7}$$
$$6\frac{4}{7}$$

9. $3\frac{1}{6}$

$$6\frac{1}{2} \ \Rightarrow \ 6\frac{3}{6}$$
$$-3\frac{1}{3} \Rightarrow - \ 3\frac{2}{6}$$
$$3\frac{1}{6}$$

4. $6\frac{7}{8}$

$$1\frac{1}{2} \ \Rightarrow \ \ 1\frac{4}{8}$$
$$+ \ 5\frac{3}{8} \Rightarrow + \ 5\frac{3}{8}$$
$$6\frac{7}{8}$$

10. $7\frac{17}{20}$

$$11\frac{1}{4} \ \Rightarrow \ 11\frac{5}{20} \Rightarrow 10\frac{25}{20}$$
$$- \ 3\frac{2}{5} \ \Rightarrow \ - \ 3\frac{8}{20} \Rightarrow - 3\frac{8}{20}$$
$$7\frac{17}{20}$$

5. $8\frac{1}{12}$

$$3\frac{3}{4} \ \Rightarrow \ \ \ 3\frac{9}{12}$$
$$+ \ 4\frac{1}{3} \ \Rightarrow \ + \ 4\frac{4}{12}$$
$$7\frac{13}{12} \Rightarrow 8\frac{1}{12}$$

11. $9\frac{5}{12}$

$$5\frac{2}{3} \ \Rightarrow \ 5\frac{8}{12}$$
$$+ \ 3\frac{3}{4} \ \Rightarrow \ +3\frac{9}{12}$$
$$8\frac{17}{12} \Rightarrow 9\frac{5}{12}$$

6. $11\frac{2}{3}$

$$20\frac{1}{3} \ \Rightarrow \ 19\frac{4}{3}$$
$$- \ 8\frac{2}{3} \ \Rightarrow \ - \ 8\frac{2}{3}$$
$$11\frac{2}{3}$$

12. $7\frac{16}{21}$

$$25\frac{1}{3} \ \Rightarrow \ 25\frac{7}{21} \ \Rightarrow \ 24\frac{28}{21}$$
$$- 17\frac{4}{7} \ \Rightarrow \ -17\frac{12}{21} \ \Rightarrow \ -17\frac{12}{21}$$
$$7\frac{16}{21}$$

7. $6\frac{7}{12}$

$$8\frac{5}{6} \ \Rightarrow \ \ 8\frac{10}{12}$$
$$- 2\frac{1}{4} \ \Rightarrow \ - \ 2\frac{3}{12}$$
$$6\frac{7}{12}$$

13. $24\frac{7}{8}$

$$40\frac{3}{4} \ \Rightarrow \ 40\frac{6}{8} \ \Rightarrow \ 39\frac{14}{8}$$
$$-15\frac{7}{8} \ \Rightarrow \ -15\frac{7}{8} \ \Rightarrow \ -15\frac{7}{8}$$
$$24\frac{7}{8}$$

8. $12\frac{1}{10}$

$$2\frac{3}{10} \ \Rightarrow \ 2\frac{3}{10}$$
$$+ \ 9\frac{4}{5} \ \Rightarrow \ + \ 9\frac{8}{10}$$
$$11\frac{11}{10} \ \Rightarrow 12\frac{1}{10}$$

14. $2\frac{11}{12}$

$$6\frac{2}{3} \ \Rightarrow \ \ 6\frac{8}{12} \ \Rightarrow \ 5\frac{20}{12}$$
$$- 3\frac{3}{4} \ \Rightarrow \ -3\frac{9}{12} \ \Rightarrow \ -3\frac{9}{12}$$
$$2\frac{11}{12}$$

Lesson 2.5—ACCUPLACER Challenge (page 57)

1. **(D)** To find the total amount used, we add using equivalent fractions with a common denominator. $2\frac{3}{4} + \frac{1}{2} + 3\frac{1}{2} = 2\frac{3}{4} + \frac{2}{4} + 3\frac{2}{4} = 5\frac{7}{4}$. We must regroup the $\frac{7}{4}$, because it's improper. $5\frac{7}{4} = 6\frac{3}{4}$.

2. **(A)** To find the remaining piece, we subtract the part, $6\frac{3}{4}$ feet, from the whole 16–foot pipe. To do so, we must borrow 1 from the 16 to have a fraction from which to subtract the $\frac{3}{4}$. $16 - 6\frac{3}{4} = 15\frac{4}{4} - 6\frac{3}{4} = 9\frac{1}{4}$ feet.

3. **(A)** She weighs $5\frac{3}{4}$ pounds less in December than she did in October, so we must subtract $5\frac{3}{4}$ pounds from her October weight of $158\frac{1}{4}$ pounds.

$$158\frac{1}{4} - 5\frac{3}{4} = 157\frac{5}{4} - 5\frac{3}{4} = 152\frac{2}{4} = 152\frac{1}{2} \text{ pounds}$$

Lesson 2.6—Skills Check (page 61)

1. $\frac{8}{15}$ $\qquad \frac{2}{3} \times \frac{4}{5} = \frac{2 \times 4}{3 \times 5} = \frac{8}{15}$

2. $\frac{3}{14}$ $\qquad \frac{3}{4} \times \frac{2}{7} = \frac{3 \times 2}{4 \times 7} = \frac{6}{28} = \frac{3}{14}$

3. $1\frac{1}{20}$ $\qquad \frac{7}{5} \times \frac{3}{4} = \frac{7 \times 3}{5 \times 4} = \frac{21}{20} = 1\frac{1}{20}$

4. $\frac{2}{3}$ $\qquad \frac{8}{15} \times \frac{5}{4} = \frac{8 \times 5}{15 \times 4} = \frac{40}{60} = \frac{2}{3}$

5. 1 $\qquad \frac{5}{2} \times \frac{2}{5} = \frac{5 \times 2}{2 \times 5} = \frac{10}{10} = 1$

6. $5\frac{19}{20}$ $\qquad 3\frac{2}{5} \times 1\frac{3}{4} = \frac{17}{5} \times \frac{7}{4} = \frac{119}{20} = 5\frac{19}{20}$

7. 6 $\qquad 2\frac{2}{5} \times 2\frac{1}{2} = \frac{12}{5} \times \frac{5}{2} = \frac{60}{10} = 6$

8. $16\frac{1}{2}$ $\qquad 2\frac{3}{4} \times 6 = 2\frac{3}{4} \times \frac{6}{1} = \frac{11}{4} \times \frac{6}{1} = \frac{66}{4} = 16\frac{2}{4} = 16\frac{1}{2}$

9. 7 $\qquad 3\frac{1}{9} \times 2\frac{1}{4} = \frac{28}{9} \times \frac{9}{4} = \frac{252}{36} = 7$

10. $\frac{4}{5}$ $\qquad \frac{4}{7}$ of $1\frac{2}{5} = \frac{4}{7} \times 1\frac{2}{5} = \frac{4}{7} \times \frac{7}{5} = \frac{28}{35} = \frac{4}{5}$

Lesson 2.6—ACCUPLACER Challenge (pages 61–62)

1. **(B)** $\frac{1}{6}$ of the 500 questionnaires were returned. "Of" translates to multiply in this context, so $\frac{1}{6}$ of $500 = \frac{1}{6} \times 500 = \frac{1}{6} \times \frac{500}{1} = \frac{500}{6} = 83.\overline{3}$. The question asks approximately how many. The closest answer is 80.

2. **(D)** If approximately $\frac{1}{5}$ of Javier's income was withheld for federal taxes and approxi-

mately $\frac{1}{8}$ of his income was withheld for state taxes, then $\frac{1}{5} + \frac{1}{8}$ of his income was

withheld in total. $\frac{1}{5} + \frac{1}{8} = \frac{8}{40} + \frac{5}{40} = \frac{13}{40}$. To find $\frac{13}{40}$ of \$80,000, his income, we can

multiply:

$$\frac{13}{\overset{}{\underset{1}{40}}} \times \frac{\overset{2,000}{\cancel{80,000}}}{1} = \frac{13 \times 2,000}{1 \times 1} = \frac{26,000}{1} = \$26,000$$

3. **(B)** If Erin used $\frac{2}{3}$ of the fabric to make the curtains, then she has $\frac{1}{3}$ of the fabric

left over. She does not have $\frac{1}{3}$ yards left over, though, so be careful not to pick choice

A here. We need to find $\frac{1}{3}$ of $5\frac{1}{2}$ yards to figure out how much fabric she has left. We

multiply $\frac{1}{3} \times 5\frac{1}{2} = \frac{1}{3} \times \frac{11}{2} = \frac{11}{6} = 1\frac{5}{6}$ yards of fabric left over.

Lesson 2.7—Skills Check (page 66)

1. **1** $\frac{2}{3} \div \frac{2}{3} = \frac{2}{3} \times \frac{3}{2} = \frac{6}{6} = 1$

2. **$\frac{3}{4}$** $\frac{1}{2} \div \frac{2}{3} = \frac{1}{2} \times \frac{3}{2} = \frac{3}{4}$

3. **$3\frac{1}{3}$** $\frac{5}{7} \div \frac{3}{14} = \frac{5}{7} \times \frac{14}{3} = \frac{70}{21} = 3\frac{7}{21} = 3\frac{1}{3}$

4. **7** $\frac{7}{8} \div \frac{1}{8} = \frac{7}{8} \times \frac{8}{1} = \frac{56}{8} = 7$

5. **$\frac{1}{3}$** $\frac{14}{48} \div \frac{7}{8} = \frac{14}{48} \times \frac{8}{7} = \frac{112}{336} = \frac{1}{3}$

6. **$\frac{21}{31}$** $2\frac{1}{3} \div 3\frac{4}{9} = \frac{7}{3} \div \frac{31}{9} = \frac{7}{3} \times \frac{9}{31} = \frac{63}{93} = \frac{21}{31}$

7. **$2\frac{1}{4}$** $3\frac{3}{4} \div 1\frac{2}{3} = \frac{15}{4} \div \frac{5}{3} = \frac{15}{4} \times \frac{3}{5} = \frac{45}{20} = \frac{9}{4} = 2\frac{1}{4}$

8. **20** $16 \div \frac{4}{5} = \frac{16}{1} \times \frac{5}{4} = \frac{80}{4} = 20$

9. **36** $4\frac{1}{2} \div \frac{1}{8} = \frac{9}{2} \times \frac{8}{1} = \frac{72}{2} = 36$

10. **$2\frac{14}{27}$** $\frac{2}{3} \div 1\frac{1}{2} \times 5\frac{2}{3} = \frac{2}{3} \div \frac{3}{2} \times \frac{17}{3} = \frac{2}{3} \times \frac{2}{3} \times \frac{17}{3} = \frac{4}{9} \times \frac{17}{3} = \frac{68}{27} = 2\frac{14}{27}$

Lesson 2.7—ACCUPLACER Challenge (pages 66-67)

1. **(C)** $\frac{8}{9} \div \frac{4}{3}$ is equal to $\frac{8}{9} \times \frac{3}{4} = \frac{8 \times 3}{9 \times 4}$.

2. **(D)** This question can be answered by dividing: $6 \div \frac{2}{3} = \frac{6}{1} \times \frac{3}{2} = \frac{18}{2} = 9$.

3. **(A)** The tutor needs $1\frac{2}{3}$ hours per student, including the time to drive to the next student. He has $5\frac{1}{2}$ hours available. We need to know how many times $1\frac{2}{3}$ goes into $5\frac{1}{2}$, so we divide. $5\frac{1}{2} \div 1\frac{2}{3} = \frac{11}{2} \div \frac{5}{3} = \frac{11}{2} \times \frac{3}{5} = \frac{33}{10} = 3\frac{3}{10}$. The tutor can only tutor whole students, not parts of students, so he can tutor only 3 students in this time.

4. **(D)** If it costs \$41.40 for $4\frac{1}{2}$ yards of material, then it costs $\$41.40 \div 4\frac{1}{2}$ yards for one yard of the material.

$$\$41.40 \div 4\frac{1}{2} = \frac{41.40}{1} \div \frac{9}{2} = \frac{41.40}{1} \times \frac{2}{9} = \frac{82.80}{9} = \$9.20$$

Decimals

3

Decimals, like fractions, allow us to work with quantities that are between whole numbers. For example, something that costs $5.75 costs more than $5, but less than $6. Decimals give us an alternative way of representing quantities that are not whole. Since our dollar is broken up into cents using decimals, many people are more comfortable working with decimals than fractions. Whichever your preference, decimals or fractions, you need to be well versed in using both systems to excel on the ACCUPLACER.

In this chapter, we will work to master

- converting fractions to decimals and decimals to fractions,
- comparing decimals,
- placing decimals in order from least to greatest or greatest to least,
- rounding decimals, and
- adding, subtracting, multiplying, and dividing with decimals.

LESSON 3.1—PLACE VALUE AND THE DECIMAL SYSTEM

Place value is how we use the digits 0–9 to make all other numbers. A change in place is also a change in value. Consider $694.75. Here, the 6 has a greater value than the 9 because of the place the digits occupy. The 6 is in the hundreds place, so it represents $600 by itself, whereas the 9 is in the tens place, so it represents $90 by itself. The digit that is furthest left of the other numbers is 6, whereas 5 is the digit that is furthest right of the other numbers. The further left a number is, the greater the place value. The further right a number is, the lesser the place value. The place a number occupies is critical to recognizing its value.

Here are some important points to review and remember:

- The decimal point separates the whole numbers, which are found to the left of the decimal point, from the numbers that are smaller than one, which are found to the right of the decimal point.
- As we move left from the ones place, each place value increases by a power of ten, or times ten.
- As we move right from the ones place, each place value decreases by a power of ten, or divide by ten.
- When a decimal point is not present, the last place is the ones. The decimal point is always immediately to the right of the ones place.

- The suffix -*ths* indicates that a place name is describing a place value that is less than the ones place, or to the right of the ones place.
- The digit furthest right is only in the ones place when there is no decimal point present. When there is a decimal point present, the ones place is immediately to the left of the decimal point. In 47.923, the 7 is in the ones place.
- There is no such thing as a *oneths* place.

Consider the number 7,392.658. We have illustrated the name of the place value for each digit in that number. Notice that as we move right and left from the ones place, the names of the places are practically the same. The major difference is that the numbers to the right of the ones place are in places that are less than one and have the suffix –*ths* added on to indicate this.

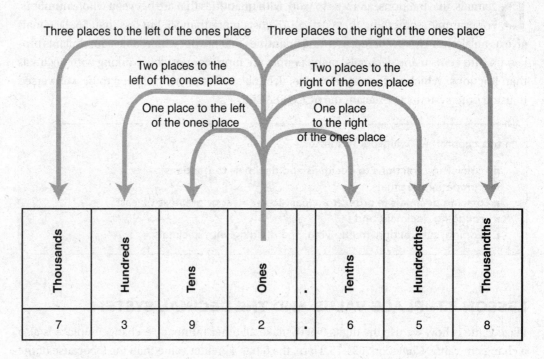

LESSON 3.2—COMPARING AND ORDERING DECIMALS

In whole numbers, longer numbers, or those with more digits, are always greater than shorter numbers, or those with fewer digits. 5,892 is greater than 589. However, this leads to a common mistake in decimals. The length of a decimal is less important than its place value. 0.6 is greater than 0.06. The further left a place is, the larger the place value it is in. In 0.6, the 6 is in the tenths place. In 0.06, the 6 is in the hundredths place. The tenths place is further left than the hundredths place. More importantly, tenths are greater than hundredths.

To Compare Decimals:

1. Write numbers above one another so that their decimal points and places are lined up.
2. Compare numbers that are written above one another working left to right.
3. Use zeros as placeholders so that numbers can have the same number of digits.

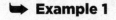 **Example 1**

Which is greater: 0.92 or 0.8547?

 TIP

These zeros were added so that both numbers have the same number of places.

$$
\begin{array}{r}
0.92\mathbf{00} \\
0.8547
\end{array}
$$

Solution

Looking in the tenths place, 9 is greater than 8, so 0.92 is greater than 0.8547. We can also say that 9200 is greater than 8547, so 0.92**00** is greater than 0.8547.

Answer: Written out using inequality symbols, 0.92 > 0.8547.

Example 2

Which is larger: 0.1033 or 0.13?

Solution

These two decimals both have the same digit in the tenths place, so we'll need to look a little more carefully. First, add zeros at the end of 0.13 so that it has the same number of digits as 0.1033.

$$
\begin{array}{r}
0.1033 \\
0.13\mathbf{00}
\end{array}
$$

Now we just focus on the digits after the decimal. We see that 1033 is smaller than 1300, and so the decimal 0.1033 is less than 0.13.

Answer: 0.1033 < 0.13

To Order More than Two Decimals:

1. Follow the same rules for comparing two decimals, but rearrange the list of decimals until you have them in the desired order of least to greatest or greatest to least.

➡ Example 3

The table below lists the finish times of four swimmers competing in the 25-meter freestyle competition.

Swimmer	Finish Time in Seconds
Tony	30.8
Miguel	30.475
Javanti	31.1
Mark	30.89

Which of the following is the correct order in which the swimmers finished the race from first to last place?

(A) Miguel, Tony, Mark, Javanti
(B) Miguel, Mark, Tony, Javanti
(C) Javanti, Miguel, Tony, Mark
(D) Tony, Miguel, Mark, Javanti

Solution

It is helpful to remember that in a race, first place goes to the person that finishes with the least amount of time. Therefore, we want to put the swimmers in order from the least amount of time to the greatest amount of time. To do this, we can rewrite the numbers with their decimal points lined up, and we can use 0s as placeholders so that the numbers have the same number of digits.

$$30.8\textbf{00}$$
$$30.475$$
$$31.1\textbf{00}$$
$$30.89\textbf{0}$$

Once we have done this, we can go down each column of digits, working left to right, comparing as we go. First, we look at the tens place. All four numbers start with a 3, so we go to the ones place. One of the numbers is a 1, and the others are 0s. 1 is larger than 0, so last place in the race is the 31.1 time. Therefore, we can rearrange the numbers like this:

$$30.8\textbf{00}$$
$$30.475$$
$$30.89\textbf{0}$$
$$31.100$$

Now that we have the numbers ordered by their whole numbers, we can look to the decimal numbers. Since we filled in missing places with 0s, we can look at all the decimal places at once. Doing so, we can see that 475 is less than 800, and 800 is less than 890. We then know that 30.475 < 30.800 < 30.890.

We now have all that we need to put the decimals in order. We can rewrite the table with the appropriate names and time orders.

Swimmer	Finish Time in Seconds
Miguel	30.475
Tony	30.800
Mark	30.890
Javanti	31.100

<u>Answer:</u> Therefore, the correct answer is choice A.

LESSON 3.2—SKILLS CHECK

Directions: Write the greater than symbol, >, or the less than symbol, <, on each line to make the statement true.

1. 32.7 _____ 43.9

2. 0.6 _____ 0.7

3. 0.08 _____ 0.04

4. 0.5 _____ 0.05

5. 0.03 _____ 0.2

6. 0.8 _____ 0.2051

7. 32.129 _____ 32.93

Directions: Put the following lists of numbers in order from least to greatest.

8. 0.43, 0.0034, 0.34, 0.403

9. 0.307, 0.037, 0.37, 0.3069

10. 5.8799, 5.903, 5.912, 5.9

(Answers are on page 99.)

LESSON 3.2—ACCUPLACER CHALLENGE

Directions: For each of the questions below, choose the best answer from the four choices given.

1. Which of the following numbers is the least?

    (A) 0.307
    (B) 0.703
    (C) 0.073
    (D) 0.37

2. Which of the following numbers is the greatest?

   (A) 0.0809
   (B) 0.089
   (C) 0.08
   (D) 0.0098

3. Which of the following places the four numbers in order from least to greatest?

   (A) 0.55, 0.511, 0.505, 0.5
   (B) 0.5, 0.505, 0.55, 0.511
   (C) 0.5, 0.55, 0.505, 0.511
   (D) 0.5, 0.505, 0.511, 0.55

4. Five gymnasts are in a gymnastics competition. Four of the gymnasts have already done their routines and received scores of 7.09, 7.18, 7.108, and 7.093. What must the fifth gymnast score in order to win the competition?

   (A) > 7.09
   (B) > 7.18
   (C) > 7.108
   (D) > 7.093

(Answers are on page 99.)

LESSON 3.3—ROUNDING DECIMALS

It sometimes happens that we don't need numbers to be so specific. For example, if a husband buys something for $206.21, he may tell his spouse that he paid $200. Situations that arise in science, finance, and mathematics of all kinds may result in numbers that are longer or more accurate than need be. For this reason, we sometimes round decimals.

To round numbers of any kind, we must first identify the place value that we wish to round to. On ACCUPLACER questions, that place value will be identified in the question. When you round, the main goal is deciding whether it is closer to round up or to round down. 26 rounded to the nearest 10 is 30 because 4 up from 26 is 30, whereas it would be 6 to round down to 20. In this case, it is closer to round up than to round down.

6 down to 20   4 up to 30

20, 21, 22, 23, 24, 25, (26), 27, 28, 29, 30

To Round Numbers:

1. Underline the place value mentioned in the question.
2. If the digit to the right of the underlined place is
   - ≥ 5, round the underlined number up one and make all the places to the right of that place 0s.
   - < 5, leave the underlined number as it was and make all the places to the right of that number 0s.
3. Drop unnecessary 0s.

   Let's take a look at rounding the following three numbers, 62.5836, 315.2384, and 96.9582, to various places.

	62.5836	**315.2384**	**96.9582**
Rounded to the nearest thousandths place	62.58<u>3</u>6 ↓ 62.584**0**	315.238**0**	96.958**0**
Rounded to the nearest hundredths place	62.5<u>8</u>36 ↓ 62.58**00**	315.24**00**	96.96**00**
Rounded to the nearest tenths place	62.<u>5</u>836 ↓ 62.6**000**	315.2**000**	97.0**000**
Rounded to the nearest units place (ones place)	6<u>2</u>.5836 ↓ 63.**0000**	315.**0000**	97.**0000**
Rounded to the nearest tens place	<u>6</u>2.5836 ↓ 60.**0000**	320.**0000**	100.**0000**

LESSON 3.3—SKILLS CHECK

> **Directions:** Round each number to the place value mentioned.

1. 145.6789 rounded to the nearest thousandths place is _____.

2. 2,305.4916 rounded to the nearest hundredths place is _____.

3. 89.961 rounded to the nearest tenths place is _____.

4. 2,004.21 rounded to the nearest units place (ones place) is _____.

5. 89.3 rounded to the nearest tens place is _____.

6. 69.294 rounded to the nearest hundreds place is _____.

7. 1,995.83 rounded to the nearest thousands place is _____.

8. 456,329 rounded to the nearest ten thousands place is _____.

(Answers are on page 100.)

LESSON 3.4—THE RELATIONSHIP BETWEEN DECIMALS AND FRACTIONS

Decimals and fractions can be used to represent the same quantity. Many of us are familiar with the fact that 0.5 and $\frac{1}{2}$ can both be used to communicate the idea of a half. Many of us also know that $\frac{3}{4} = 0.75$. But, how exactly does that work? What about other decimals? How do we write 0.27 as a fraction? How do we write $\frac{3}{8}$ as a decimal?

To Change a Decimal to a Fraction:

1. Place the numbers to the right of the decimal point in the numerator.
2. Place the name of the final decimal place (the rightmost number) as the number in the denominator. For example, if the rightmost digit was in the *thousandths* place, the denominator would be 1,000.
3. Reduce whenever possible.

 **Example 1**

Convert 0.27, 0.027, 0.105, and 56.083 to fractions.

Solution

In the number 0.27, 7 is in the hundredths place, so 100 becomes the denominator.

$$0.27 = \frac{27}{100}$$

$$0.027 = \frac{27}{1000}$$

$$0.105 = \frac{105}{1000} = \frac{21}{200}$$

$$56.083 = 56\frac{83}{1000}$$

<u>Answers:</u> 0.27 becomes $\frac{27}{100}$. 0.027 becomes $\frac{27}{1000}$. 0.105 becomes $\frac{21}{200}$. 56.083 becomes $56\frac{83}{1000}$.

Changing Fractions to Decimals

It's worth noting that some fractions are much easier to convert to decimals than others. This is because our decimal system only has place values that are powers of 10. In other words, if a fraction already has 10, 100, 1,000, or another power of 10 as a denominator, then we can more easily convert the fraction to a decimal.

Changing Fractions to Decimals When the Denominator Is a Multiple of 10

(i.e., $\frac{}{10}$, $\frac{}{100}$, $\frac{}{1000}$, etc.):

1. Write the numerator to the right of a decimal point.
2. Ensure that the last digit lands in the decimal place with the same "name" as the denominator. That is, if the denominator of the fraction is 1,000, then the rightmost digit in the decimal should be in the thousandths place.
3. Use zeros as placeholders where necessary.

EXAMPLES

$\frac{1}{10} = 0.1$	$\frac{7}{10} = 0.7$	$\frac{9}{100} = 0.09$	$\frac{107}{10,000} = 0.0107$	$1\frac{9}{10} = 1.9$
tenths · tenths place		hundredths · hundredths place		

Other fractions can be easily made into equivalent fractions with 10, 100, 1,000, or other multiples of ten as a denominator. These fractions would include those that have 4, 5, 20, and 25 as denominators.

To Change Fractions to Decimals Using Equivalent Fractions:

1. Convert fractions to an equivalent with 10, 100, 1,000, or other multiples of ten as a denominator.
2. Follow the steps for converting fractions, with a multiple of ten as a denominator, to decimals.

➥ Example 2

Change $\frac{1}{2}$ to a decimal.

Solution

$$\frac{1}{2} = \frac{1}{2} \times \frac{5}{5} = \frac{5}{10} = 0.5$$

Answer: 0.5

➥ Example 3

Change $\frac{3}{4}$ to a decimal.

Solution

$$\frac{3}{4} = \frac{3}{4} \times \frac{25}{25} = \frac{75}{100} = 0.75$$

Answer: 0.75

➥ Example 4

Change $\frac{3}{5}$ to a decimal.

Solution

$$\frac{3}{5} = \frac{3}{5} \times \frac{2}{2} = \frac{6}{10} = 0.6$$

Answer: 0.6

➥ Example 5

Change $\frac{7}{25}$ to a decimal.

Solution

$$\frac{7}{25} = \frac{7}{25} \times \frac{4}{4} = \frac{28}{100} = 0.28$$

Answer: 0.28

➥ Example 6

Change $\frac{11}{20}$ to a decimal.

Solution

$$\frac{11}{20} = \frac{11}{20} \times \frac{5}{5} = \frac{55}{100} = 0.55$$

Answer: 0.55

As mentioned earlier, some fractions are easier than others to change into a decimal. If it's not easy enough to make the denominator into a multiple of 10, then we can always use long division. This method will work for all fractions.

To Change a Fraction to a Decimal Using Long Division:

1. Divide the denominator into the numerator using long division (denominator outside the $\overline{)}$ sign, numerator under the $\overline{)}$ sign).
2. Place a decimal point to the right of the dividend (the number under the $\overline{)}$ sign). Then place another decimal point on top of the $\overline{)}$ sign straight above the first one.
3. Divide following the steps for long division.
4. Continue until you have a remainder of 0 or until you have reached a desirable place value.

➥ Example 7

Change $\frac{1}{2}$ to a decimal using long division.

Solution

First, rewrite the fraction as a long division problem.

$$2\overline{)1}$$

Since we can't immediately divide 2 into 1, we add a decimal point and 0 as a placeholder. We write another decimal point straight above 1.0 for our future result, or quotient.

$$2\overline{)1.0}^{\,\cdot}$$

We then divide as if it were 2 into 10. Make sure to put the 5 above the 0 in 1.0, which makes the result, or quotient, 0.5.

$$
\begin{array}{r}
0.5 \\
2\overline{)1.0} \\
\underline{-1.0} \\
0
\end{array}
$$

Answer: Therefore, $\frac{1}{2} = 0.5$.

➥ Example 8

Change $\frac{3}{4}$ to a decimal using long division.

Solution

$$
\begin{array}{r}
0.75 \\
4\overline{)3.00} \\
\underline{-28}\downarrow \\
20 \\
\underline{-20} \\
0
\end{array}
$$

Answer: Therefore, $\frac{3}{4} = 0.75$.

Fraction-Decimal Relationships Worth Memorizing

In addition to those already mentioned, it is useful to memorize the following fraction–decimal relationships.

$$\frac{1}{3} = 0.\overline{3} \qquad \frac{2}{3} = 0.\overline{6} \qquad \frac{1}{6} = 0.1\overline{6} \qquad \frac{1}{8} = 0.125$$

LESSON 3.4—SKILLS CHECK

Directions: Change each decimal below to a fraction in simplest form.

1. 0.3

3. 0.0309

5. 0.05

2. 0.03

4. 4.17

Directions: Change each fraction to a decimal.

6. $\frac{3}{10}$

10. $\frac{13}{50}$

14. $4\frac{3}{5}$

7. $\frac{51}{100}$

11. $\frac{9}{12}$

15. $5\frac{9}{40}$

8. $\frac{79}{1000}$

12. $\frac{4}{11}$

9. $\frac{7}{25}$

13. $\frac{7}{9}$

(Answers are on page 100.)

LESSON 3.4—ACCUPLACER CHALLENGE

Directions: For each of the questions below, choose the best answer from the four choices given.

1. $9\frac{3}{20} =$

(A) 0.9320

(B) 0.915

(C) 9.3

(D) 9.15

2. Engineers often use inches written in decimal form. How would $\frac{3}{8}$ of an inch be expressed in decimal form?

   (A) 0.375 inches
   (B) 0.38 inches
   (C) 3.8 inches
   (D) 375 inches

3. Surveyors often use inches expressed in decimal feet form. To the nearest hundredth, how many decimal feet does $4\frac{3}{8}$ inches equal?

   (A) 0.36 feet
   (B) 0.438 feet
   (C) 0.7 feet
   (D) 4.375 feet

(Answers are on page 101.)

LESSON 3.5—ADDING AND SUBTRACTING DECIMALS

If you collect $5 from one person and $10.20 from another, then you have collected $15.20 in total. After years of using money, many of us know this intuitively. We have correctly calculated the total in this type of situation because we paid attention to place value. If we strip away the $ signs and now approach this as 5 + 10.20, some people may make the mistake of finding the sum of these two numbers to be 10.25 or 60.20 or even 10.7. Why? The answer is because they did not respect the cardinal rule for adding decimals: we must line up our place values and our decimal points to make sure we are adding the correct places to one another.

To Add or Subtract Decimals:

1. Vertically line up numbers at their decimal points.
2. Use zeros as placeholders to make each number the same number of places.
3. Bring the decimal point down below.
4. Add vertically to find the sum; subtract vertically to find the difference.

➡ Example 1

Find the sum of 10.94 + 5.076.

Solution

$$
\begin{array}{r}
10.940 \\
+\ 5.076 \\
\hline
16.016
\end{array}
$$

Answer: 16.016

➥ Example 2

Find the sum of 3.8 + 47 + 1.83.

Solution

$$
\begin{array}{r}
3.80 \\
47.00 \\
+\ \ 1.83 \\
\hline
52.63
\end{array}
$$

<u>Answer:</u> 52.63

➥ Example 3

Find the difference of 5.98 – 3.47.

Solution

$$
\begin{array}{r}
5.98 \\
-\ 3.47 \\
\hline
2.51
\end{array}
$$

<u>Answer:</u> 2.51

➥ Example 4

Find the difference of 0.7 and 0.254.

Solution

$$
\begin{array}{r}
0.700 \\
-\ 0.254 \\
\hline
0.446
\end{array}
$$

<u>Answer:</u> 0.446

➥ Example 5

Boone County consists of Sparta City, which has a population of 1.37 million people, and the nearby suburbs, which have another 1.1 million people. What is the total population of Boone County?

(A)   1.48 million people
(B)   2.38 million people
(C)   2.47 million people
(D)   14.8 million people

Solution

Whenever we are given two parts, in this case the population of Sparta City and the population of the surrounding suburbs, and are asked to find the whole, in this case the *total* population of Boone County, we add. To add these two numbers together, we have to be sure to line up the decimal points.

$$
\begin{array}{r}
1.37 \\
+\ 1.10 \\
\hline
2.47
\end{array}
$$

Answer: Therefore, the correct answer is choice C.

LESSON 3.5—SKILLS CHECK

Directions: Find the sum or difference as directed.

1. $9.85 + 0.032 =$

2. $12.684 - 8.5 =$

3. $12.42 - 1.38 =$

4. $1.8 + 0.3 + 8.09 =$

5. $8.03 - 4.2 + 12.006 =$

6. $16 + 9.24 + 170.3 =$

(Answers are on page 101.)

LESSON 3.5—ACCUPLACER CHALLENGE

Directions: For each of the questions below, choose the best answer from the four choices given.

1. $1.3 - 0.05 + 14.6 =$

   (A) 15.4
   (B) 15.85
   (C) 15.95
   (D) 27.1

2. What is the total thickness of a piece of wood that is 4.9 centimeters thick with a 0.15-centimeter-thick laminate on top?

   (A) 4.105
   (B) 5.05
   (C) 5.5
   (D) 6.4

3. To pay for a new concert hall, the town of Center Bay raised $1.04 million from the state, $2.3 million from the county, and $0.05 million from local businesses. The total cost of the new concert hall is estimated to be $5 million. How much more money does the town need to raise to complete the project?

   (A) $1.04 million
   (B) $1.25 million
   (C) $1.61 million
   (D) $3.39 million

4. Nick cut two lengths of copper tubing each 1.45 meters long from a 4-meter piece of tubing. Which of the following expressions shows the length of the remaining piece?

   (A) 4 − 2 − 1.45
   (B) 2(4 − 1.45)
   (C) 4 − 2(1.45)
   (D) 4(2 − 1.45)

(Answers are on page 102.)

LESSON 3.6—MULTIPLYING DECIMALS

In the same way that it is necessary to pay attention to place value when adding and subtracting decimals, it is equally important to pay special attention to place value when multiplying decimals. We do not, however, need to line up our decimals when multiplying. We'll examine why below.

To Multiply Decimals:

1. Write the multiplication problem vertically with the longer number (that contains more digits) on top.
2. Line the top and bottom number up to the right (the decimal points do not need to be lined up).
3. Ignore the decimal point for now and multiply as you would normally.
4. Count the total number of decimal places found to the right of both decimal points in the two original numbers.
5. Place the decimal point in the product so that the product has the same *total* number of decimal places as the two original numbers.

➡ Example 1

Find the product of 2.12 and 0.2.

Solution

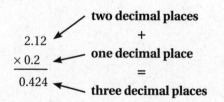

Answer: 0.424

➡ Example 2

Find the product of 2.12 and 0.02.

Solution

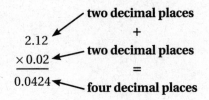

$$
\begin{array}{r}
2.12 \\
\times\, 0.02 \\
\hline
0.0424
\end{array}
$$

two decimal places
+
two decimal places
=
four decimal places

<u>Answer:</u> 0.0424

➡ Example 3

Find the product of 0.212 and 0.002.

Solution

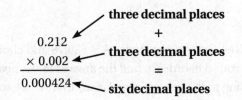

$$
\begin{array}{r}
0.212 \\
\times\, 0.002 \\
\hline
0.000424
\end{array}
$$

three decimal places
+
three decimal places
=
six decimal places

<u>Answer:</u> 0.000424

Decimal points affect the place value of the product of two numbers, but not the digits in the product—that is if we don't count the zeros we use as placeholders. All of the answers to these examples had 424 as part of the answer, but only the place values of those digits changed.

Notice the similarities and differences below:

$$283 \times 10 = 2{,}830$$
$$28.3 \times 10 = 283$$
$$2.83 \times 10 = 28.3$$
$$0.283 \times 10 = 2.83$$
$$0.0283 \times 10 = 0.283$$
$$2.83 \times 0.1 = 0.283$$
$$2.83 \times 0.01 = 0.0283$$
$$2.83 \times 10{,}000 = 28{,}300$$

➡ Example 4

Lenny bought 12 gallons of gasoline at $2.399 per gallon. To the nearest cent, how much did he pay for the gasoline in total?

(A) $28.78
(B) $28.79
(C) $28.80
(D) $35.99

Solution

We know that each gallon of gasoline costs $2.399. We also know that Lenny bought 12 of those gallons, so we need to multiply the cost times 12.

$$
\begin{array}{r}
2.399 \\
\times\ \ \ 12 \\
\hline
4798 \\
+23990 \\
\hline
28.788
\end{array}
$$

Looking at the answer choices, we see that choice A is $28.78 and choice B is $28.79. Our answer is $28.788, so we must round in order to find the answer to the nearest cent. There are 100 cents in a dollar, so rounding to the nearest cent means that we are to round our decimal to the nearest hundredth. In our answer of $28.788, 8 is in the hundredths place, followed by another 8 to its right.

Answer: Therefore, we round up and the correct answer is choice B.

LESSON 3.6—SKILLS CHECK

Directions: Find the product for each problem below.

1. 0.5×0.7

2. 0.05×0.7

3. 0.0005×7

4. $(34.1)(2.4)$

5. 1.2×23.6

(Answers are on page 103.)

LESSON 3.6—ACCUPLACER CHALLENGE

> **Directions:** For each of the questions below, choose the best answer from the four choices given.

1. Jermaine bought 8 gallons of gasoline at $2.199 per gallon. To the nearest cent, how much did he pay for the gasoline in total?

   (A) $2.20
   (B) $17.59
   (C) $17.60
   (D) $25.59

2. Find the cost of 1.5 pounds of Swiss cheese on sale for $4.50 per pound.

   (A) $3.00
   (B) $6.75
   (C) $67.50
   (D) $6.00

3. The North American Plate moves westward at an average rate of about 0.034 inches per month. On average, how far does the North American Plate move in a year?

   (A) 0.102 inch
   (B) 0.34 inch
   (C) 0.408 inch
   (D) 12.034 inches

(Answers are on page 103.)

LESSON 3.7—DIVIDING DECIMALS

How many quarters are there in three dollars? Many people would think to themselves that there are four quarters in a dollar, so there must be twelve quarters in three dollars. Mathematically that could look like $3 \times 4 = 12$. This is absolutely correct. There are twelve quarters in three dollars.

However, a question asked in the form of "How many a are in b?" could also be interpreted as a division problem that could be written as $b \div a$. For example, the answer to the question, "How many fives are in thirty?" can be solved as $30 \div 5 = 6$. When a question asks how many quarters there are in three dollars, we can also represent this scenario with the division problem, $\$3.00 \div \$0.25$, remembering that a quarter = $0.25. We already know the answer is 12, but just how do we calculate $3 \div 0.25$ and get 12 using long division?

To Divide a Number by a Decimal:

1. Change the divisor (the number outside the $\overline{)}$) to a whole number by moving the decimal point to the right as many places as it takes to make a whole number.
2. Move the decimal point of the dividend (the number under the $\overline{)}$) the same number of spaces to the right that you moved the divisor's decimal point.
3. Complete the long division problem.

Note: We always move the decimal point of both the divisor (outside the $\rceil$) and the dividend (under the $\rceil$) the same number of places. If the dividend becomes a whole number too, then that's great. If the dividend does not become a whole number, then that is fine too. We are only concerned with making the divisor a whole number and then staying consistent with how we move the decimal point of the dividend, wherever it may land.

To demonstrate, let's go back to $3.00 ÷ $0.25.

➡ Example 1

Find the quotient of 3 ÷ 0.25.

Solution

First, write the problem as long division.

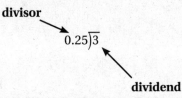

Then make 0.25 a whole number by moving the decimal point twice to the right. Then do the same to 3 by writing 3.00, and then moving the decimal point twice to the right there also.

Then complete the long division problem.

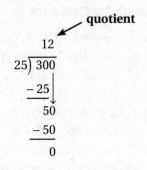

Answer: The quotient of 3 ÷ 0.25 is 12.

Now let's look at a trickier example.

➡ **Example 2**

Find the quotient of $\dfrac{8.2}{0.24}$.

Solution

Remember, fractions can be thought of as division.

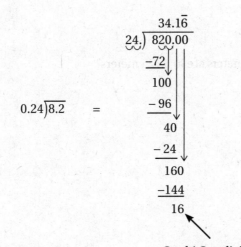

$$0.24\overline{)8.2} \quad = \quad \begin{array}{r} 34.1\overline{6} \\ 24\overline{)820.00} \\ -72 \\ \hline 100 \\ -96 \\ \hline 40 \\ -24 \\ \hline 160 \\ -144 \\ \hline 16 \end{array}$$

Ouch! Our division continues forever! We end up with a *repeating decimal*. The bar above the 6 indicates that the 6 repeats forever, as in 34.166666666…

Answer: The quotient of $\dfrac{8.2}{0.24}$ is $34.1\overline{6}$.

LESSON 3.7—SKILLS CHECK

Directions: Find the quotient for each problem.

1. 4.8 ÷ 0.3

2. 0.56 ÷ 0.8

3. $\dfrac{0.56}{0.008}$

4. 9.9 ÷ 0.44

5. $\dfrac{0.6695}{0.0206}$

6. 15.34 ÷ 0.236

(Answers are on page 103.)

LESSON 3.7—ACCUPLACER CHALLENGE

Directions: For each of the questions below, choose the best answer from the four choices given.

1. $\dfrac{3}{20} =$

   (A) 0.15
   (B) 1.5
   (C) 3.20
   (D) 20.3

2. How many 0.25 meters are in 5.125 meters?

   (A) 1
   (B) 4.875
   (C) 5.375
   (D) 20.5

3. $0.42 \div 1.8 =$

   (A) $0.2\overline{3}$
   (B) 0.236
   (C) $2.\overline{3}$
   (D) 4.286

(Answers are on page 104.)

To complete a set of practice questions that reviews all of Chapter 3, go to:
barronsbooks.com/TP/accuplacer/math29qw/

Lesson 3.2—Skills Check (page 81)

1. **32.7 < 43.9**   Working left to right, compare digits. 32.7 has a 3 in the tens place. 43.9 has a 4 in the tens place. 3 < 4, so 32.7 < 43.9.

2. **0.6 < 0.7**   Working left to right, compare digits. 0.6 has a 6 in the tenths place. 0.7 has a 7 in the tenths place. 6 < 7, so 0.6 < 0.7.

3. **0.08 > 0.04**   Working left to right, compare digits. 0.08 has an 8 in the hundredths place. 0.04 has a 4 in the hundredths place. 8 > 4, so 0.08 > 0.04.

4. **0.5 > 0.05**   Working left to right, compare digits. 0.5 has a 5 in the tenths place. 0.05 has a 0 in the tenths place. 5 > 0, so 0.5 > 0.05.

5. **0.03 < 0.2**   Working left to right, compare digits. 0.03 has a 0 in the tenths place. 0.2 has a 2 in the tenths place. 0 < 2, so 0.03 < 0.2.

6. **0.8 > 0.2051**   Working left to right, compare digits. 0.8 has an 8 in the tenths place. 0.2051 has a 2 in the tenths place. 8 > 2, so 0.8 > 0.2051.

7. **32.129 < 32.93**   Working left to right, compare digits. The first two digits are the same in both numbers. 32.129 has a 1 in the tenths place. 32.93 has a 9 in the tenths place. 1 < 9, so 32.129 < 32.93.

8. **0.0034, 0.34, 0.403, 0.43**   Use zeros as placeholders to compare the numbers as decimals with the same number of places: 0.0034, 0.34**00**, 0.403**0**, 0.43**00**.

9. **0.037, 0.3069, 0.307, 0.37**   Use zeros as placeholders to compare the numbers as decimals with the same number of places: 0.037**0**, 0.3069, 0.307**0**, 0.37**00**.

10. **5.8799, 5.9, 5.903, 5.912**   Use zeros as placeholders to compare the numbers as decimals with the same number of places: 5.8799, 5.9**000**, 5.903**0**, 5.912**0**.

Lesson 3.2—ACCUPLACER Challenge (pages 81–82)

1. **(C)** Use zeros as placeholders to compare the answer choices as decimals with the same number of places. After doing so, we can see that 0.073, 0.307, 0.37**0**, 0.703 are ordered from least to greatest because 73, 307, 370, 703 are in order. Therefore, 0.073 is the least.

2. **(B)** Use zeros as placeholders to compare the answer choices as decimals with the same number of places. After doing so, we can see that 0.0098, 0.08**00**, 0.0809, 0.089**0** are ordered from least to greatest. Ignoring the decimal point, 0890 is the greatest of the four digit numbers as compared to 0098, 0800, and 0809.

3. **(D)** We can make all of the numbers have the same number of decimal places by adding zeros as placeholders. After doing so, we can see that 0.5**00**, 0.505, 0.511, 0.55**0** is the correct order because 500, 505, 511, 550 is ordered from least to greatest.

4. **(B)** Using zeros as placeholders, we can compare and order each of the four scores as decimals with the same number of places from least to greatest as 7.090, 7.093, 7.108, 7.18**0**. Since 7.18 is the highest score so far, the fifth gymnast must receive a score greater than this, or > 7.18.

Lesson 3.3—Skills Check (page 84)

1. **145.679**   Underlined, the thousandths place is 145.678̲9. To the right of 8 is 9, so we round up to 145.679.

2. **2,305.49**   Underlined, the hundredths place is 2,305.49̲16. To the right of 9 is 1, so we round down to 2,305.49.

3. **90.0**   Underlined, the tenths place is 89.9̲61. To the right of 9 is 6, so we round up to 90.0.

4. **2,004**   Underlined, the units place is 2,004̲.21. To the right of 4 is 2, so we round down to 2,004.

5. **90**   Underlined, the tens place is 8̲9.3. To the right of 8 is 9, so we round up to 90.

6. **100**   Underlined, the hundreds place is 0̲69.294. To the right of 0 is 6, so we round up to 100.

7. **2,000**   Underlined, the thousands place is 1̲,995.83. To the right of 1 is 9, so we round up to 2,000.

8. **460,000**   Underlined, the ten thousands place is 45̲6,329. To the right of 5 is 6, so we round up to 460,000.

Lesson 3.4—Skills Check (page 88)

1. $\frac{3}{10}$ $0.3 = \frac{3}{10}$ because the 3 is in the tenths place.

2. $\frac{3}{100}$ $0.03 = \frac{3}{100}$ because the 3 is in the hundredths place.

3. $\frac{309}{10,000}$ $0.0309 = \frac{309}{10,000}$ because the 9, the digit furthest to the right, is in the ten thousandths place.

4. $4\frac{17}{100}$ Since 4.17 has a 4 in the ones place, this number will become a mixed number with a 4 in front. To the right of the decimal point is 17, so 17 becomes the numerator. The 7, the number furthest to the right, is in the hundredths place, so 100 becomes the denominator. $4.17 = 4\frac{17}{100}$.

5. $\frac{1}{20}$ $0.05 = \frac{5}{100}$ because the 5 is in the hundredths place. Since 5 and 100 share a common factor, we must reduce: $\frac{5}{100} = \frac{1}{20}$.

6. **0.3** $\frac{3}{10} = 0.3$ because the 3 must be in the tenths place.

7. **0.51** $\frac{51}{100} = 0.51$ because the 1, the last digit in the numerator, must be in the hundredths place.

8. **0.079** $\frac{79}{1000} = 0.079$ because the 9, the last digit in the numerator, must be in the thousandths place.

9. **0.28** Convert the fraction to an equivalent over 100 by multiplying: $\frac{7}{25} \times \frac{4}{4} = \frac{28}{100}$.

   Then, write 28 as a decimal with the 8 in the hundredths place. $\frac{28}{100} = 0.28$.

10. **0.26** Convert the fraction to an equivalent over 100 by multiplying: $\frac{13}{50} \times \frac{2}{2} = \frac{26}{100}$.

    Then, write 26 as a decimal with the 6 in the hundredths place. $\frac{26}{100} = 0.26$.

11. **0.75** Divide using long division. $\frac{9}{12} = 9 \div 12 = 0.75$.

12. **$0.\overline{36}$** Divide using long division. $\frac{4}{11} = 4 \div 11 = 0.\overline{36}$.

13. **$0.\overline{7}$** Divide using long division. $\frac{7}{9} = 7 \div 9 = 0.\overline{7}$.

14. **4.6** Convert the fraction to an equivalent over 10 by multiplying: $4\frac{3}{5} \times \frac{2}{2} = 4\frac{6}{10}$. Then, write a decimal with the 6 in the tenths place and 4, the whole number, in the ones place. $4\frac{6}{10} = 4.6$.

15. **5.225** Divide using long division. $9 \div 40 = 0.225$. Then, write 5, the whole number, in the ones place. $5\frac{9}{40} = 5.225$.

Lesson 3.4—ACCUPLACER Challenge (pages 88–89)

1. **(D)** $9\frac{3}{20} = 9\frac{3}{20} \times \frac{5}{5} = 9\frac{15}{100} = 9.15$

2. **(A)** $\frac{3}{8} = 3 \div 8 = 0.375$ inches

3. **(A)** As a decimal, $4\frac{3}{8}$ inches is $4 + (3 \div 8) = 4.375$ inches. However, the question asks how many decimal feet $4\frac{3}{8}$ inches equals. To convert inches to feet, we divide by 12 because there are 12 inches in a foot. $4.375 \div 12 = 0.364583$. Rounded to the nearest hundredth, $4\frac{3}{8}$ inches $= 0.36$ feet.

Lesson 3.5—Skills Check (page 91)

1. **9.882**

$$
\begin{array}{r}
9.850 \\
+\,0.032 \\
\hline
9.882
\end{array}
$$

2. **4.184**

$$
\begin{array}{r}
\overset{1}{\cancel{1}}2.684 \\
-\,8.500 \\
\hline
4.184
\end{array}
$$

3. 11.04

$$\begin{array}{r} \overset{3}{\cancel{1}}\,\overset{1}{}\\ 12.\cancel{4}2 \\ -\ 1.38 \\ \hline 11.04 \end{array}$$

5. 15.836

$$\begin{array}{r} \overset{7}{\cancel{8}}\,\overset{1}{}.03 \\ -\ 4.20 \\ \hline 3.83 \end{array} \qquad \begin{array}{r} 3.830 \\ +12.006 \\ \hline 15.836 \end{array}$$

4. 10.19

$$\begin{array}{r} 1 \\ 1.80 \\ 0.30 \\ +\ 8.09 \\ \hline 10.19 \end{array}$$

6. 195.54

$$\begin{array}{r} 1 \\ 16.00 \\ 9.24 \\ +170.30 \\ \hline 195.54 \end{array}$$

Lesson 3.5—ACCUPLACER Challenge (pages 91–92)

1. **(B)** Since this problem involves subtraction, then addition, evaluate in two steps.

$$\begin{array}{r} \overset{2}{\cancel{1}}\,\overset{1}{}\\ 1.\cancel{3}0 \\ -\ 0.05 \\ \hline 1.25 \end{array} \qquad \begin{array}{r} 1.25 \\ +14.60 \\ \hline 15.85 \end{array}$$

2. **(B)** The total thickness is found by adding the 4.9 centimeters to the 0.15 centimeters.

$$\begin{array}{r} 1 \\ 4.90 \\ +\ 0.15 \\ \hline 5.05 \end{array}$$

3. **(C)** We first need to calculate how much money, in millions of dollars, has been raised so far by adding $1.04 + $2.3 + $0.05 = $3.39 million. The question, however, asks how much *more* money the town needs to raise to complete the project, so we must subtract the $3.39 million they have already raised from the total cost of $5 million. Subtracting, $5.00 million – $3.39 million = $1.61 million.

4. **(C)** Since Nick cut two lengths of pipe that were each 1.45 meters long from the total 4-meter pipe, we need to subtract 1.45 meters from 4 meters twice. That would look like 4 – 1.45 – 1.45. However, that is not a choice. An equivalent answer is 4 – 2(1.45), which is 4 meters, the whole pipe, minus two times 1.45. We can also figure out how much pipe Nick would have left by subtracting: 4 – 1.45 – 1.45 = 1.1 meters. Then, we need to evaluate each of the four choices. Choice A = 4 – 2 – 1.45 = 0.55. Choice B = 2(4 – 1.45) = 5.1. Choice C = 4 – 2(1.45) = 1.1. Choice D = 4(2 – 1.45) = 2.2. Only choice C gives the correct amount of pipe that would be remaining.

Lesson 3.6—Skills Check (page 94)

1. **0.35**

$$\begin{array}{r} \overset{3}{0.5} \\ \times 0.7 \\ \hline 0.35 \end{array}$$

2. **0.035**

$$\begin{array}{r} \overset{3}{0.05} \\ \times\ 0.7 \\ \hline 0.035 \end{array}$$

3. **0.0035**

$$\begin{array}{r} \overset{3}{0.0005} \\ \times\qquad 7 \\ \hline 0.0035 \end{array}$$

4. **81.84**

$$\begin{array}{r} \overset{1}{34.1} \\ \times\ \ 2.4 \\ \hline 1364 \\ +6820 \\ \hline 81.84 \end{array}$$

5. **28.32**

$$\begin{array}{r} \overset{1}{23.6} \\ \times\ 1.2 \\ \hline 472 \\ +2360 \\ \hline 28.32 \end{array}$$

Lesson 3.6—ACCUPLACER Challenge (page 95)

1. **(B)** Jermaine paid $2.199 per gallon for 8 gallons. In total, he spent $8 \times \$2.199 = \$17.592$. To the nearest cent, that is $17.59.

2. **(B)** At $4.50 per pound, 1.5 pounds of Swiss cheese costs $\$4.50 \times \$1.50 = \$6.75$.

3. **(C)** The North American Plate moves westward at an average rate of about 0.034 inches per month. There are 12 months in a year. Therefore, multiply 0.034 by 12 to get 0.408, which is the number of inches that the North American Plate moves in a year on average.

Lesson 3.7—Skills Check (page 97)

1. **16** $4.8 \div 0.3 = 16$

2. **0.7** Moving both decimal points one place to the right, $0.56 \div 0.8$ becomes $5.6 \div 8$. Using long division, we get 0.7 as shown below.

$$\begin{array}{r} 0.7 \\ 8\overline{)5.6} \\ -5.6 \\ \hline 0 \end{array}$$

3. **70** Moving both decimal points three places to the right, $\dfrac{0.56}{0.008}$ becomes $560 \div 8 = 70$.

4. **22.5** Moving both decimal points two places to the right, $9.9 \div 0.44$ becomes $990 \div 44$. Using long division, we get 22.5 as shown below.

$$
\begin{array}{r}
022.5 \\
44\overline{)990.0} \\
-88 \\
\hline
110 \\
-88 \\
\hline
220 \\
-220 \\
\hline
0
\end{array}
$$

5. **32.5** Moving both decimal points four places to the right, $\dfrac{0.6695}{0.0206}$ becomes

   $6{,}695 \div 206$. Using long division, we get 32.5.

6. **65** Moving both decimal points three places to the right, $15.34 \div 0.236$ becomes $15{,}340 \div 236$. Using long division, we get 65 as shown below.

$$
\begin{array}{r}
00065 \\
236\overline{)15340} \\
-1416 \\
\hline
1180 \\
-1180 \\
\hline
0
\end{array}
$$

Lesson 3.7—ACCUPLACER Challenge (page 98)

1. **(A)** $\dfrac{3}{20} = 3 \div 20$. Using long division, we get 0.15 as shown below.

$$
\begin{array}{r}
0.15 \\
20\overline{)3.00} \\
-20 \\
\hline
100 \\
-100 \\
\hline
0
\end{array}
$$

2. **(D)** To find how many 0.25 meters are in 5.125 meters, we divide 5.125 by 0.25. Moving both decimal points two places to the right, we change the problem to $512.5 \div 25$. Using long division, we get 20.5 as shown below.

$$
\begin{array}{r}
20.5 \\
25\overline{)512.5} \\
-50 \\
\hline
12 \\
-0 \\
\hline
125 \\
-125 \\
\hline
0
\end{array}
$$

3. **(A)** Moving both decimal points one place to the right, $0.42 \div 1.8$ becomes $4.2 \div 18$. Using long division, we get $0.2\overline{3}$ as shown below.

$$
\begin{array}{r}
0.2\overline{3} \\
18{\overline{\smash{\big)}\,4.20}} \\
-36 \downarrow \\
\hline
60 \\
-54 \\
\hline
6
\end{array}
$$

Ratios and Proportions 4

Often in life—either in the workplace, at home, or in school—we need to use information from a small sample to make predictions or draw conclusions about a larger sample. In fact, we do this all the time without really thinking about it. Consider this scenario: You need 1 gallon of paint to cover 400 square feet of wall space. How many gallons would you need to paint a house with 1,600 square feet of wall space? You may be able to do this one pretty quickly. If 1 gallon covers 400 square feet, then 2 gallons would cover 800 square feet. 3 gallons would cover 1,200 square feet, which means that you would need 4 gallons to cover 1,600 square feet. By solving this problem, you are already applying the concepts of ratios and proportions, even if you didn't realize it! This example may seem simple, but it won't always be this straightforward.

As you work your way through this chapter, you'll develop the skills you need to solve problems like this one, along with others that require a deep understanding of ratios and proportions. Since the basic ideas of ratios and proportions will show up again in algebra and geometry topics, a clear understanding of these key concepts is essential to doing well on the ACCUPLACER.

In this chapter, we will work to master

- creating ratios based on word problems,
- writing ratios three different ways,
- calculating the unknown value in a proportion,
- solving proportion word problems, and
- establishing whether two ratios are proportional.

LESSON 4.1—RATIOS

A **ratio** is a comparison of two quantities that is usually expressed as a fraction. Before we get started working with ratios, it's important to understand the concept of quantity. In mathematics, a **quantity** is anything that can be measured or counted. A few examples of quantities are inches, pizzas, people, points, liters, and days. Notice that we can associate numbers with any one of these quantities; that is, we can measure an object as being 36 inches long, or we can tell someone that there are 18 people in a math class. In both of these examples, we have a numerical **value** and **units**.

Quantity of students = 18 (value) people (units)

value   units

The quantity of students in a math class is 18 people.

When we work with ratios, it is important that we pay careful attention to the quantities involved, not just the numbers. The biggest challenge in setting up and solving ratio questions is keeping track of the information you are given and organizing it correctly.

➡ Example 1

In a math class of 18 students, 12 students are girls. What is the ratio of girls to total students in the class?

Solution

The key word in ratio questions is the word *to*. The quantity that comes before the word *to* becomes the numerator in our ratio. The quantity that follows *to* is the denominator.

In this problem, the quantity before *to* is girls, and the quantity after *to* is total students, so we set up the fraction $\dfrac{\text{girls}}{\text{total students}}$.

Now we add the values of these quantities. The fraction becomes $\dfrac{12 \text{ girls}}{18 \text{ total students}}$.

With ratios, as with all fractions, we always have to reduce our answer. Looking at the fraction $\dfrac{12}{18}$, we see that the greatest common factor is 6, and we reduce as follows: $\dfrac{12}{18} \div \dfrac{6}{6} = \dfrac{2}{3}$.

Answer: The ratio of girls to total students in the class is 2 to 3. Notice that we read the ratio $\dfrac{2 \text{ girls}}{3 \text{ total students}}$ as "2 to 3."

ACCUPLACER TIPS

When you're working on the problems in this chapter, circle the word *to*. The quantity that comes before the word *to* should be the numerator in your ratio, and the quantity that follows *to* should be the denominator.

Ways of Writing Ratios

In Example 1, we saw that a ratio can be written as a fraction. This is the most common way to set up ratios and reduce them. The ratio "two to three" can be written out in three different ways, and you should be prepared to see any or all of these on the ACCUPLACER.

- **As a fraction:** $\frac{2}{3}$
- **With a colon:** $2:3$
- **With the word *to*:** 2 to 3

➡ Example 2

A math class has 18 students, 12 of whom are girls. What is the ratio of girls to boys in the class?

Solution

Be careful! This question looks similar to the one we just did, but there's actually a subtle difference. The quantities we are looking at in this question are girls and boys. Since the word *girls* comes before the word *to*, it will be in the numerator of our ratio. This means that the quantity *boys* will be in the denominator.

$$\frac{\text{girls}}{\text{boys}}$$

This ratio question requires an additional first step. We know the total number of people in the class, and we know the number of girls, so we need to subtract to find the number of boys. We calculate that $18 - 12 = 6$ boys.

Our ratio becomes $\frac{12 \text{ girls}}{6 \text{ boys}}$. Now, we look for a common denominator so that we can reduce the ratio.

$$\frac{12 \text{ girls}}{6 \text{ boys}} \div \frac{6}{6} = \frac{2}{1}$$

Normally, we would say that $\frac{2}{1}$ is just 2. With ratio questions, however, we have to give our answer as the relationship between two quantities. Therefore, the 1 in the denominator is crucial to our answer.

<u>Answer:</u> The ratio of girls to boys in the class is $2:1$.

You're probably noticing that the process for reducing ratios is the same as it is for reducing fractions. Generally speaking, it is. The big difference is that if you reduce a ratio, and it ends up with a 1 in the denominator, you need to make sure to leave it there. Going back to our definition, a ratio is a comparison of two quantities.

LESSON 4.1—SKILLS CHECK

Directions: Write the following ratios as fractions, and simplify them.

1. 36 to 40

2. 3 : 39

3. 15 : 45

4. 35 to 70

5. 60 : 60

6. 75 to 25

7. 56 to 42

8. 72 : 2

9. 64 to 256

10. 1.9 to 3.8

(Answers are on page 117.)

LESSON 4.1—ACCUPLACER CHALLENGE

Directions: For each of the questions below, choose the best answer from the four choices given.

1. Last season, a softball team played 18 games. The team won 15 of these games. What is the ratio of the softball team's wins to its total number of games played?

   (A) 1 to 5
   (B) 1 to 6
   (C) 5 to 6
   (D) 6 to 5

2. A standard deck of 52 cards is divided equally into four suits: hearts, diamonds, spades, and clubs. What is the ratio of hearts to clubs?

   (A) 1 : 52
   (B) 1 : 1
   (C) 4 : 1
   (D) 1 : 4

3. What is the ratio of the length of any square's side to its perimeter?

   (A) 1 : 8
   (B) 1 : 4
   (C) 1 : 2
   (D) 1 : 1

4. Eli's total monthly expenses include $180 for his utility bills, $155 for his groceries, and $835 for his house payment. What is the ratio of the cost of his utility bills to his total monthly expenses?

   (A) 31 to 234
   (B) 11 to 13
   (C) 4 to 13
   (D) 2 to 13

(Answers are on page 117.)

LESSON 4.2—PROPORTIONS

On the ACCUPLACER, you will often need to set up ratios to construct proportions and solve for an unknown quantity. Not only is this a valuable ACCUPLACER skill, but it's also something that can help you outside of the classroom. Let's start with a straightforward definition of a proportion: A **proportion** represents equality between two ratios. Proportions look like this:

$$\frac{a}{b} = \frac{c}{d}$$

On ACCUPLACER problems involving proportions, you will often be given three of the quantities needed to make a proportion. You will then need to solve for the missing item.

To see how proportions work, let's revisit the example of students in a math class.

➡ Example 1

On a given night, out of every 6 students who are enrolled in a math class, only 5 students attend class. If there are 24 students enrolled in a class, how many can we expect to attend?

Solution

Look carefully at the question. At the beginning of the prompt, we are given a ratio:

$$\frac{5 \text{ students attend class}}{6 \text{ students enrolled in the class}}$$

Based on this small amount of information, we can determine how many students would attend if the class had 12, 18, or even 600 students.

The second part of the question, though, is what determines how we should set up our proportion. It asks how many students will attend class if 24 students are enrolled in the class. We need to use this information to set up a ratio. The quantities we will use are *attend* and

enrolled, so we can write the ratio as $\frac{\text{attend}}{\text{enrolled}}$.

Both sides of the equation need to have the same relationship between the quantities. Since the question asked for the ratio of those who attend to those enrolled, the other side should have the same relationship. It will look like this:

$$\frac{\text{attend}}{\text{enrolled}} = \frac{\text{attend}}{\text{enrolled}}$$

Now we need to attach values to each quantity. Begin with the part of the problem that asked the question. We will fill in this information on the right side of the equal sign. Since we don't know how many will attend, we'll put a question mark there. You can also use a letter, like x or n, to represent this unknown part. (If you're not comfortable with using letters just yet, don't worry—we'll talk about this in Unit 2.)

$$\frac{\text{attend}}{\text{enrolled}} = \frac{?\ \text{attend}}{24\ \text{enrolled}}$$

Since the right side is finished, we now need to fill in the information on the left side. To do this, we revisit the ratio given in the first part of the question. Notice that the numbers are smaller here. It's a good idea—and common practice when using proportions—to have the smaller ratio on the left and the bigger ratio on the right.

$$\frac{5\ \text{attend}}{6\ \text{enrolled}} = \frac{?\ \text{attend}}{24\ \text{enrolled}}$$

Now our proportion is complete, and all we have to do from here is solve. To solve any proportion problem, you will follow two steps: First cross-multiply and then divide. For now, we can't do anything with the question mark. Notice that 5 and 24 are diagonal from each other, so you multiply these. After you do this, the 6 is the only number you haven't used yet. You can then divide 5×24 by 6. The process looks like this:

$$\frac{5 \times 24}{6} = 20$$

This works because the cross products are equal in a proportion. Therefore, if we find the product of the two values that are diagonal from one another and then divide by whatever is left, we can isolate the unknown part. This will make more sense when we start working on algebra later in this book.

Answer: If there are 24 students in a class, we could expect that 20 students would attend the class on a given night.

In all proportion questions, the most important thing to do is to make sure that the units in your ratios are the same on both sides. Since we placed attend over enrolled on one side, we have to have the exact same thing on the other side. If these do not match, you will get the wrong answer every time—and these wrong answers will almost certainly appear as multiple choices on the ACCUPLACER! Be careful and make sure to take the extra few seconds to add labels. It's also important to note that your answer to a proportion question could be a fraction, a decimal, or a whole number.

➡ Example 2

A softball team wins 3 games for every 2 games that they lose. If the team plays 20 games in a season, how many can they expect to win?

Solution

As we did with the last example, start with the part of the problem that poses a question. Notice that this question states that the team *plays* 20 games, and it asks us to find out how many games the team would *win*. We could think of the number of games that the team plays as the total, so let's set up the proportion as

$$\frac{\text{wins}}{\text{total}} = \frac{\text{wins}}{\text{total}}$$

Now we fill in the blanks. On the right side, we use the information from the question, putting a question mark in place of *wins* and 20 in place of *total*. Now let's fill in the left side. The prompt tells us that the team wins 3 games and loses 2, which means that they played 5 in total. Therefore, we fill in the proportion as:

$$\frac{3 \text{ wins}}{5 \text{ total}} = \frac{? \text{ wins}}{20 \text{ total}}$$

Cross-multiply and divide.

$$\frac{3 \times 20}{5} = \frac{60}{5} = 12$$

<u>Answer:</u> The softball team could expect to win 12 out of 20 games.

Let's look at one more common application of a proportion.

➡ Example 3

You are planning a road trip, and you consult a map. When you measure the distance between two cities, you see that they are 3 inches apart. The map scale says that $\frac{1}{4}$ of an inch is equal to 120 actual miles. How many miles apart are the two cities?

Solution

Here, we need to use the information from the map to draw a conclusion about the actual distance between the two cities. There are a few different ratios we could set up to get started, but it makes the most sense to start with the scale that we are given. We are told that $\frac{1}{4}$ of an inch on the map is equal to 120 miles. Therefore, we start with the ratio:

$$\frac{\frac{1}{4} \text{ inches}}{120 \text{ miles}}$$

It's worth noting that if we set the proportion up as $\frac{\text{miles}}{\text{inches}}$, we would still get the correct answer, as long as both sides of the proportion are set up the same way.

Now we know that the other ratio needs to be set up with inches in the numerator and miles in the denominator. We can fill out the rest using the other piece of information that we were given in the prompt.

$$\frac{\frac{1}{4} \text{ inches}}{120 \text{ miles}} = \frac{3 \text{ inches}}{? \text{ miles}}$$

The last step is to cross-multiply and divide. $120 \times 3 = 360$, and $360 \div \frac{1}{4} = 1,440$ miles.

<u>Answer:</u> The actual distance is 1,440 miles apart.

Look at the proportion examples we've done so far. One thing you should notice is that the unknown, or variable, can be in a different place each time, but the process used to solve each proportion is the same. To reiterate, that process is:

1. Set up a proportion.
2. Multiply the two numerical quantities that are diagonal from one another.
3. Divide by the remaining numerical quantity.

How to Determine if Two Ratios Are Proportional

In some math problems, you will need to check to see if two ratios are proportional to one another. Remember that **proportional** means that the ratios are equivalent. One very simple way of doing this is to set the two ratios up as fractions and then, through the processes of raising or reducing, check to see if the fractions are equivalent.

➡ Example 4

Is the ratio 4 : 5 proportional to the ratio 75 : 100?

Solution

You might notice right away that these ratios aren't proportional because they aren't equivalent fractions! The ratio $\frac{4}{5}$ is already in its simplest form, and the ratio $\frac{75}{100}$ simplifies to $\frac{3}{4}$.

Since $\frac{4}{5}$ is not equal to $\frac{3}{4}$, the ratios aren't proportional. Let's look at another way of determining proportionality—one that might come in handy when the ratios you're working with don't reduce nicely or contain fractions or decimals.

To get started, let's assume that the ratios are in fact proportional. You'll write out the proportion like this:

$$\frac{4}{5} = \frac{75}{100}$$

If the two ratios are proportional, then their **cross products** will be equal. This means that when we multiply the two sets of diagonal numbers, they should be equal.

The first cross product is $4 \times 100 = 400$. The other cross product is $5 \times 75 = 375$.

Answer: Since 375 is not equal to 400, the two ratios are not proportional.

➡ Example 5

Consider two rectangles. The first rectangle has a length of 8 inches and a width of 28 inches. The second rectangle has a length of 12 inches and a width of 42 inches. Are the two rectangles proportional to one another?

Solution

It's clear from the question prompt that the two rectangles are not equal in size, but they could still be proportional. This means that they share a similar relationship between their lengths and widths.

Let's begin by setting up our proportion. Notice that the question mentions the quantities of length and width. A good place to start would be writing a proportion where the quantities associated with the first rectangle are on the left, and the quantities associated with the second rectangle are on the right.

$$\frac{8 \text{ length}}{28 \text{ width}} = \frac{12 \text{ length}}{42 \text{ width}}$$

To check for proportionality, we need to cross-multiply. The first cross product is $8 \times 42 = 336$, and the second cross product is $28 \times 12 = 336$.

Answer: Since the two cross products are equal, we can conclude that the two rectangles are proportional.

LESSON 4.2—SKILLS CHECK

Directions: Solve for the unknown in each of the following proportions.

1. $\dfrac{9}{?} = \dfrac{18}{36}$

2. $\dfrac{?}{15} = \dfrac{40}{150}$

3. $\dfrac{7}{9} = \dfrac{49}{?}$

4. $\dfrac{2}{3} = \dfrac{?}{10}$

5. $\dfrac{1}{12} = \dfrac{?}{30}$

6. $\dfrac{2}{9} = \dfrac{80}{?}$

(Answers are on page 118.)

LESSON 4.2—ACCUPLACER CHALLENGE

Directions: For each of the questions below, choose the best answer from the four choices given.

1. After looking over your finances, you have determined that you spend $9 for every $16 that you make. If you made $40,000 in a given year, how much did you spend that year?

   (A) $17,500
   (B) $22,500
   (C) $24,875
   (D) $29,325

2. Mr. Appleton teaches mathematics at a high school in New York City. He says that on a given day, he can expect 4 students to be on time for every 2 students who are late. If there are 24 students in his class, how many can he expect to be on time?

   (A) 16
   (B) 18
   (C) 20
   (D) 22

3. A paint mixture calls for 2 parts white paint and 3 parts blue paint. How much white paint would you need in order to produce 17.5 gallons of the paint mixture?

(A) 5 gallons
(B) 5.5 gallons
(C) 6.5 gallons
(D) 7 gallons

4. One triangle has a height of 15 feet and a base length of 8 feet. It is proportional to a second triangle that has a height of 37.5 feet. What is the length of the second triangle's base?

(A) 20 feet
(B) 28 feet
(C) 30 feet
(D) 70 feet

5. Each day, 375 flights leave the Rapid Metropolis International Airport. The airport's leadership team has determined that 3 flights out of every 15 are delayed. How many flights leaving Rapid Metropolis International Airport could we expect to be on time?

(A) 63
(B) 75
(C) 250
(D) 300

6. Consider the diagram below.

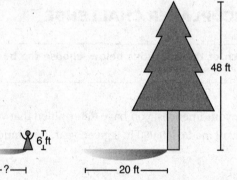

What would be the length of the woman's shadow in the diagram?

(A) 3 feet, 2 inches
(B) 2 feet, 10 inches
(C) 2 feet, 8 inches
(D) 2 feet, 6 inches

(Answers are on page 118.)

To complete a set of practice questions that reviews all of Chapter 4, go to:
barronsbooks.com/TP/accuplacer/math29qw/

Lesson 4.1—Skills Check (page 110)

1. $\dfrac{9}{10}$ 36 to 40 $= \dfrac{36}{40} \div \dfrac{4}{4} = \dfrac{9}{10}$

2. $\dfrac{1}{13}$ $3:39 = \dfrac{3}{39} \div \dfrac{3}{3} = \dfrac{1}{13}$

3. $\dfrac{1}{3}$ $15:45 = \dfrac{15}{45} \div \dfrac{15}{15} = \dfrac{1}{3}$

4. $\dfrac{1}{2}$ 35 to 70 $= \dfrac{35}{70} \div \dfrac{35}{35} = \dfrac{1}{2}$

5. $\dfrac{1}{1}$ $60:60 = \dfrac{60}{60} \div \dfrac{60}{60} = \dfrac{1}{1}$. Remember that a ratio is a comparison of two numbers,

   so we need to have a 1 in the numerator and a 1 in the denominator. It reads "one to one."

6. $\dfrac{3}{1}$ 75 to 25 $= \dfrac{75}{25} \div \dfrac{25}{25} = \dfrac{3}{1}$

7. $\dfrac{4}{3}$ 56 to 42 $= \dfrac{56}{42} \div \dfrac{14}{14} = \dfrac{4}{3}$

8. $\dfrac{36}{1}$ $72:2 = \dfrac{72}{2} \div \dfrac{2}{2} = \dfrac{36}{1}$

9. $\dfrac{1}{4}$ 64 to 256 $= \dfrac{64}{256} \div \dfrac{64}{64} = \dfrac{1}{4}$

10. $\dfrac{1}{2}$ 1.9 to 3.8 $= \dfrac{1.9}{3.8} \div \dfrac{1.9}{1.9} = \dfrac{1}{2}$

Lesson 4.1—ACCUPLACER Challenge (pages 110–111)

1. **(C)** $\dfrac{15 \text{ wins}}{18 \text{ total games}} \div \dfrac{3}{3} = \dfrac{5}{6}$. The ratio of wins to total games played is 5 to 6.

2. **(B)** If there are an equal number of cards in each suit, then the number of hearts is $52 \div 4 = 13$. There must also be 13 clubs. Therefore, the ratio of hearts to clubs is $\dfrac{13 \text{ hearts}}{13 \text{ clubs}}$. This ratio is equal to $1:1$.

3. **(B)** A square has four equal sides, and a square's perimeter is calculated by adding the lengths of all four sides, or by multiplying the length of one side by 4. Therefore, the ratio of a square's side to its perimeter is always $1:4$, no matter the size of the square.

4. **(D)** Eli's total monthly expenses are $\$180 + \$155 + \$835 = \$1{,}170$. To find the ratio of his utility bills to his total monthly expenses, write:

$$\frac{180 \text{ dollars in utilities}}{1170 \text{ total monthly expenses}} \div \frac{10}{10} = \frac{18}{117} \div \frac{9}{9} = \frac{2}{13}$$

Lesson 4.2—Skills Check (page 115)

1. **18** $\dfrac{9}{?} = \dfrac{18}{36}$

    $\dfrac{9 \times 36}{18} = \dfrac{324}{18} = 18$

2. **4** $\dfrac{?}{15} = \dfrac{40}{150}$

    $\dfrac{15 \times 40}{150} = \dfrac{600}{150} = 4$

3. **63** $\dfrac{7}{9} = \dfrac{49}{?}$

    $\dfrac{9 \times 49}{7} = \dfrac{441}{7} = 63$

4. **$6\dfrac{2}{3}$** $\dfrac{2}{3} = \dfrac{?}{10}$

    $\dfrac{2 \times 10}{3} = \dfrac{20}{3} = 6\dfrac{2}{3}$

5. **$2\dfrac{1}{2}$** $\dfrac{1}{12} = \dfrac{?}{30}$

    $\dfrac{1 \times 30}{12} = \dfrac{30}{12} = 2\dfrac{1}{2}$

6. **360** $\dfrac{2}{9} = \dfrac{80}{?}$

    $\dfrac{9 \times 80}{2} = \dfrac{720}{2} = 360$

Lesson 4.2—ACCUPLACER Challenge (pages 115–116)

1. **(B)** $\dfrac{9 \text{ spent}}{16 \text{ made}} = \dfrac{? \text{ spent}}{40{,}000 \text{ made}}$

    $\dfrac{9 \times 40{,}000}{16} = \dfrac{360{,}000}{16} = \$22{,}500$

2. **(A)** $\dfrac{4 \text{ on time}}{6 \text{ total}} = \dfrac{? \text{ on time}}{24 \text{ total}}$

    $\dfrac{4 \times 24}{6} = \dfrac{96}{6} = 16$

3. **(D)** $\dfrac{2 \text{ white}}{5 \text{ mixture}} = \dfrac{? \text{ white}}{17.5 \text{ mixture}}$

    $\dfrac{2 \times 17.5}{5} = \dfrac{35}{5} = 7$

4. **(A)** $\dfrac{15 \text{ height}}{8 \text{ base}} = \dfrac{37.5 \text{ height}}{? \text{ base}}$

    $\dfrac{8 \times 37.5}{15} = \dfrac{300}{15} = 20$

5. **(D)** If 3 out of every 15 flights are delayed, then 12 flights must depart on time.

    $\dfrac{12 \text{ on time}}{15 \text{ total}} = \dfrac{? \text{ on time}}{375 \text{ total}}$

    $\dfrac{12 \times 375}{15} = \dfrac{4500}{15} = 300$

6. **(D)** $\dfrac{6 \text{ height}}{? \text{ shadow}} = \dfrac{48 \text{ height}}{20 \text{ shadow}}$

    $\dfrac{6 \times 20}{48} = \dfrac{120}{48} = 2\dfrac{1}{2}$

    $2\dfrac{1}{2}$ feet is equal to 2 feet, 6 inches.

Percentages

5

In the previous chapter, we explored proportions and proportional relationships. One of the most common, everyday applications of proportions is the use of percentages. Having a strong knowledge of this critical life skill is also essential to scoring well on the Arithmetic test of the ACCUPLACER.

> **In this chapter, we will work to master**
>
> - identifying equivalent fractions, decimals, and percentages,
> - solving the three types of percentage word problems using proportions,
> - solving "is/of" percentage problems, and
> - solving multistep percentage word problems using proportions.

The Meaning of Percent

Centum means one hundred in Latin; as such, words that contain *cent* often relate to the number 100. For example, a century is 100 years and a centimeter is $\frac{1}{100}$ of a meter. Naturally, **percent** also relates to 100. Both the word percent and the % symbol mean "for every 100."

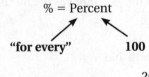

$$20\% = 20 \text{ for every } 100 = \frac{20}{100}$$

LESSON 5.1—FRACTION, DECIMAL, AND PERCENTAGE CONVERSIONS

On the ACCUPLACER, you will be required to demonstrate your mastery of representing percentages as equivalent fractions or decimals. Any percent can be converted into a fraction or a decimal. Becoming familiar with the process for making these conversions can help you solve a variety of problems involving percentages.

Take a look at Example 1, which asks you to relate a percentage to fractions and a decimal.

➡ Example 1

All of the following are equivalent ways to write 40% of *W* EXCEPT

(A) $\dfrac{40W}{100}$

(B) $\dfrac{1}{40}W$

(C) $0.4W$

(D) $\dfrac{2}{5}W$

Solution

First, let's recognize that "40% of *W*" can be translated into 40% × *W*, since the word *of* means multiply in this context. All of the answer choices have *W* times some quantity, but only one of these quantities is NOT equal to 40%.

Throughout this chapter, we will review how to change a percentage to a fraction or to a decimal. After reviewing this chapter, you will easily be able to determine that:

$$40\% = \frac{40}{100} = \frac{2}{5} = 0.40 = 0.4$$

Looking at the answer choices, we can determine that only choice B is NOT equivalent because $\dfrac{1}{40} \neq 40\%$.

"not equal" symbol

Answer: Therefore, the correct answer is choice B.

Now, let's review how to make those conversions.

To Change a Percentage to a Fraction:

1. Write the number in front of the % symbol as the numerator over a fraction bar.
2. Write 100 as the denominator.
3. Discard the % symbol.
4. Reduce fractions whenever possible.

➡ Example 2

Convert 23% to a fraction.

Solution

$$23\% = \frac{23}{100}$$

Answer: $\dfrac{23}{100}$

➡ Example 3

Convert 8% to a fraction.

Solution

$$8\% = \frac{8}{100} = \frac{2}{25}$$

Answer: $\dfrac{8}{100}$ or $\dfrac{2}{25}$

➡ Example 4

Convert 32.9% to a fraction.

Solution

$$32.9\% = \frac{32.9}{100} = \frac{329}{1000}$$

Answer: $\frac{32.9}{100}$ or $\frac{329}{1000}$

➡ Example 5

Convert 360% to a fraction.

Solution

$$360\% = \frac{360}{100} = \frac{18}{5} = 3\frac{3}{5}$$

Answer: $\frac{360}{100}$ or $\frac{18}{5}$ or $3\frac{3}{5}$

To Change a Decimal to a Percentage:

1. Move the decimal point two places to the right.
2. Add a % symbol.

➡ Example 6

Convert 0.45 to a percentage.

Solution

$$0.45 = 45\%$$

Answer: 45%

➡ Example 7

Convert 1.283 to a percentage.

Solution

$$1.283 = 128.3\%$$

Answer: 128.3%

➡ Example 8

Convert 0.03 to a percentage.

Solution

$$0.03 = 3\%$$

Answer: 3%

➡ Example 9

Convert 8 to a percentage.

Solution

$$8.00 = 800\%$$

Answer: 800%

➡ Example 10

Convert 0.5 to a percentage.

Solution

$$0.50 = 50\%$$

Answer: 50%

To Change a Percentage to a Decimal:

1. Move the decimal point two places to the left.
2. Discard the % symbol.

➡ Example 11

Convert 21% to a decimal.

Solution

$$2\,1\% = 0.21$$

Answer: 0.21

➡ Example 12

Convert 37% to a decimal.

Solution

$$3\,7\% = 0.37$$

Answer: 0.37

➡ Example 13

Convert 147% to a decimal.

Solution

$$1\,4\,7\% = 1.47$$

Answer: 1.47

➡ Example 14

Convert 83.5% to a decimal.

Solution

$$8\,3\,.\,5\% = 0.835$$

Answer: 0.835

➡ Example 15

Convert 5% to a decimal.

Solution

$$5\% = 0.05$$

Answer: 0.05

To Change a Fraction to a Percentage:

1. Change the fraction to a decimal by dividing the numerator by the denominator.
2. Next, change the decimal to a percentage by moving the decimal two places to the right.
3. Add a % symbol.

➡ Example 16

Convert $\frac{9}{50}$ to a percentage.

Solution

$$\frac{9}{50} = 9 \div 50 = 0.18 = 18\%$$

Answer: 18%

➡ Example 17

Convert $\frac{3}{10}$ to a percentage.

Solution

$$\frac{3}{10} = 3 \div 10 = 0.30 = 30\%$$

Answer: 30%

Example 18

Convert $\frac{4}{5}$ to a percentage.

Solution

$$\frac{4}{5} = 4 \div 5 = 0.80 = 80\%$$

Answer: 80%

Example 19

Convert $\frac{2}{500}$ to a percentage.

Solution

$$\frac{2}{500} = 2 \div 500 = 0.004 = 0.4\%$$

Answer: 0.4%

Example 20

Convert $\frac{247}{100}$ to a percentage.

Solution

$$\frac{247}{100} = 247 \div 100 = 2.47 = 247\%$$

Answer: 247%

Example 21

Convert $2\frac{3}{10}$ to a percentage.

Solution

$$2\frac{3}{10} = \frac{23}{10} = 23 \div 10 = 2.30 = 230\%$$

Answer: 230%

ACCUPLACER TIPS

A handy and interesting thing to remember is that the U.S. dollar is broken into 100 pieces called cents. Percentages, decimals, and fractions have the exact same relationship as dollars and cents!

1 cent = \$0.01 = $\frac{1}{100}$ of a dollar = 1 percent of a dollar *or* 1% of \$1.00

5 cents = \$0.05 = $\frac{5}{100}$ of a dollar = 5 percent of a dollar *or* 5% of \$1.00

10 cents = \$0.10 = $\frac{10}{100}$ of a dollar = 10 percent of a dollar *or* 10% of \$1.00

25 cents = \$0.25 = $\frac{25}{100}$ of a dollar = 25 percent of a dollar *or* 25% of \$1.00

50 cents = \$0.50 = $\frac{50}{100}$ of a dollar = 50 percent of a dollar *or* 50% of \$1.00

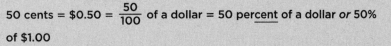

100 cents = \$1.00 = $\frac{100}{100}$ of a dollar = 100 percent of a dollar *or* 100% of \$1.00

LESSON 5.1—SKILLS CHECK

> **Directions:** Change each percentage to a fraction that is reduced to its simplest form.

1. 47% =

2. 6% =

3. 120% =

4. 2.5% =

5. 0.6% =

> **Directions:** Change each decimal to a percentage.

6. 0.92 =

7. 3.287 =

8. 0.08 =

9. 0.3 =

10. 6 =

> **Directions:** Change each percentage to a decimal.

11. 43% =

12. 76% =

13. 125% =

14. 45.5% =

15. 8% =

> **Directions:** Change each fraction to a percentage.

16. $\dfrac{3}{25}$ =

17. $\dfrac{7}{10}$ =

18. $\dfrac{2}{5}$ =

19. $\dfrac{7}{250}$ =

20. $\dfrac{180}{100}$ =

(Answers are on page 139.)

LESSON 5.1—ACCUPLACER CHALLENGE

> **Directions:** For each of the questions below, choose the best answer from the four choices given.

1. All of the following are equivalent ways to write 80% of P EXCEPT

    (A) $80P \div 100$

    (B) $\dfrac{80}{100}P$

    (C) $0.08P$

    (D) $\dfrac{4}{5}P$

2. Which of the following lists of numbers is in order from least to greatest?

    (A) $6\%, 0.4, \dfrac{1}{2}, 0.32$

    (B) $0.32, 0.4, \dfrac{1}{2}, 6\%$

    (C) $0.4, 0.32, \dfrac{1}{2}, 6\%$

    (D) $6\%, 0.32, 0.4, \dfrac{1}{2}$

3. 250% is equivalent to

    (A) 0.25

    (B) $2\dfrac{1}{2}$

    (C) 25

    (D) 250

4. 0.4% is equivalent to

    (A) $\dfrac{1}{2500}$

    (B) $\dfrac{1}{250}$

    (C) $\dfrac{1}{25}$

    (D) 40

(Answers are on page 139.)

LESSON 5.2—SOLVING THE THREE TYPES OF PERCENTAGE WORD PROBLEMS

On the ACCUPLACER, you will find percentage application problems, or word problems. Loosely speaking, there are only three general types of percentage word problems. This stems from the fact that we can always represent a percentage using the following proportion:

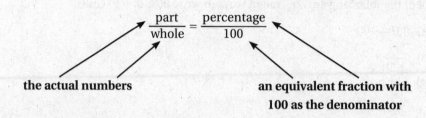

$$\frac{\text{part}}{\text{whole}} = \frac{\text{percentage}}{100}$$

the actual numbers   an equivalent fraction with
                                      100 as the denominator

Notice that there will always be four values in the proportion. We know that, in a percentage problem, one of the numbers is always 100. That leaves only three other quantities to worry about—hence the three types of percentage problems.

Consider the following three problems:

- Filomena spent 15% of her paycheck this week on groceries. How much did she spend on groceries if her paycheck was $1,200?
- Ralph spent $320, or 20%, of his paycheck on groceries. How much was his paycheck?
- Gina spent $150 of her $2,000 paycheck on groceries. What percent of her paycheck did she spend on groceries?

On the surface, these three problems are very similar, but take a minute to notice what the question is actually asking in each question. In the first question, we're asked to find the amount spent on groceries; in the second, we need to find the total amount of the paycheck; and in the third, we have to find a missing percentage. Now let's look at how the solution methods for each of these three problems differ.

To Solve a Percentage Word Problem Using a Proportion:

1. Identify the part, the whole, and the percentage mentioned in the problem.
2. Set up a proportion using the relationship $\frac{\text{part}}{\text{whole}} = \frac{\text{percentage}}{100}$.
3. Cross-multiply and divide.

➥ Example 1

Finding the Part in a Percentage Word Problem

Filomena spent 15% of her paycheck this week on groceries. How much did she spend on groceries if her paycheck was $1,200?

Solution

First, read the problem carefully. Note that we're being asked to find the amount spent on groceries. Then go back, and identify the part, the whole, and the percentage. The quantity that the question is asking us to find is the unknown, so we will replace that quantity with a question mark.

- The part = the amount that Filomena spent on groceries this week = ?
- The whole = her paycheck = $1,200
- The percentage = the percentage of her paycheck that she spent on groceries = 15%

Next, write your proportion using the relationship $\frac{part}{whole} = \frac{percentage}{100}$.

$$\frac{?}{1200} = \frac{15}{100}$$

Use the cross-multiplication skills you practiced in Chapter 4 to solve for the unknown quantity. Remember that we multiply the two numbers that are diagonal from one another, and then divide by the remaining quantity. In this case, first multiply $1,200 by 15, and then divide by 100.

$$\frac{1200 \times 15}{100} = \frac{18,000}{100} = 180$$

Answer: $180

Go back and make sure that your answer appropriately answers the question and that it makes sense. In this case, if she spends 15% of her paycheck on groceries, then $180 is a good answer.

➡ Example 2

Finding the Whole in a Percentage Word Problem

Ralph spent $320, or 20%, of his paycheck on groceries. How much was his paycheck?

Solution

First, read the problem carefully. Notice that we're being asked to calculate the total amount of Ralph's paycheck. Now identify the part, the whole, and the percentage. The quantity that the question is asking us to find is the unknown, so we will replace that quantity with a question mark.

- The part = the amount that Ralph spent on groceries this week = $320
- The whole = his paycheck = ?
- The percentage = the percent of his paycheck that he spent on groceries = 20%

Next, write your proportion using the relationship $\frac{part}{whole} = \frac{percentage}{100}$.

$$\frac{320}{?} = \frac{20}{100}$$

(Note that the question mark is in a different position than it was in Example 1. This is because we're trying to find the whole rather than the part.)

Now, cross-multiply and divide.

$$\frac{320 \times 100}{20} = \frac{32,000}{20} = 1,600$$

Answer: $1,600

Let's check to see if this answer makes sense. We know that the amount of Ralph's paycheck must have been quite a bit larger than the amount that he spent on groceries, so it makes sense that Ralph's paycheck would be $1,600.

➡ **Example 3**

Finding the Percentage in a Percentage Word Problem

Gina spent $150 of her $2,000 paycheck on groceries. What percent of her paycheck did she spend on groceries?

Solution

This time around, we're being asked to find the missing percentage. Next, identify the three parts of the proportion that we need to fill in.

- The part = the amount that Gina spent on groceries this week = $150
- The whole = her paycheck = $2,000
- The percentage = the percent of her paycheck that she spent on groceries = ?

Next, write your proportion using the relationship $\frac{part}{whole} = \frac{percentage}{100}$.

$$\frac{150}{2000} = \frac{?}{100}$$

(Note that here the question mark is above the 100. This is where it will always be when we need to find a missing percentage.)

Now, cross-multiply and divide.

$$\frac{150 \times 100}{2000} = \frac{15,000}{2000} = 7.5$$

Answer: 7.5%

Go back and make sure that your answer appropriately answers the question. Relative to her income, Gina only spent a small amount of her paycheck on groceries. Therefore, it makes sense that this would amount to only 7.5%.

LESSON 5.2—ACCUPLACER CHALLENGE

Directions: For each of the questions below, choose the best answer from the four choices given.

1. The Tornados won 12 games out of the 25 games that they played. What percentage of their games did the Tornados win?

   (A) 12%
   (B) 21%
   (C) 46%
   (D) 48%

2. On a test, a student got 80% of the questions correct. If the student got 32 questions correct, how many questions were on the test?

   (A) 25
   (B) 40
   (C) 48
   (D) 80

3. Sean earns $1,200 a week. How much money is deducted from Sean's weekly pay for taxes and Social Security if his employer deducts 26% of his earnings for taxes and Social Security?

   (A) $12.26
   (B) $300.00
   (C) $312.00
   (D) $888.00

4. After a car crash, Stan's insurance company paid 85% of the cost of getting his car fixed. If the repair bill was $725, how much did his insurance company pay?

   (A) $108.75
   (B) $329.50
   (C) $616.25
   (D) $640.00

5. Got-To-Go Delivery Company employs 100 drivers. These drivers make up 80% of the company. How many employees work for Got-To-Go Delivery Company?

   (A) 80
   (B) 120
   (C) 125
   (D) 180

6. Fiona makes $2,419 a month and pays $595 for rent. Rent is approximately what percent of Fiona's income?

   (A) 25%
   (B) 30%
   (C) 33%
   (D) 40%

(Answers are on page 140.)

LESSON 5.3—IS/OF PERCENTAGE PROBLEMS

On the ACCUPLACER, you can expect to see the classic "is/of" percentage problems. These work exactly the same as the other word problems we already looked at, but there are fewer words that tell us what we need to do.

For example, consider the following three problems:

- What is 15% of 240?
- 20 is 40 percent of what number?
- 3 is what percent of 24?

To make sense of these problems, let's first look at a true statement involving "is" and "of" percentage.

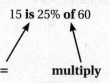

Changing the percent into a decimal, we can translate this statement into:

We can also use the proportion $\dfrac{\text{part}}{\text{whole}} = \dfrac{\text{percentage}}{100}$ to solve problems like this.

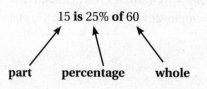

We have all of the pieces, so there isn't anything to solve for here. By cross-multiplying and dividing, we can see that our proportion fits the "is/of" type of percentage problem nicely.

$$\frac{15}{60} = \frac{25}{100}$$

$$(15)(100) = (25)(60)$$

$$1,500 = 1,500 \checkmark$$

Now let's take a closer look at how we can use this proportion to solve our example problems.

To Solve "Is/Of" Percentage Word Problems Using a Proportion:

1. If necessary, rewrite the problem in the form: The *part* is a *percentage* of the *whole*.

2. Set up a proportion using the relationship $\frac{part}{whole} = \frac{percentage}{100}$.

3. Cross-multiply and divide.

➡ Example 1

Finding the "Part" in an "Is/Of" Percentage Word Problem

What is 15% of 240?

Solution

First, rewrite the sentence in the form: The *part* is a *percentage* of the *whole*. In this problem, the word *What* is where the *part* needs to go. Since we don't know the value of the part, we use a question mark.

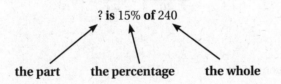

? **is** 15% **of** 240

the part   the percentage   the whole

Next, write your proportion using the relationship $\frac{part}{whole} = \frac{percentage}{100}$.

$$\frac{?}{240} = \frac{15}{100}$$

Then, cross-multiply and divide.

$$\frac{240 \times 15}{100} = \frac{3600}{100} = 36$$

Answer: 36 is 15% of 240.

Check your answer.

$$36 = (0.15)(240)$$

$$36 = 36 \quad ✔$$

➡ Example 2

Finding the "Whole" in an "Is/Of" Percentage Word Problem

20 is 40 percent of what number?

Solution

First, rewrite the sentence in the form: The *part* is a *percentage* of the *whole*. In this problem, notice how the phrase "what number" comes after the word "of." This tells us that the number we're looking for is the whole.

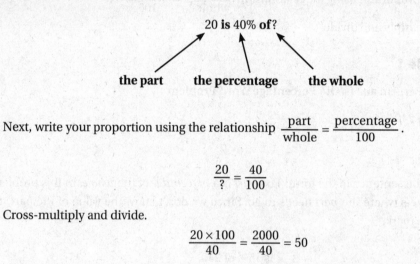

Next, write your proportion using the relationship $\dfrac{\text{part}}{\text{whole}} = \dfrac{\text{percentage}}{100}$.

$$\frac{20}{?} = \frac{40}{100}$$

Cross-multiply and divide.

$$\frac{20 \times 100}{40} = \frac{2000}{40} = 50$$

Answer: 20 is 40 percent of 50.

Check your answer.

$$20 = (0.4)(50)$$

$$20 = 20 \quad ✔$$

➡ Example 3

Finding the "Percentage" in an "Is/Of" Percentage Word Problem

3 is what percent of 24?

Solution

First, rewrite the sentence in the form: The *part* is a *percentage* of the *whole*. The phrase "what percent" tells us that we need to find a missing value for the percentage.

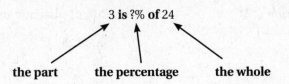

$$3 \text{ is } ?\% \text{ of } 24$$

the part   the percentage   the whole

Next, write your proportion using the relationship $\dfrac{\text{part}}{\text{whole}} = \dfrac{\text{percentage}}{100}$.

$$\frac{3}{24} = \frac{?}{100}$$

Cross-multiply and divide.

$$\frac{3 \times 100}{24} = \frac{300}{24} = 12.5$$

Answer: 3 is 12.5% of 24.

Check your answer. Remember that to do this, we should convert 12.5% back into a decimal.

$$3 = (0.125)(24)$$

$$3 = 3 \ \checkmark$$

> ## ACCUPLACER TIPS
>
> Another way to find a "% of" something is to change the percentage to a decimal and then multiply. This can be handy for problems such as
>
> ### What is 20% of 14?
>
> To find a "% of" a whole using multiplication, change the percentage to a decimal by moving the decimal point two places to the left. Then multiply the decimal times the whole.
>
> ### 20% of 14 becomes 0.20 × 14 = 2.8

LESSON 5.3—ACCUPLACER CHALLENGE

> **Directions:** For each of the questions below, choose the best answer from the four choices given.

1. 10 is 20% of what number?

    (A) 0.2
    (B) 2
    (C) 50
    (D) 500

2. What number is 15% of 70?

    (A) 4.7
    (B) 10.5
    (C) 467
    (D) 1,050

3.   20 is what percent of 5?
   (A) 4
   (B) 25
   (C) 250
   (D) 400

4.   What percent of 120 is 18?

   (A) 15
   (B) 67
   (C) 150
   (D) 670

5.   20% of what number is 5?

   (A) 0.25
   (B) 1
   (C) 25
   (D) 100

6.   150% of 50 is what number?

   (A) 3
   (B) 7.5
   (C) 33.3
   (D) 75

(Answers are on page 141.)

LESSON 5.4—MULTISTEP PERCENTAGE WORD PROBLEMS

All percentage relationships can be written using the proportion $\dfrac{\text{part}}{\text{whole}} = \dfrac{\text{percentage}}{100}$.
However, not all problems can be directly answered using only this proportion. In some cases, there may be an extra step or steps you have to take in order to answer the question.
   Consider the following three examples:

- Wendy bought a $120 dress on sale for 40% off. How much did she pay for the dress?
- John spends $10 on materials for every toy car that he makes and sells. How much should he charge for each toy car if he wants to make a 15% profit?
- This year, 20% more students attend the Mountain View School than last year. How many students attended the school last year if 72 students attend the school this year?

You have already practiced all of the skills you need to solve these problems successfully. You just need to make sure that you read the problems carefully and that your answer is appropriate for the question asked.

➥ Example 1

Wendy bought a $120 dress on sale for 40% off. How much did she pay for the dress?

Solution

Here, $120 is the regular price of the dress. This makes it the whole. The part is the amount that the dress is discounted.

- The part = ?
- The whole = $120
- The percentage = 40%

Next, write your proportion using the relationship $\dfrac{\text{part}}{\text{whole}} = \dfrac{\text{percentage}}{100}$.

$$\frac{?}{120} = \frac{40}{100}$$

Then cross-multiply and divide.

$$\frac{120 \times 40}{100} = \frac{4800}{100} = 48$$

We found the part, which is $48, but that does not answer the question. The question is asking how much Wendy paid for the dress. If the dress was originally $120, and it is on sale for 40% off, Wendy must pay at least $60, since that would be 50% off. She's only getting 40% off, so she must pay more than $60 for the dress.

$48 is not how much she pays; it's how much is taken off of the price of the dress because the dress is on sale. To find out how much Wendy pays in the end, we need to subtract the discount of $48 from the original full price amount of $120.

$$\$120 - \$48 = \$72$$

Answer: Therefore, Wendy pays $72 for the dress.

Remember, whenever you think you have completed a problem, stop and ask yourself, "Did I answer the question that was asked and does my answer make sense?"

➡ Example 2

John spends $10 on materials for every toy car that he makes and sells. How much should he charge for each toy car if he wants to make a 15% profit?

Solution

This one is a little bit tricky. John wants to make 15% more than he spends on materials; that's the idea of profit. This suggests that the total cost of materials should be the whole.

- The part = ?
- The whole = $10
- The percentage = 15%

Next, write your proportion using the relationship $\frac{\text{part}}{\text{whole}} = \frac{\text{percentage}}{100}$.

$$\frac{?}{10} = \frac{15}{100}$$

Then cross-multiply and divide.

$$\frac{10 \times 15}{100} = \frac{150}{100} = 1.5$$

The final answer cannot be $1.50, since John needs to make a profit. Therefore, he must charge more than $10. What we have found is that $1.50 is 15% of $10. This is how much John must charge on top of the original $10 so that he can make a profit. Therefore, we need to add the 15%, or $1.50, to the original amount.

$$\$10 + \$1.50 = \$11.50$$

Answer: John must charge $11.50 for each toy car.

➡ Example 3

This year, 20% more students attend the Mountain View School than last year. How many students attended the school last year if 72 students attend the school this year?

Solution

This problem is talking about a change in percent, and more specifically about a percent increase. We know this because the problem states "20% more students." That indicates that there was an original number of students, which was the number of students who attended Mountain View School last year, and 20% was added to that amount to bring us to this year's 72 students.

One way to set up percent increase problems is to use the following proportion:

$$\frac{\text{original amount} + \text{increase}}{\text{original amount}} = \frac{100 + \text{percent increase}}{100}$$

We begin by defining the pieces of the proportion. We don't know the number of students who attended Mountain View School last year before the increase happened, so we'll use a question mark in our proportion.

- Original amount + increase = the number of students who attend Mountain View School this year after the increase has already happened = 72
- Percent increase = + 20, as stated by "20% more"
- 100 + percent increase = 100 + 20 = 120

Using these facts, we can write our proportion and then cross-multiply and divide to find the solution.

$$\frac{72}{?} = \frac{120}{100}$$

$$\frac{72 \times 100}{120} = \frac{7200}{120} = 60$$

Answer: Therefore, there were 60 students at Mountain View School last year.

To check, we need to make sure that 20% more than 60 is equal to 72.

$$20\% \text{ of } 60 = 12$$

$$60 + 12 = 72$$

Since 60 + (20% of 60) = 72, we know that we have the correct answer.

ACCUPLACER TIPS

Whenever a word problem uses words such as "30% more," it's an indication that a quantity is being increased by 30%. Other words such as "an additional 10%" or "40% greater than" also indicate a percent increase problem.

However, percentages can also decrease. Words such as "30% less" or "decreased by 10%" or "40% fewer than" indicate a percent decrease problem.

To solve a percent decrease problem, we would take the same steps as we did to solve a percent increase problem, but we would use the following proportion instead:

$$\frac{\text{original amount} - \text{decrease}}{\text{original amount}} = \frac{100 - \text{percent decrease}}{100}$$

It's critical to your success on the ACCUPLACER that you pay close attention to the words used in the problem to make sure that you are answering the exact question that you are being asked.

LESSON 5.4—ACCUPLACER CHALLENGE

Directions: For each of the questions below, choose the best answer from the four choices given.

1. Phil wants a pair of $380 high-top sneakers that are on sale at Ted's Shoes for 20% off. What is the sale price?

   (A) $76.00
   (B) $304.00
   (C) $360.00
   (D) $456.00

2. Last year, 750 people ran in the Lamb Harbor 10K race. This year, there were 60% more runners. How many people ran in the Lamb Harbor 10K race this year?

   (A) 450
   (B) 810
   (C) 1,200
   (D) 1,250

3. The owner of Lester's Shoes pays his supplier $25 for a pair of boots. He puts a 30% markup on each pair. How much do boots cost at Lester's Shoes?

   (A) $17.50
   (B) $28.00
   (C) $32.50
   (D) $42.50

4. Joe Sullivan makes $31,200 a year. He spends 25% of his income on mortgage payments. How much does Joe spend on his mortgage each month?

    (A) $650
    (B) $7,800
    (C) $23,400
    (D) $39,000

5. During the last Chester County election, 5,400 people went to the polls and voted. This represents 60% of registered voters. How many registered voters did not vote in the last Chester County election?

    (A) 3,240
    (B) 3,600
    (C) 5,400
    (D) 9,000

6. The list price of Fran's new furniture was $1,700. She chose to pay 20% down and $45 a month for 36 months. Find the total amount she paid for the furniture.

    (A) $340
    (B) $1,360
    (C) $1,620
    (D) $1,960

7. Sam weighed 200 pounds last year. After a diet, he now weighs 160 pounds. What percent of his weight did Sam lose?

    (A) 20%
    (B) 25%
    (C) 40%
    (D) 80%

8. Jerry pays $840 for a flat screen TV, including a 5% sales tax. What is the price of the TV before the tax is added?

    (A) $798
    (B) $800
    (C) $835
    (D) $882

(Answers are on page 142.)

To complete a set of practice questions that reviews all of Chapter 5, go to: barronsbooks.com/TP/accuplacer/math29qw/

Lesson 5.1—Skills Check (page 124)

1. $\dfrac{47}{100}$ $47\% = \dfrac{47}{100}$

2. $\dfrac{3}{50}$ $6\% = \dfrac{6}{100} = \dfrac{3}{50}$

3. $1\dfrac{1}{5}$ $120\% = \dfrac{120}{100} = 1\dfrac{20}{100} = 1\dfrac{1}{5}$

4. $\dfrac{1}{40}$ $2.5\% = \dfrac{2.5}{100} = \dfrac{25}{1000} = \dfrac{1}{40}$

5. $\dfrac{3}{500}$ $0.6\% = \dfrac{0.6}{100} = \dfrac{6}{1000} = \dfrac{3}{500}$

6. **92%** $0.92 = 92\%$

7. **328.7%** $3.287 = 328.7\%$

8. **8%** $0.08 = 8\%$

9. **30%** $0.3 = 30\%$

10. **600%** $6 = 600\%$

11. **0.43** $43\% = 0.43$

12. **0.76** $76\% = 0.76$

13. **1.25** $125\% = 1.25$

14. **0.455** $45.5\% = 0.455$

15. **0.08** $8\% = 0.08$

16. **12%** $\dfrac{3}{25} = \dfrac{3}{25} \times \dfrac{4}{4} = \dfrac{12}{100} = 12\%$

17. **70%** $\dfrac{7}{10} = \dfrac{7}{10} \times \dfrac{10}{10} = \dfrac{70}{100} = 70\%$

18. **40%** $\dfrac{2}{5} = \dfrac{2}{5} \times \dfrac{20}{20} = \dfrac{40}{100} = 40\%$

19. **2.8%** $\dfrac{7}{250} = \dfrac{7}{250} \times \dfrac{4}{4} = \dfrac{28}{1000} = \dfrac{2.8}{100} = 2.8\%$

20. **180%** $\dfrac{180}{100} = 180\%$

Lesson 5.1—ACCUPLACER Challenge (page 125)

1. **(C)** 80% of P can be translated to $80\% \times P$. It can also be written as $\dfrac{80}{100} \times P = \dfrac{80}{100}P$, which is choice B. $\dfrac{80}{100}P$ can be reduced to $\dfrac{4}{5}P$, which is choice D. $\dfrac{80}{100}P$ can also be written as $\dfrac{80P}{100} = 80P \div 100$, which is choice A. Choice C, $0.08P$, is the only choice that is not equivalent to 80% of P.

2. **(D)** If we change all four numbers to percentages, we have 6%, 0.32 = 32%, 0.4 = 40%, $\frac{1}{2}$ = 50%, in order from least to greatest.

3. **(B)** $250\% = \frac{250}{100} = 2\frac{50}{100} = 2\frac{1}{2}$

4. **(B)** $0.4\% = \frac{0.4}{100} = \frac{4}{1000} = \frac{1}{250}$

Lesson 5.2—ACCUPLACER Challenge (pages 129–130)

1. **(D)** The Tornados won $\frac{12}{25} = \frac{?}{100}$ of their games.

$$\frac{(12)(100)}{25} = \frac{1200}{25} = 48$$

The Tornados won 48% of their games.

2. **(B)** The 32 items correct is part of the whole test. The length of the test, or the whole, is what we need to find. We can set up the proportion $\frac{32}{?} = \frac{80}{100}$, and then cross-multiply and divide.

$$\frac{(100)(32)}{80} = \frac{3200}{80} = 40$$

There were 40 questions on this test.

3. **(C)** $1,200 is Sean's whole pay. We can set up the proportion $\frac{?}{1200} = \frac{26}{100}$, and then cross-multiply and divide.

$$\frac{(26)(1200)}{100} = \frac{31,200}{100} = 312$$

The amount of money that is deducted from Sean's weekly pay for taxes and Social Security is $312.00.

4. **(C)** $725 is the whole cost of the repair, and 85% of that is what Stan's insurance company paid. We can set up the proportion $\frac{?}{725} = \frac{85}{100}$, and then cross-multiply and divide.

$$\frac{(725)(85)}{100} = \frac{61,625}{100} = \$616.25$$

The amount of money that Stan's insurance company paid is $616.25.

5. **(C)** There are 100 drivers in the company, but not all of the employees are drivers. We must find the number of employees in the whole company. We can set up the proportion $\frac{100}{?} = \frac{80}{100}$, and then cross-multiply and divide.

$$\frac{(100)(100)}{80} = \frac{10,000}{80} = 125$$

The number of employees who work for the Got-to-Go Delivery Company is 125 employees.

6. **(A)** The amount that Fiona spends on rent is a part of her income. We can set up the proportion $\frac{595}{2419} = \frac{?}{100}$, and then cross-multiply and divide.

$$\frac{(595)(100)}{2419} = \frac{59,500}{2419} = 24.597$$

Therefore, she spends approximately 25% of her income on rent.

Lesson 5.3—ACCUPLACER Challenge (pages 133–134)

1. **(C)** 10 is 20% of some number. We can set up the proportion $\frac{10}{?} = \frac{20}{100}$, and then cross-multiply and divide.

$$\frac{(10)(100)}{20} = \frac{1000}{20} = 50$$

10 is 20% of 50.

2. **(B)** A number is 15% of 70. We can set up the proportion $\frac{?}{70} = \frac{15}{100}$, and then cross-multiply and divide.

$$\frac{(70)(15)}{100} = \frac{1050}{100} = 10.5$$

10.5 is 15% of 70.

3. **(D)** 20 is some percent of 5. We can set up the proportion $\frac{20}{5} = \frac{?}{100}$, and then cross-multiply and divide.

$$\frac{(20)(100)}{5} = \frac{2000}{5} = 400$$

20 is 400% of 5.

4. **(A)** 18 is some percent of 120. We can set up the proportion $\frac{18}{120} = \frac{?}{100}$, and then cross-multiply and divide.

$$\frac{(18)(100)}{120} = \frac{1800}{120} = 15$$

18 is 15% of 120.

5. **(C)** 5 is 20% of a number. We can set up the proportion $\frac{5}{?} = \frac{20}{100}$, and then cross-multiply and divide.

$$\frac{(5)(100)}{20} = \frac{500}{20} = 25$$

20% of 25 is 5.

6. **(D)** A number is 150% of 50. We can set up the proportion $\frac{?}{50} = \frac{150}{100}$, and then cross-multiply and divide.

$$\frac{(50)(150)}{100} = \frac{7500}{100} = 75$$

150% of 50 is 75.

Lesson 5.4—ACCUPLACER Challenge (pages 137–138)

1. **(B)** The shoes are 20% off, so Phil must pay 100% – 20% = 80 % of the cost of the shoes. We can set up the proportion $\frac{?}{380} = \frac{80}{100}$, reduce to $\frac{?}{380} = \frac{4}{5}$, and then cross-multiply and divide.

$$\frac{380 \times 4}{5} = \frac{1520}{5} = \$304.00$$

The sale price is $304.00.

2. **(C)** To find out how many more runners there are, we need to find 60% of 750 by setting up the proportion $\frac{60}{100} = \frac{?}{750}$. Cross-multiply and divide.

$$\frac{750 \times 60}{100} = \frac{45,000}{100} = 450 \text{ new runners}$$

Since there were 750 runners last year and 450 new runners this year, there is a total of 750 + 450 = 1,200 people in the race this year.

3. **(C)** The owner must sell his boots for more than $25 in order to make a profit. He wants to mark up the price by 30%, so we first set up the proportion $\frac{?}{25} = \frac{30}{100}$. Cross-multiply and divide.

$$\frac{25 \times 30}{100} = \frac{750}{100} = 7.5$$

Since the owner wants to profit this much, he must sell the boots for

$$\$25 + \$7.50 = \$32.50$$

4. **(A)** If Joe earns $31,200 a year and 25% of his income goes to mortgage payments, we can use the proportion $\frac{?}{31,200} = \frac{25}{100}$ to find out how much Joe spends on mortgage payments in a year. Cross-multiply and divide.

$$\frac{31,200 \times 25}{100} = \frac{780,000}{100} = 7,800$$

Be careful! The question asks how much Joe spends on his mortgage each *month*, so we must divide by 12. $\frac{7800}{12} = \$650$

5. **(B)** Knowing that 5,400 people, or 60% of registered voters, voted, we can use the proportion $\frac{5400}{?} = \frac{60}{100}$ to find out how many registered voters there were in Chester

County during the last election. We find that $\frac{5400 \times 100}{60} = \frac{540,000}{60} = 9,000$ registered voters. To find out how many voters did not vote, we must subtract:

$$9,000 - 5,400 = 3,600$$

There were 3,600 registered voters who did not vote.

6. **(D)** Fran paid both 20% of $1,700 down and $45 a month for 36 months. To find out how much she paid in total, we can evaluate the following expression:

$$(0.20)(\$1,700) + (\$45)(36) = \$340 + \$1,620 = \$1,960$$

7. **(A)** Sam lost $200 - 160 = 40$ pounds. To calculate the percent of his weight that he lost, we need to compare the 40 pounds that he lost to the 200 pounds he weighed last year using the proportion $\frac{40}{200} = \frac{?}{100}$. We could cross-multiply and then divide, but we could also reduce by dividing $\frac{40}{200} \div \frac{2}{2} = \frac{20}{100}$, which is equal to 20%.

8. **(B)** Since the $840 price already includes the 5% sales tax, we consider this price to be the original 100% of the cost plus 5% sales tax equals 105% of the original amount. Knowing that, we can use the proportion $\frac{\$840}{?} = \frac{105}{100}$ to find the price of the TV before the tax was added. Cross-multiply and divide.

$$\frac{840 \times 100}{105} = \frac{84,000}{105} = 800$$

The original price of the TV before the tax was added was $800.

Mean and Median

6

◤◤◤◤◤◤◤◤◤◤◤◤◤◤◤◤◤◤

Imagine that you are conducting a study about the hourly wages that students at your school are paid. You interview 120 of the students who are working, collect the data, and begin to prepare a brief report of your findings. In your report, you wouldn't want to include every respondent's income—there are just too many people and too many numbers, and they would be hard to interpret quickly. Instead, you would want to provide just a few numerical figures that capture the big picture about the hourly wages earned by students at your school. That number should fall somewhere near the middle of all the responses you received, but how do you determine where the middle is?

This is where the concepts of mean and median come in. They give us two different ways of measuring the "center" of a data set, and each is useful in a unique way. You can definitely expect to see questions about these concepts on the Arithmetic test of the ACCUPLACER, but a deep understanding of mean and median will help you on the Elementary Algebra test as well.

In this chapter, we will work to master

- understanding the difference between mean and median,
- calculating the mean of a set of numbers,
- finding a missing value when the mean is given, and
- finding the median of a set of numbers.

LESSON 6.1—FINDING THE MEAN OF A SET OF NUMBERS

Perhaps the most commonly used measure of the center of a set of data is the **mean**, or **average**. We hear about this all the time. A grade of C is average, Brian is of average height, Tina is an above average math student, and so on. In each of these cases, the word **average** is used to characterize a typical member of a group. When you are working with a collection of numbers, either from a word problem, a table, or a number set, the **mean** value will always fall somewhere between the lowest value and the highest value. For example, if we consider the set of numbers {3, 5, 10, 14}, we know that the mean will be somewhere between 3 and 14 because 3 is the smallest value and 14 is the largest value.

To Calculate the Mean When a Set of Data Is Given:

1. Find the sum of all the numbers in the set.
2. Divide by the total number of items in the set.

Represented algebraically, the process looks like this:

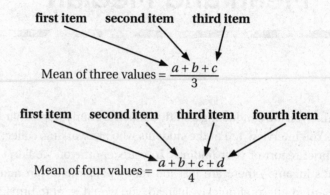

You should notice that in both of these examples, the value in the denominator is equal to the total number of items in the numerator. Therefore, if we were to find the mean of 12 numbers, we would need to divide by 12.

➡ Example 1

Find the mean of the following set of numbers: {3, 5, 10, 14}

Solution

The first step is to write the sum of all the numbers in the set as the numerator. Then use the number of items in the set as the denominator.

$$\frac{3 + 5 + 10 + 14}{4}$$

Next, add all of the numbers in the numerator. Then divide by the total number of items in the set. Since there are four items, divide by four.

$$\frac{32}{4} = 8$$

<u>Answer:</u> The mean of the four numbers in this set is 8.

Notice that our mean, 8, falls between the smallest value in the set and the largest value in the set, just as we predicted that it would earlier.

It's possible that the ACCUPLACER will just give you a set of numbers and ask you to calculate the mean, but you should also expect to answer word problems that require a deep understanding of the concept of mean.

➥ Example 2

Five employees work at a small, local restaurant. Their ages are 22, 33, 19, 46, and 35. What is the average age of the employees?

(A) 25

(B) 31

(C) 33

(D) 37

Solution

It's always a good idea to first estimate what the mean will be. This will help us determine whether our answer is accurate. Just like before, the mean will need to be between the smallest number and the largest number, which in this case is between 19 and 46. A good guess would be that the mean is somewhere around 30. Just looking at the answer choices, choices B and C are the closest to our guess.

Next, we will find the sum of the numbers and divide it by 5 because there are five employees. The expression will look like this:

$$\text{Mean} = \frac{22 + 33 + 19 + 46 + 35}{5}$$

$$\text{Mean} = \frac{155}{5} = 31$$

Answer: We get a mean age of 31, so the correct answer is choice B. Look how close this is to our initial guess of 30! This means that our answer is a good one.

ACCUPLACER TIPS

In the study of statistics, we often use the mean to generalize about a population. We might say, for example, that the average household has 1.7 children. It obviously isn't true that any family could have 1.7 children; we all have 0, 1, 2, 3, and so on. The mean of 1.7 simply shows that, on average, most households have 1 or 2 children. It's perfectly fine for the mean of a set of numbers to be a fraction or a decimal.

We should add one other quick—but important—point about calculating the mean. Sometimes, you'll be given a set of numbers in which one or more of the values are repeated, like this one: {4, 7, 7, 12, 15, 15}.

In calculating the mean of these numbers, we have to add all six values exactly as they appear—in other words, we shouldn't use the 7 and the 15 only once. The correct calculation is:

$$\frac{4 + 7 + 7 + 12 + 15 + 15}{6} = 10$$

The same rule applies for sets that contain the number 0. You have to count it as an item, even though it doesn't change the sum of your scores. Have you ever missed a test, a quiz, or a homework assignment and had to take a score of 0 on it? It really drags down your overall average! That's because the zero doesn't increase the sum of your scores at all, but it does increase the total number of items in your class average! To see just how much a 0 can hurt your average, try question 3 in the next set of ACCUPLACER Challenge questions.

LESSON 6.1—SKILLS CHECK

Directions: Find the mean of each set of numbers below.

1. 13, 51, 24, 36

2. 8.9, 3.4, 6.6

3. 5, 6, 7, 8, 9, 10, 11

4. $2\frac{1}{2}, 7\frac{3}{4}, 10\frac{3}{4}$

(Answers are on page 156.)

LESSON 6.1—ACCUPLACER CHALLENGE

Directions: For each of the questions below, choose the best answer from the four choices given.

1. Three friends are discussing how much they pay in rent. Luis pays $950, Eleanor pays $625, and Neil pays $1,035. What is their mean monthly rent?

    (A) $830
    (B) $870
    (C) $950
    (D) $1,305

2. Five friends attending the same college are discussing how many credit hours they are each taking this semester. They produce the following responses: 12, 15, 16, 16, and 13. What is the mean number of credit hours for the five students?

    (A) 13
    (B) 14.4
    (C) 15.5
    (D) 16

3. Alvin's math teacher gave four tests last semester. On the first test, Alvin got a score of 90. On the second test, he got a score of 78. On the third test, he got a score of 96. He skipped class on the day of the fourth test and received a score of 0. What was Alvin's average score for these four tests?

    (A) 66
    (B) 72
    (C) 88
    (D) 90

4. The table below shows the hourly wages received by 12 high school students. What is the mean wage for these students?

Hourly Wage	Number of Students
$9	7
$10	2
$11	2
$12	1

(A) $9.25
(B) $9.75
(C) $10.50
(D) $29.25

5. A small business has eight employees. Three of them work 32 hours per week, four of them work 28 hours per week, and one of them works 40 hours per week. What is the average number of hours worked by an employee at this company?

(A) 12 hours, 30 minutes
(B) 29 hours, 40 minutes
(C) 31 hours
(D) 33 hours, 20 minutes

(Answers are on page 156.)

LESSON 6.2—FINDING MISSING DATA WHEN THE MEAN IS GIVEN

In the examples in Lesson 6.1, we found the mean of a set of numbers when all the data in the set was given. Sometimes, however, the mean of the set is already provided, and we are asked to find a missing piece of data that belongs in the set. Let's take a look at an example.

➡ Example 1

Four students took a math test, and their average score was 78 percent. Three of the students remember their scores. One got a 93 percent, one got a 67 percent, and one got a 72 percent. What was the fourth student's score?

(A) 58 percent
(B) 77 percent
(C) 80 percent
(D) 84 percent

Solution

This time, both the question and the solution process are very different, even though this question still involves the concept of mean. If we are given the mean of a set of numbers but are missing one of the numbers in the set, we begin by multiplying the mean score by the number of students who took the test.

The average for the students would be calculated like this, but notice that there is a question mark in place of the fourth student's score:

$$\frac{93 + 67 + 72 + ?}{4} = 78$$

We calculate $78 \times 4 = 312$. Think about it this way. Since we're essentially working backwards, we take the steps that are opposite from the ones that we took in the earlier examples of Lesson 6.1; therefore, we multiply instead of divide. This tells us that the 312 in the previous step is the sum of the four students' scores. We can then subtract the scores that we already know from the sum of 312. After we have done this, we will be left with the fourth student's score.

$$312 - 93 - 67 - 72 = 80$$

Answer: In order for the average score to be a 78, the fourth student must have received a score of 80 percent.

ACCUPLACER TIPS

Taking tests, like the ACCUPLACER, is stressful. We've been there, and we know what it's like. Sometimes, even when you think you remember a process perfectly, it just slips out of your mind at the last minute. Let's say that this happens to you on a problem like the one we just looked at. If you don't remember this process, don't panic! Use the answer choices and work backwards, using the general formula for mean that we looked at earlier. For example, in the problem in Example 1, you would first plug in the answer for choice A, which is 58, and check to see if it produces a mean of 78.

$$\frac{93 + 67 + 72 + 58}{4} = 72.5$$

This doesn't work, so try choice B.

$$\frac{93 + 67 + 72 + 77}{4} = 77.25$$

We're a lot closer this time, but don't assume that 77.25 is close enough. Try choice C.

$$\frac{93 + 67 + 72 + 80}{4} = 78$$

Choice C produces a mean of exactly 78, so this is the correct answer. Even though it's undoubtedly important to have deep knowledge of the content on the ACCUPLACER, it's just as important to become a smart test taker by using strategies like this one.

➡ Example 2

The four students in the previous example timed themselves when they took the math test. They discussed it after the test and determined that their average testing time was 27 minutes. The times for three of the students are as follows: 19 minutes, 28 minutes, and 31 minutes. How long did it take the fourth student to complete the math test?

(A) 20 minutes
(B) 26 minutes
(C) 28 minutes
(D) 30 minutes

Solution

As in the previous example, the mean is given, but one number in the set is missing. We begin by multiplying the mean score by the total number of people who tested.

$$27 \times 4 = 108$$

Now we need to subtract each student's time from the total time of 108 minutes.

$$108 - 19 - 28 - 31 = 30$$

It took the fourth student 30 minutes to complete the exam, so the correct answer is D. We could also solve this problem by working backwards. Each of the answer choices represents a possible time for the fourth student. We're looking to find which answer choice, when plugged into the mean formula along with the three other times that we already know, would produce a mean time of 27 minutes. The only answer choice that would work is choice D.

Answer: The correct answer is choice D.

LESSON 6.2—SKILLS CHECK

Directions: Use the process outlined in this section to find the missing data.

1. Three numbers have a mean of 16. The first number is 11, and the second number is 9. What is the third number?

2. Four numbers have a mean of 20. The first number is 19, the second is 28, and the third is 23. What is the fourth number?

3. Five numbers have a mean of 50. The first number is 73, the second is 22, the third is 37, and the fourth is 68. What is the fifth number?

(Answers are on page 156.)

LESSON 6.2—ACCUPLACER CHALLENGE

Directions: For each of the questions below, choose the best answer from the four choices given.

1. My goal for the first three weeks of July is to have an average income of $400 per week. During the first week, I made $475, and during the second week I made $310. How much would I need to make in the third week to have a mean income of $400?

    (A) $392.50
    (B) $395
    (C) $415
    (D) $450

2. Ivan had four of his musical instruments appraised at a local shop, and the shopkeeper told him that the mean value of the instruments was $750. He knows that his keyboard is worth $825, his drum kit is worth $615, and his bass guitar is worth $360, but he doesn't know how much his electric guitar is worth. Since the mean value of the four instruments is $750, how much was the electric guitar worth?

    (A) $600
    (B) $637.50
    (C) $950
    (D) $1,200

3. Ryan runs five days per week, and he tries to run an average of 6 miles each time. Based on the table below, how many miles would he need to run on the fifth day to make sure that he meets his average of 6 miles?

Day	Number of Miles
1	6
2	4
3	9
4	3
5	?

    (A) 5
    (B) 6
    (C) 7
    (D) 8

(Answers are on page 157.)

LESSON 6.3—FINDING THE MEDIAN

Another way of describing the center of a set of numbers is to use the **median**, or middle number. The median provides us with an alternative way of characterizing a set of data that can sometimes seem more accurate, or appropriate, than the mean.

Consider a company that has several employees. Five of those employees were asked about their annual salaries, and this is how they responded:

Employee 1: $41,000
Employee 2: $24,000
Employee 3: $36,000
Employee 4: $2,450,000
Employee 5: $52,000

We want to choose a number that is an accurate representation of the typical salary at that company. When we calculate the mean by adding all the numbers and dividing by the 5 employees, we see that the "average" is $520,600! It's clear that Employee 4's gigantic salary is pulling the mean up much higher—so high, in fact, that it doesn't give us a clear picture of what an employee at this company could expect to make. Therefore, we should use the median as an alternative.

To Find the Median:

1. Arrange the numbers from smallest to largest.
2. If the set contains an *odd* number of items, there will only be one number in the middle. This number is the median.
3. If the set contains an *even* number of items, there will be two numbers in the middle. You will then need to calculate the mean of these two numbers. The mean of those two middle numbers will be the median.

To find the median of these five salaries, first put them in order from smallest to largest.

$24,000 $36,000 $41,000 $52,000 $2,450,000

There is only one number in the middle, $41,000, which is the median. The median of $41,000 is a much more accurate representation of the typical salary than the mean of $520,600. Let's try a couple of examples.

➡ Example 1

Find the median of the following set of numbers: {16, 11, 45, 34, 23}

Solution

First, rewrite the numbers in order from smallest to largest:

11   16   23   34   45

Since this set contains five numbers (an odd number), we know that there will only be one number in the middle. We can cross off one number at the beginning of the set and one number at the end of the set. Then we can cross off the next number at the beginning and the next number at the end until there is only one number left.

~~11~~   ~~16~~   23   ~~34~~   ~~45~~

Answer: The only number that hasn't been crossed out is 23. This is the median.

➥ Example 2

Rachel wrote four emails this afternoon, and she counted the number of words in each email. The four emails contained the following number of words: 103, 81, 65, and 210. Find the median number of words she wrote in her four emails.

(A) 65
(B) 73
(C) 81
(D) 92

Solution

First, put the numbers in order from smallest to largest.

65   81   103   210

This time there is an even number of items. We can cross off the one at the beginning and the one at the end. We're then left with the two numbers in the middle.

~~65~~   81   103   ~~210~~

To get the median, we still have to find the mean of 81 and 103. Remember how to do this? You'll need to find the sum of these two numbers, and then divide by two: $81 + 103 = 184$, and $184 \div 2 = 92$.

Answer: The median number of words she wrote is choice D, 92 words.

LESSON 6.3—SKILLS CHECK

Directions: Find the median for each set of numbers below.

1. 24, 16, 11, 9, 19

2. 45, 58, 29, 97, 34, 52, 101

3. 3.65, 6.35, 3.6, 5.3, 5.36

4. 9, 25, 8, 16

5. 90, 42, 39, 76, 42, 46

6. $6\frac{2}{5}, 8\frac{1}{10}, 11\frac{3}{4}, 4\frac{7}{8}$

(Answers are on page 157.)

LESSON 6.3—ACCUPLACER CHALLENGE

> **Directions:** For each of the questions below, choose the best answer from the four choices given.

1. Gary played six basketball games last month and recorded the number of points he scored in each: 27, 11, 15, 19, 6, and 20. What is the median number of points he scored for these six games?

   (A) 15
   (B) 17
   (C) 19
   (D) 20

2. Five consecutive numbers have a mean of 11. What is the median of the five numbers?

   (A) 55
   (B) 16
   (C) 11
   (D) 5.5

3. Four numbers have a mean of 36. Three of the numbers are 18, 45, and 60. What is the median of the four numbers?

   (A) 21
   (B) 33
   (C) 40
   (D) 45

4. What is the median of the first 10 prime numbers? (*Hint: A number is prime if it has exactly two factors—1 and itself. The number 1 is not prime, as it only has one factor.*)

   (A) 8
   (B) 10
   (C) 12
   (D) 14

(Answers are on page 158.)

> To complete a set of practice questions that reviews all of Chapter 6, go to:
> barronsbooks.com/TP/accuplacer/math29qw/

Lesson 6.1—Skills Check (page 148)

1. **31** $\dfrac{13+51+24+36}{4} = \dfrac{124}{4} = 31$

2. **6.3** $\dfrac{8.9+3.4+6.6}{3} = \dfrac{18.9}{3} = 6.3$

3. **8** $\dfrac{5+6+7+8+9+10+11}{7} = \dfrac{56}{7} = 8$

4. **7** $\dfrac{2\frac{1}{2}+7\frac{3}{4}+10\frac{3}{4}}{3} = \dfrac{21}{3} = 7$

Lesson 6.1—ACCUPLACER Challenge (pages 148–149)

1. **(B)** $\dfrac{950+625+1035}{3} = \dfrac{2610}{3} = \$870$

2. **(B)** $\dfrac{12+15+16+16+13}{5} = \dfrac{72}{5} = 14.4$ credit hours

3. **(A)** $\dfrac{90+78+96+0}{4} = \dfrac{264}{4} = 66$

4. **(B)** Even though only four different wages are listed in the table, we have to account for all 12 students.

$$\dfrac{9(7)+10(2)+11(2)+12(1)}{12} = \dfrac{63+20+22+12}{12} = \dfrac{117}{12} = \$9.75$$

5. **(C)** $\dfrac{32(3)+28(4)+40(1)}{8} = \dfrac{96+112+40}{8} = \dfrac{248}{8} = 31$ hours

Lesson 6.2—Skills Check (page 151)

1. **28** The mean is 16, so multiply it by 3 and then subtract the two known numbers in the set.

$$16 \times 3 = 48$$
$$48 - 11 - 9 = 28$$

2. **10** The mean is 20, so multiply it by 4 and then subtract the known numbers.

$$20 \times 4 = 80$$
$$80 - 19 - 28 - 23 = 10$$

3. **50** The mean is 50, so multiply it by 5 and then subtract the known numbers.

$$50 \times 5 = 250$$

$$250 - 73 - 22 - 37 - 68 = 50$$

Lesson 6.2—ACCUPLACER Challenge (page 152)

1. **(C)**

$$\$400 \times 3 = \$1,200$$

$$\$1,200 - \$475 - \$310 = \$415$$

2. **(D)**

$$\$750 \times 4 = \$3,000$$

$$\$3,000 - \$825 - \$615 - \$360 = \$1,200$$

3. **(D)**

$$6 \times 5 = 30$$

$$30 - 6 - 4 - 9 - 3 = 8 \text{ miles}$$

Lesson 6.3—Skills Check (page 154)

1. **16**

24, 16, 11, 9, 19

9, ~~11~~, 16, ~~19~~, ~~24~~

The median is 16.

2. **52**

45, 58, 29, 97, 34, 52, 101

~~29~~, ~~34~~, ~~45~~, 52, ~~58~~, ~~97~~, ~~101~~

The median is 52.

3. **5.3**

3.65, 6.35, 3.6, 5.3, 5.36

~~3.6~~, ~~3.65~~, 5.3, ~~5.36~~, ~~6.35~~

The median is 5.3.

4. **12.5 *or* $12\frac{1}{2}$**

9, 25, 8, 16

~~8~~, 9, 16, ~~25~~

$$\frac{9 + 16}{2} = \frac{25}{2} = 12.5 \text{ or } 12\frac{1}{2}$$

5. **44**

90, 42, 39, 76, 42, 46

~~39~~, ~~42~~, 42, 46, ~~76~~, ~~90~~

$$\frac{42 + 46}{2} = \frac{88}{2} = 44$$

6. $7\frac{1}{4}$

$$6\frac{2}{5}, 8\frac{1}{10}, 11\frac{3}{4}, 4\frac{7}{8} \qquad 6\frac{2}{5} + 8\frac{1}{10} = 14\frac{1}{2}$$

$$4\frac{7}{8}, 6\frac{2}{5}, 8\frac{1}{10}, 11\frac{3}{4} \qquad 14\frac{1}{2} \div 2 = 7\frac{1}{4}$$

Lesson 6.3—ACCUPLACER Challenge (page 155)

1. **(B)**

$$27, 11, 15, 19, 6, 20$$

$$6, \cancel{11}, 15, 19, \cancel{20}, \cancel{27}$$

$$\frac{15 + 19}{2} = \frac{34}{2} = 17 \text{ points}$$

2. **(C)** The five numbers that have a mean of 11 are 9, 10, 11, 12, and 13.

$$\frac{9 + 10 + 11 + 12 + 13}{5} = \frac{55}{5} = 11$$

$$9, \cancel{10}, 11, \cancel{12}, \cancel{13}$$

The median is also 11.

3. **(B)** First, find the missing number.

$$36 \times 4 = 144$$

$$144 - 18 - 45 - 60 = 21$$

The four numbers are 18, 21, 45, and 60.

$$\cancel{18}, 21, 45, \cancel{60}$$

$$\frac{21 + 45}{2} = \frac{66}{2} = 33$$

4. **(C)** The first ten prime numbers are 2, 3, 5, 7, 11, 13, 17, 19, 23, and 29. Find the median.

$$2, 3, 5, 7, 11, 13, \cancel{17}, \cancel{19}, \cancel{23}, \cancel{29}$$

$$\frac{11 + 13}{2} = \frac{24}{2} = 12$$

Exponents, Scientific Notation, and Square Roots

7

The examples and exercises in this chapter straddle the line between two of the ACCUPLACER math tests: Arithmetic and Elementary Algebra. You should expect to see any or all of these three topics—exponents, scientific notation, and square roots—on both tests, depending on the type of questions that you're asked to answer.

> **In this chapter, we will work to master**
>
> - performing calculations with exponents,
> - converting numbers from scientific notation to standard notation, and vice-versa,
> - finding the square root of perfect squares,
> - estimating the values of square roots other than perfect squares, and
> - simplifying radical expressions.

LESSON 7.1—EXPONENTS

Let's start our discussion of exponents with some terminology you've probably heard before—**squaring** a number—and then build up from there. To put it simply, when we square a number, we multiply the number by itself. For example, $5 \times 5 = 25$ and $8 \times 8 = 64$. You can even think about squaring a number in terms of finding the area of a square, which we calculate by multiplying the length of the square by its width. Let's think about 5 squared and 8 squared in terms of the area of a square, as exemplified in the following illustrations.

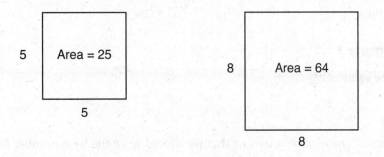

We can therefore refer to the numbers 25 and 64 as perfect squares. A **perfect square** is the result we get when we multiply a whole number by itself. The perfect squares are commonly occurring numbers in mathematics, and it's important to know them well. Knowing your perfect squares will save you lots of calculation time on the ACCUPLACER, and perfect squares will be even more important later on in the Elementary Algebra unit. Let's look at some basic terminology.

We write 8 squared like this:

$$8^2$$

base   exponent

In this example, the number 8 is called the **base**, and the 2 is called the **exponent**. The exponent tells us how many times we need to write the base number in order to set up a multiplication problem. In this case, since the exponent is a 2, we write the base number, 8, twice. Then we put a multiplication sign in between.

$$8 \times 8$$

The first twelve perfect squares are the ones that you're likely to see the most often, so we've provided them here for your reference. To calculate each of these, we have multiplied the base number times itself.

$$1^2 = 1 \times 1 = 1 \qquad\qquad 7^2 = 7 \times 7 = 49$$
$$2^2 = 2 \times 2 = 4 \qquad\qquad 8^2 = 8 \times 8 = 64$$
$$3^2 = 3 \times 3 = 9 \qquad\qquad 9^2 = 9 \times 9 = 81$$
$$4^2 = 4 \times 4 = 16 \qquad\qquad 10^2 = 10 \times 10 = 100$$
$$5^2 = 5 \times 5 = 25 \qquad\qquad 11^2 = 11 \times 11 = 121$$
$$6^2 = 6 \times 6 = 36 \qquad\qquad 12^2 = 12 \times 12 = 144$$

Often, the number in the exponent position is a number other than 2. It could be a 3, 4, 10, or any other number. When we see the mathematical expression 2^4, we read it as "two to the fourth power." The number 2 is the base, and the raised 4 is the exponent, or power. Exponents tell us the number of times that we should write the base number to set up a multiplication problem and solve.

To Calculate the Value of a Number Raised to a Power:

1. Look at the exponent, and then write the base number that many times.
2. Put multiplication signs in between each of the base numbers.
3. Multiply.

➡ Example 1

What is the value of 2^4?

Solution

To evaluate 2^4, the exponent tells us that we should write the base number four times and then multiply.

$$2 \times 2 \times 2 \times 2$$

Answer: 16

➧ Example 2

Find the value of $\left(\frac{2}{5}\right)^3$.

Solution: The exponent in this case is 3, so we need to write the base number out three times and put multiplication signs between them. Even though we're working with a fraction in this example, the process is the same.

$$\frac{2}{5} \times \frac{2}{5} \times \frac{2}{5}$$

To evaluate, we multiply all the numerators, and then multiply all the denominators, which produces the fraction below.

$$\frac{2}{5} \times \frac{2}{5} \times \frac{2}{5} = \frac{8}{125}$$

Answer: The correct answer is $\frac{8}{125}$.

You might notice that, in the numerator, we calculated 2^3, and in the denominator, we calculated 5^3. This reveals something interesting about raising fractions to a particular power. We can rewrite $\left(\frac{2}{5}\right)^3$ as $\left(\frac{2^3}{5^3}\right)$. To put it more generally,

$$\left(\frac{a}{b}\right)^n = \left(\frac{a^n}{b^n}\right)$$

Special Cases

While most exponents can be calculated using the process outlined above, there are a couple of special cases you should know.

Whenever a number is raised to the first power, the answer to the expression is the same as the base number. For example, $9^1 = 9$. This is true for any number, whether it is a fraction or a decimal. Therefore, $\left(\frac{1}{2}\right)^1 = \frac{1}{2}$ and $9.875^1 = 9.875$. The 1 in the exponent tells you to write the base number once. Since it's only written once, there is nowhere to put a multiplication sign in between!

This next special case is a little stranger. If any number except zero is raised to the zero power, the answer to the expression is always 1. Therefore, $15^0 = 1$, and $3.14^0 = 1$. This is something you definitely need to memorize. Note that 0^0 is undefined.

LESSON 7.1—SKILLS CHECK

> **Directions:** Solve each of the following problems.

1. $9^2 =$

2. $18^2 =$

3. $\left(\dfrac{3}{4}\right)^2 =$

4. $1^0 =$

5. $3^3 =$

6. $4^4 =$

7. $15^0 =$

8. $3^3 + 4^3 =$

9. $1^{10} =$

10. $0^3 =$

11. $\left(\dfrac{2}{3}\right)^4 =$

12. $10^5 =$

13. $0.25^2 =$

14. $1.5^3 =$

(Answers are on page 178.)

LESSON 7.1—ACCUPLACER CHALLENGE

> **Directions:** For each of the questions below, choose the best answer from the four choices given.

1. Find the difference of 2^6 and 8^2.

   (A) 0
   (B) 4
   (C) 32
   (D) 60

2. $\dfrac{3}{5}$ squared =

   (A) 1.2
   (B) 0.925
   (C) 0.36
   (D) 0.25

3. Which of the following examples, when raised to a positive power, will produce a smaller number than the base number?

   (A) 1.009
   (B) $\dfrac{8}{9}$
   (C) 0
   (D) $\dfrac{9}{8}$

4. $(3^3)^2 =$

   (A) 18
   (B) 81
   (C) 243
   (D) 729

(Answers are on page 178.)

LESSON 7.2—USING SCIENTIFIC NOTATION FOR VERY LARGE NUMBERS

Scientists estimate that there are about 37,200,000,000,000 cells in the average human body. That's a huge number! When we write out a number in this way, it is referred to as **standard notation**, because this is the way that numbers are usually written. In math and science, however, we often use something called **scientific notation** to express very large numbers, like this one, so that we don't have to write out all those zeros. Scientific notation is also used when we talk about things like the number of stars in the known universe or the distance between galaxies. Scientific notation uses multiplication and the powers of 10 to estimate the value of these very large numbers. Here's what 37,200,000,000,000 looks like in scientific notation:

$$3.72 \times 10^{13}$$

Remember that 10^{13} means that you would write the number 10 thirteen times with multiplication signs between each of the tens. It's clear that this would produce a huge number. We've listed the first few powers of 10 in the following table so that you can see how the numbers grow as the power increases. On the Arithmetic test of the ACCUPLACER, you'll need to be able to convert between standard and scientific notation, and fortunately there are some easy steps you can follow for making these conversions.

$10^1 = 10$
$10^2 = 100$
$10^3 = 1,000$
$10^4 = 10,000$
$10^5 = 100,000$
$10^6 = 1,000,000$

To Convert a Number in Scientific Notation to Standard Notation:

1. Write the first number in the expression.
2. Move the decimal point the same number of places to the right as the exponent above the 10.
3. Add zeros for each place you move past the end of the original decimal numbers.

➡ Example 1

Convert 1.56×10^7 to standard notation.

Solution

First, write the number 1.56. The exponent 7 tells us that the decimal point needs to be moved seven places to the right. Next, put your pen or pencil on the decimal point, and then move seven places to the right. We recommend that you use arrows, as shown below, to do this so that you don't lose track of how many places you've moved.

$$1.56 \overset{\frown}{}\overset{\frown}{}\overset{\frown}{}\overset{\frown}{}\overset{\frown}{}\overset{\frown}{}\overset{\frown}{}$$
$$1 \quad 2 \quad 3 \quad 4 \quad 5 \quad 6 \quad 7$$

In each of the "valleys" you drew, insert zeros, as shown below.

$$1\ 5\ 6\ 0\ 0\ 0\ 0\ 0.$$
$$1\ 2\ 3\ 4\ 5\ 6\ 7$$

<u>Answer:</u> Adding commas back into the number, we have 15,600,000, or fifteen million six hundred thousand.

Notice that, in this example, the power of 10, or exponent, did not tell us how many zeros to add but rather how many places to move the decimal point. The power was 7, but we only had to add 5 zeros in order to move the decimal point 7 places.

ACCUPLACER TIPS

Don't confuse commas and decimal points! They serve very different purposes. A decimal indicates a break between the whole number and a part of the whole, so the placement of the decimal has a very real impact on the total value of the number. Commas, on the other hand, are only used to make it easier for us to read numbers. For example, it's easier to read 15,600,000 than it is to read 15600000.

To Convert Large Numbers in Standard Notation to Scientific Notation:

1. Put your pen or pencil on the decimal point, which is found at the end of whole numbers.
2. Move the decimal point to the left until there is only one digit to the left of the decimal.
3. Write out the decimal number without any zeros.
4. Count the number of places you moved the decimal, and make it the power of 10.

➡ Example 2

Convert 679,000,000,000 to scientific notation.

Solution

Begin by putting your pen or pencil down after the last zero in the number. Next, move the decimal point to the left until it ends up between the 6 and the 7. When writing a number in scientific notation, there should always be one nonzero digit to the left of the decimal point.

$$6\ 7\ 9\ 0\ 0\ 0\ 0\ 0\ 0\ 0\ 0\ 0$$

Next, count the number of times you moved the decimal point. This number is going to be the exponent above the 10.

$$6\ 7\ 9\ 0\ 0\ 0\ 0\ 0\ 0\ 0\ 0\ 0$$
$$11\ 10\ 9\ 8\ 7\ 6\ 5\ 4\ 3\ 2\ 1$$

Answer: Write out the decimal number, 6.79, followed by a times sign and 10 to the 11th power, like this:

$$6.79 \times 10^{11}$$

LESSON 7.2—SKILLS CHECK

Directions: Convert each number in scientific notation to standard notation.

1. 1.14×10^5

2. 9×10^8

3. 4.595×10^{10}

4. 7.3×10^6

Directions: Convert each number in standard notation to scientific notation.

5. 401,000

6. 1,700,000,000

7. 68,250,000,000

8. 1,990,000,000,000,000

(Answers are on page 179.)

LESSON 7.2—ACCUPLACER CHALLENGE

Directions: For each of the questions below, choose the best answer from the four choices given.

1. The approximate distance from Earth to the sun is 93 million miles. How would this number be expressed in scientific notation?

   (A) 930×10^5 miles
   (B) 93×10^6 miles
   (C) 9.3×10^7 miles
   (D) 0.93×10^8 miles

2. As of December 2015, Earth's population totaled about 7.3×10^9 people. How would this number be expressed in standard notation?

   (A) 73,000,000
   (B) 730,000,000
   (C) 7,300,000,000
   (D) 73,000,000,000

3. Find the sum of 2.5×10^3 and 3.01×10^4.

   (A) 3,260,000

   (B) 32,600

   (C) 551,000

   (D) 5,510,000

4. What is the product of $75,000 \times 300$?

   (A) 2.25×10^7

   (B) 2.25×10^5

   (C) 225×10^7

   (D) 2.25×10^8

(Answers are on page 179.)

LESSON 7.3—USING SCIENTIFIC NOTATION FOR VERY SMALL NUMBERS

In the examples from Lesson 7.2, we looked at how scientific notation can be used to write very large numbers so that we don't have to write out all of the zeros. Scientific notation can also be used to write out very small numbers, and you should expect to see an example of this on the ACCUPLACER. Consider a red blood cell. Cells are very tiny and can only be seen with the help of high-powered microscopes. One red blood cell measures about 0.0000075 meters in length. There are lots of other examples like this in nature. We use scientific notation to write out very small numbers like the weight or length of a proton, the diameter of a DNA helix, or the size of an average bacterium.

How can we write numbers like these in scientific notation? Well, we do it more or less the same way as before, only this time we use negative powers of 10. We'll discuss negative numbers and powers more later on, but for now it's important to know that when you raise ten to a negative power, the decimal point moves to the left instead of to the right. The table that follows lists the first few negative powers of 10. As you can see, the decimal point moves further and further to the left as we move down the table, which means that the numbers are getting smaller!

$10^{-1} = 0.1$
$10^{-2} = 0.01$
$10^{-3} = 0.001$
$10^{-4} = 0.0001$
$10^{-5} = 0.00001$
$10^{-6} = 0.000001$

To Convert a Very Small Number in Standard Notation to Scientific Notation:

1. Move the decimal point to the right until the first nonzero digit is in the ones place.
2. Count the number of places you moved the decimal point.
3. Write this number as a negative power of 10.

➡ Example 1

A human red blood cell measures about 0.0000075 meters in length. How would this number be expressed in scientific notation?

Solution

First, place your pencil on the decimal point, and begin moving it to the right until it ends up between the 7 and the 5. Count the number of places you moved the decimal—you will need this number for your power of 10.

$$0 \underset{1}{.} \underset{2}{0} \underset{3}{0} \underset{4}{0} \underset{5}{0} \underset{6}{0} \, 7 \; 5$$

Write the 7.5. Now, notice that you moved the decimal point six places to the right. This means that you will need to multiply 7.5 by 10 to the negative 6th power.

Answer: Therefore, we would write 0.0000075 in scientific notation as 7.5×10^{-6}.

Notice that the steps for converting very large and very small numbers into scientific notation are mostly similar! The big difference is the direction that the decimal point is moved toward, and you need to remember that negative exponents are used for very small numbers.

To Convert a Very Small Number in Scientific Notation to Standard Notation:

1. Put your pencil on the decimal point.
2. Look at the power of 10 to determine how many places to the left to move the decimal point.
3. Add zeros before the beginning of the decimal number as needed.

➡ Example 2

Convert 3.6×10^{-4} to standard notation.

Solution

First, write out the decimal number. Next, look at the power of 10. In this case, it's −4, so you will need to move the decimal place four places to the left. The negative exponent tells us that, in standard notation, this number will be a very small decimal number.

$$\underset{4}{0} \, \underset{3}{0} \, \underset{2}{0} \, \underset{1}{3} \, . \, 6$$

Answer: After moving the decimal point four places to the left and including the three additional zeros between the decimal point and the three, we see that 3.6×10^{-4} is equal to 0.00036.

LESSON 7.3—SKILLS CHECK

> **Directions:** Convert each number in scientific notation to standard notation.

1. 5.89×10^{-3}

3. 9.375×10^{-4}

2. 7×10^{-5}

4. 2.22×10^{-7}

> **Directions:** Convert each number in standard notation to scientific notation.

5. 0.00045

7. 0.002

6. 0.000699

8. 0.00001002

(Answers are on page 180.)

LESSON 7.3—ACCUPLACER CHALLENGE

> **Directions:** For each of the questions below, choose the best answer from the four choices given.

1. $5.7 \times 10^{-6} =$

    (A) 57,000,000
    (B) 5,700,000
    (C) 0.0000057
    (D) 0.00000057

2. $0.00054 - 0.0000319 =$

    (A) 5.081×10^{-4}
    (B) 5.81×10^{-5}
    (C) 5.719×10^{-4}
    (D) 2.65×10^{-3}

3. Find the product of 3.5×10^{-4} and 2×10^{-3}.

    (A) 0.0007
    (B) 0.00007
    (C) 0.000007
    (D) 0.0000007

4. Which of the following is greater than 0.00405?

    (A) 4.05×10^{-4}
    (B) 4.5×10^{-4}
    (C) 4.55×10^{-5}
    (D) 4.5×10^{-3}

(Answers are on page 180.)

LESSON 7.4—SQUARE ROOTS

Now that we've spent some time looking at exponents—and squaring numbers in particular—we can review how to calculate roots. Remember that when we square a number, or raise it to the second power, we multiply the number by itself to get a new number. The answer you get when doing this is called a perfect square. Finding the **square root** of a number can be thought of as the inverse of squaring a number. As you'll see, perfect squares are the only numbers that will have whole number square roots.

The symbol that indicates that we need to calculate a square root is called a **radical sign**. A radical sign looks like a check mark with a line at the top that goes over the number we need to find the square root of. It looks like this:

$$\sqrt{}$$

When you calculate a square root, you need to ask yourself this question: "What number times itself is equal to the number under the radical sign?" Even when you might not be able to figure out the *exact* square root of a number, asking yourself this question will help you make a close estimate.

➡ Example 1

Find the square root of 49.

Solution

In mathematical notation, we would write this as $\sqrt{49}$. Since the number 49 is under the radical sign, we need to ask ourselves: "What number times itself is equal to 49?" The answer is 7 because 7 times itself is equal to 49.

Answer: $\sqrt{49} = 7$

➡ Example 2

Find the square root of $\sqrt{100}$.

Solution

You should remember from earlier in this unit that 100 is a perfect square. This means that the square root of 100 will be a whole number. Next, we ask ourselves: "What number times itself is equal to 100?" Going through our times tables, we know that $10 \times 10 = 100$.

Answer: $\sqrt{100} = 10$

Estimating Square Roots

Very frequently, we will not come up with a whole number answer when calculating a square root. The square root of two, for example, is an irrational number, which means that it will have an infinite number of nonrepeating decimal places. On the ACCUPLACER, you most likely won't have access to a calculator for solving complicated square roots. Despite not having access to a calculator, it is possible to estimate what a square root will be by comparing it to other square roots that *can* be calculated easily.

➡ Example 3

Estimate the value of $\sqrt{72}$.

Solution

Just as before, you should first ask yourself: "What number times itself is equal to 72?" Unfortunately, there is no whole number that can be multiplied by itself to get a result of 72. This means that the answer will fall somewhere between two whole numbers; in other words, it will include a decimal.

Since we know that our answer won't be a whole number, we need to think about the other perfect squares that are close to 72. We know that $8 \times 8 =$ the perfect square 64, and $9 \times 9 =$ the perfect square 81. These are the two perfect squares that are closest to 72. One of them is less than 72, and the other is greater than 72.

$$\sqrt{64} = 8$$
$$\sqrt{72} = ?$$
$$\sqrt{81} = 9$$

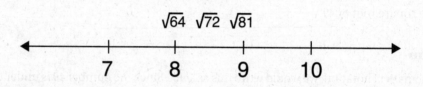

<u>Answer:</u> Looking at it this way, we can conclude that the square root of 72 will fall somewhere between the whole numbers 8 and 9.

➡ Example 4

Which two consecutive whole numbers will $\sqrt{10}$ fall between?

Solution

Since there is no whole number that can be multiplied by itself to get 10, we need to estimate the value of $\sqrt{10}$ by finding other perfect squares that are close to 10. The perfect square $3 \times 3 = 9$ is close to 10, as is the perfect square $4 \times 4 = 16$.

$$\sqrt{9} = 3$$
$$\sqrt{10} = ?$$
$$\sqrt{16} = 4$$

<u>Answer:</u> The square root of 10 will fall between 3 and 4. Which value will it be closer to? Since 10 is closer to 9 than 16, the square root of 10 must be fairly close to 3. (It is actually about 3.1622776.)

LESSON 7.4—SKILLS CHECK

Directions: Solve for the following.

1. $\sqrt{4} =$

2. $\sqrt{36} =$

3. $\sqrt{1} =$

4. $\sqrt{144} =$

5. $\sqrt{0} =$

6. $\sqrt{400} =$

7. $\sqrt{225} =$

8. $\sqrt{\dfrac{1}{4}} =$

(Answers are on page 180.)

LESSON 7.4—ACCUPLACER CHALLENGE

Directions: For each of the questions below, choose the best answer from the four choices given.

1. $\sqrt{324} =$
   - (A) 16
   - (B) 17
   - (C) 18
   - (D) 19

2. $\sqrt{1369} =$
   - (A) 17
   - (B) 29
   - (C) 31
   - (D) 37

3. Which two consecutive whole numbers will $\sqrt{200}$ fall between?
   - (A) 13 and 14
   - (B) 14 and 15
   - (C) 15 and 16
   - (D) 16 and 17

4. Rounded to the nearest whole number, what would be the value of $\sqrt{80}$?
   - (A) 7
   - (B) 8
   - (C) 9
   - (D) 40

5. $\sqrt{5^2} =$
   - (A) 2
   - (B) 5
   - (C) 10
   - (D) 25

(Answers are on page 181.)

LESSON 7.5—SIMPLIFYING RADICALS AND OPERATIONS ON RADICALS

Some problems on the ACCUPLACER will ask you to simplify radical expressions, either through factoring or through performing a series of operations on radicands (numbers under a radical sign). We have already looked at how to calculate the square root of a perfect square and how to estimate the square root of some other numbers that are not perfect squares. Now we will look at some ACCUPLACER problems that actually ask you to leave a radical sign in your answer. You should be prepared to see problems like this on all three ACCUPLACER tests.

Let's look back at an example from the previous section in which we estimated the value of $\sqrt{72}$. What would you do if the problem and the answer choices looked like this?

$$\sqrt{72} =$$

(A) $2\sqrt{6}$

(B) $6\sqrt{3}$

(C) $3\sqrt{6}$

(D) $6\sqrt{2}$

The answer choices are simplified versions of $\sqrt{72}$. Here's a quick process you can use to solve problems like this one.

To Simplify a Square Root:

1. Find the largest possible perfect square factor of the number under the radical sign.
2. Separate the original number into two factors, one of which is the perfect square.
3. Find the square root of the perfect square factor, and leave the other factor under the radical sign.
4. Check to make sure that the number you left under the radical sign doesn't have any perfect square factors. If it does, go through the process again.

Now that we know the steps, let's try to solve this question. First of all, it's difficult to eliminate any of these answer choices without the aid of a calculator. Quick mental math won't allow us to rule out any choices, so we're going to have to work the problem out carefully.

 Example 1

$\sqrt{72} =$

(A) $2\sqrt{6}$

(B) $6\sqrt{3}$

(C) $3\sqrt{6}$

(D) $6\sqrt{2}$

Solution

The first step is to find the largest possible perfect square factor of 72. The perfect squares that we could consider are 1, 4, 9, 16, 25, 36, 49, and 64. Using 1 isn't going to help us out much, but we could consider 4, 9, and 36, since all three are factors of 72. The number 36 is the largest perfect square of 72, so we should use that one. Next, rewrite the radical expression as:

$$\sqrt{72} = \sqrt{36} \times \sqrt{2}$$

We just discovered something interesting here, and it's worth making note of. When you multiply or divide two numbers under separate radical signs, their product or quotient can also be written under a radical sign. For example, $\sqrt{36} \times \sqrt{2} = \sqrt{72}$. We'll come back to this later. The next step is to calculate $\sqrt{36}$.

$$\sqrt{72} = \sqrt{36} \times \sqrt{2}$$
$$\sqrt{72} = 6 \times \sqrt{2}$$
$$\sqrt{72} = 6\sqrt{2}$$

<u>Answer</u>: Since $\sqrt{2}$ can't be simplified, that means that our work is finished and choice D is correct. The expression $6\sqrt{2}$ really means "6 times the square root of 2," but we often read it as "6 square root of 2."

ACCUPLACER TIPS

You might not notice the biggest perfect square factor right away, but don't worry! You can still arrive at the same answer, although through a slightly longer path. You might choose to factor out 9 instead of 36, since 9 is also a perfect square factor. Let's see how that will work out.

$$\sqrt{72} = \sqrt{9} \times \sqrt{8}$$
$$\sqrt{72} = 3 \times \sqrt{8}$$

Notice that $\sqrt{8}$ has a perfect square factor of 4. Therefore, we'll have to do a little more work.

$$\sqrt{72} = 3 \times \sqrt{8}$$
$$\sqrt{72} = 3 \times \sqrt{4} \times \sqrt{2}$$
$$\sqrt{72} = 3 \times 2 \times \sqrt{2}$$
$$\sqrt{72} = 6\sqrt{2}$$

Problems like this will often require you to simplify the number under the radical sign more than once. Just keep your work organized, and you'll be fine!

Example 2

$\sqrt{48} = ?$

Solution

Since 48 has two perfect square factors (other than 1), let's look at two different ways of solving the problem. Either method will give the same answer!

Method 1	Method 2
$\sqrt{48} = \sqrt{16} \times \sqrt{3}$	$\sqrt{48} = \sqrt{4} \times \sqrt{12}$
$\sqrt{48} = 4 \times \sqrt{3}$	$\sqrt{48} = 2 \times \sqrt{12}$
$\sqrt{48} = 4\sqrt{3}$	$\sqrt{48} = 2 \times \sqrt{4} \times \sqrt{3}$
	$\sqrt{48} = 2 \times 2 \times \sqrt{3}$
	$\sqrt{48} = 4\sqrt{3}$

<u>Answer:</u> No matter which of the two methods you use, the solution is $4\sqrt{3}$.

Multiplying and Dividing Square Roots

When you multiply two numbers that are each under a radical sign, the product of the numbers will also be under a radical sign. The same concept applies for division. On the ACCUPLACER, you should be prepared to perform operations like multiplication or division before simplifying a radical expression. Consider the following example.

Example 3

$\sqrt{6} \times \sqrt{10} =$

(A) $3\sqrt{5}$

(B) $2\sqrt{15}$

(C) $15\sqrt{2}$

(D) 60

Solution

We can estimate that $\sqrt{6}$ falls between 2 and 3, and $\sqrt{10}$ falls between 3 and 4. Therefore, we can estimate that $\sqrt{6} \times \sqrt{10}$ must be bigger than $2 \times 3 = 6$ and certainly not more than $3 \times 4 = 12$. Choices C and D are already bigger than our estimate, so we're left with only choices A and B.

First, multiply the two numbers under the radical sign.

$$\sqrt{6} \times \sqrt{10} = \sqrt{60}$$

Next, follow the steps we outlined earlier to simplify $\sqrt{60}$.

$$\sqrt{60} = \sqrt{4} \times \sqrt{15}$$
$$\sqrt{60} = 2 \times \sqrt{15}$$
$$\sqrt{60} = 2\sqrt{15}$$

<u>Answer</u>: Since 15 doesn't have any perfect square factors, it can't be simplified further. Choice B is correct.

➡ Example 4

Simplify $\dfrac{\sqrt{45}}{\sqrt{5}}$.

Solution

In the last problem, we multiplied first and then simplified. This time we should divide first and then simplify.

$$\frac{\sqrt{45}}{\sqrt{5}} = \sqrt{9} = 3$$

<u>Answer</u>: The correct answer is 3. This one works out nicely because the square root of 9 is equal to 3!

To reiterate, when multiplying or dividing two or more numbers that are each under a radical sign, you should:

1. First, multiply or divide the numbers under the radical signs.
2. Then, follow the steps for simplifying the product or the quotient.

Adding and Subtracting Square Roots

This process is a little trickier because you can't add two square root expressions unless the number under the radical sign is the same for both terms. In other words, you may need to simplify the radical expressions as much as possible before adding or subtracting. Let's look at a couple of examples.

➡ Example 5

Simplify $4\sqrt{3} + 5\sqrt{3}$.

Solution

Both terms have $\sqrt{3}$. This means that they are like terms, so we can go ahead and add them. To do so, we add or subtract the numbers in front of the radical signs, and keep the numbers under the radical signs the same. Think about it this way. If you have 4 square roots of 3 and get 5 more square roots of 3, then you have 9 square roots of 3 altogether.

<u>Answer</u>: $4\sqrt{3} + 5\sqrt{3} = 9\sqrt{3}$

➥ Example 6

Simplify $\sqrt{18} + \sqrt{8}$.

Solution

We can't add $\sqrt{18}$ and $\sqrt{8}$ initially because they have different numbers under the radical signs. First, we'll have to simplify both of these square roots, and then we can add the two if they have the same number under the radical signs after being simplified. We need to simplify $\sqrt{18}$ and $\sqrt{8}$ separately.

$$\sqrt{18} = \sqrt{9} \times \sqrt{2} \qquad \sqrt{8} = \sqrt{4} \times \sqrt{2}$$
$$\sqrt{18} = 3\sqrt{2} \qquad \sqrt{8} = 2\sqrt{2}$$

Now that both are simplified, we can rewrite the original expression as

$$3\sqrt{2} + 2\sqrt{2}$$

This expression states that we are adding 3 square roots of 2 to 2 square roots of 2, which means that we will have 5 square roots of 2.

Answer: $\sqrt{18} + \sqrt{8} = 3\sqrt{2} + 2\sqrt{2} = 5\sqrt{2}$

If the problem you are asked to solve involves subtraction instead of addition, you should follow the same steps, which are:

1. First, simplify all the square roots in the expression.
2. Then, add or subtract the numbers in front of the radical signs, keeping the number under the radical signs the same.

LESSON 7.5—SKILLS CHECK

Directions: Simplify each square root.

1. $\sqrt{20} =$ 5. $\sqrt{50} =$ 9. $\sqrt{32} =$

2. $\sqrt{12} =$ 6. $\sqrt{24} =$ 10. $\sqrt{45} =$

3. $\sqrt{28} =$ 7. $\sqrt{200} =$ 11. $\sqrt{27} =$

4. $\sqrt{54} =$ 8. $\sqrt{90} =$ 12. $\sqrt{800} =$

(Answers are on page 181.)

LESSON 7.5—ACCUPLACER CHALLENGE

Directions: For each of the questions below, choose the best answer from the four choices given.

1. $5\sqrt{6} + 11\sqrt{6} =$

   (A) $16\sqrt{6}$

   (B) $11\sqrt{12}$

   (C) 22

   (D) $48\sqrt{2}$

2. $\sqrt{90} - \sqrt{40} =$

   (A) $5\sqrt{2}$

   (B) $5\sqrt{10}$

   (C) $\sqrt{10}$

   (D) $2\sqrt{5}$

3. $\sqrt{5} \times \sqrt{20} =$

   (A) $\sqrt{10}$

   (B) 10

   (C) 20

   (D) 100

4. $\sqrt{27} + \sqrt{75} =$

   (A) $2\sqrt{3}$

   (B) $12\sqrt{2}$

   (C) $\sqrt{102}$

   (D) $8\sqrt{3}$

5. $\sqrt{120} \div \sqrt{5} =$

   (A) $2\sqrt{6}$

   (B) $6\sqrt{2}$

   (C) $2\sqrt{15}$

   (D) $2\sqrt{5}$

6. $\sqrt{28} - \sqrt{63} =$

   (A) $9\sqrt{5}$

   (B) $-\sqrt{7}$

   (C) $5\sqrt{7}$

   (D) $13\sqrt{7}$

(Answers are on page 181.)

> To complete a set of practice questions that reviews all of Chapter 7, and to complete a set of practice questions that covers all of Unit 1, go to:
> *barronsbooks.com/TP/accuplacer/math29qw/*

Lesson 7.1—Skills Check (page 162)

1. **81** $9^2 = 9 \times 9 = 81$

2. **324** $18^2 = 18 \times 18 = 324$

3. $\dfrac{9}{16}$ $\left(\dfrac{3}{4}\right)^2 = \dfrac{3}{4} \times \dfrac{3}{4} = \dfrac{9}{16}$

4. **1**   Any number raised to the zero power has a value of 1.

5. **27** $3^3 = 3 \times 3 \times 3 = 27$

6. **256** $4^4 = 4 \times 4 \times 4 \times 4 = 256$

7. **1**   Any number raised to the zero power has a value of 1.

8. **91** $3^3 = 3 \times 3 \times 3 = 27$
   $4^3 = 4 \times 4 \times 4 = 64$
   $27 + 64 = 91$

9. **1** $1^{10} = 1 \times 1 \times 1 \times 1 \times 1 \times 1 \times 1 \times 1 \times 1 \times 1 = 1$

10. **0** $0^3 = 0 \times 0 \times 0 = 0$
    Zero raised to any nonzero power has a value of zero.

11. $\dfrac{16}{81}$ $\left(\dfrac{2}{3}\right)^4 = \dfrac{2}{3} \times \dfrac{2}{3} \times \dfrac{2}{3} \times \dfrac{2}{3} = \dfrac{16}{81}$

12. **100,000** $10^5 = 10 \times 10 \times 10 \times 10 \times 10 = 100,000$

13. **0.0625** $0.25^2 = 0.25 \times 0.25 = 0.0625$

14. **3.375** $1.5^3 = 1.5 \times 1.5 \times 1.5 = 3.375$

Lesson 7.1—ACCUPLACER Challenge (page 162)

1. **(A)** Calculate the value of 2^6 and 8^2, and then find the difference of the two.

$$2^6 = 2 \times 2 \times 2 \times 2 \times 2 \times 2 = 64$$
$$8^2 = 8 \times 8 = 64$$
$$64 - 64 = 0$$

2. **(C)**

$$\left(\frac{3}{5}\right)^2 = \frac{3}{5} \times \frac{3}{5} = \frac{9}{25} \times \frac{4}{4} = \frac{36}{100} = 0.36$$

3. **(B)** Choices A and D will increase in value when raised to a power. Zero will have the same value, 0, when raised to any nonzero power.

$$\left(\frac{8}{9}\right)^2 = \frac{64}{81}$$
$$\frac{64}{81} < \frac{8}{9}$$

4. **(D)** First calculate the value of the expression within the parentheses.

$$3^3 = 3 \times 3 \times 3 = 27$$

Then square 27.

$$27^2 = 27 \times 27 = 729$$

Lesson 7.2—Skills Check (page 165)

1. **114,000** $1.14 \times 10^5 = 1.1\,4\,0\,0\,0 = 114{,}000$

2. **900,000,000** $9 \times 10^8 = 9.0\,0\,0\,0\,0\,0\,0\,0 = 900{,}000{,}000$

3. **45,950,000,000** $4.595 \times 10^{10} = 4.5\,9\,5\,0\,0\,0\,0\,0\,0\,0 = 45{,}950{,}000{,}000$

4. **7,300,000** $7.3 \times 10^6 = 7.3\,0\,0\,0\,0\,0 = 7{,}300{,}000$

5. **4.01×10^5** $401{,}000 = 4\,0\,1\,0\,0\,0 = 4.01 \times 10^5$

6. **1.7×10^9** $1{,}700{,}000{,}000 = 1\,7\,0\,0\,0\,0\,0\,0\,0\,0 = 1.7 \times 10^9$

7. **6.825×10^{10}** $68{,}250{,}000{,}000 = 6\,8\,2\,5\,0\,0\,0\,0\,0\,0\,0 = 6.825 \times 10^{10}$

8. **1.99×10^{15}** $1{,}990{,}000{,}000{,}000{,}000 = 1\,9\,9\,0\,0\,0\,0\,0\,0\,0\,0\,0\,0\,0\,0\,0$
$$= 1.99 \times 10^{15}$$

Lesson 7.2—ACCUPLACER Challenge (pages 165–166)

1. **(C)** Remember that a number in scientific notation should only have a nonzero digit in the ones place.

$$93 \text{ million} = 93{,}000{,}000$$
$$9\,3\,0\,0\,0\,0\,0\,0 = 9.3 \times 10^7 \text{ miles}$$

2. **(C)** $7.3 \times 10^9 = 7\,3\,0\,0\,0\,0\,0\,0\,0\,0 = 7{,}300{,}000{,}000$

3. **(B)**

$$2.5 \times 10^3 = 2.5\,0\,0 = 2{,}500$$

$$3.01 \times 10^4 = 3.0\,1\,0\,0 = 30{,}100$$

$$2{,}500 + 30{,}100 = 32{,}600$$

4. **(A)** $75{,}000 \times 300 = 22{,}500{,}000$

$$22{,}500{,}000 = 2\,2\,5\,0\,0\,0\,0 = 2.25 \times 10^7$$

Lesson 7.3—Skills Check (page 168)

1. **0.00589** $5.89 \times 10^{-3} = 0\ 0\ 5.89 = 0.00589$

2. **0.00007** $7 \times 10^{-5} = 0\ 0\ 0\ 0\ 7. = 0.00007$

3. **0.0009375** $9.375 \times 10^{-4} = 0\ 0\ 0\ 9.375 = 0.0009375$

4. **0.000000222** $2.22 \times 10^{-7} = 0\ 0\ 0\ 0\ 0\ 0\ 2.22 = 0.000000222$

5. **4.5×10^{-4}** $0.00045 = 0.0\ 0\ 0\ 0\ 4\ 5 = 4.5 \times 10^{-4}$

6. **6.99×10^{-4}** $0.000699 = 0.0\ 0\ 0\ 6\ 99 = 6.99 \times 10^{-4}$

7. **2×10^{-3}** $0.002 = 0.0\ 0\ 2 = 2 \times 10^{-3}$

8. **1.002×10^{-5}** $0.00001002 = 0.0\ 0\ 0\ 0\ 1\ 002 = 1.002 \times 10^{-5}$

Lesson 7.3—ACCUPLACER Challenge (page 168)

1. **(C)** $5.7 \times 10^{-6} = 0\ 0\ 0\ 0\ 0\ 5.7 = 0.0000057$

2. **(A)** $0.00054 - 0.0000319 = 0.0005081 = 0.0\ 0\ 0\ 5\ 081 = 5.081 \times 10^{-4}$

3. **(D)**

$$3.5 \times 10^{-4} = 0\ 0\ 0\ 3.5 = 0.00035$$

$$2 \times 10^{-3} = 0\ 0\ 2. = 0.002$$

$$0.00035 \times 0.002 = 0.0000007$$

4. **(D)**

$$4.5 \times 10^{-3} = 0\ 0\ 4.5 = 0.0045$$

$$0.0045 > 0.00405$$

Lesson 7.4—Skills Check (page 171)

1. **2** $2 \times 2 = 4$, so $\sqrt{4} = 2$.

2. **6** $6 \times 6 = 36$, so $\sqrt{36} = 6$.

3. **1** $1 \times 1 = 1$, so $\sqrt{1} = 1$.

4. **12** $12 \times 12 = 144$, so $\sqrt{144} = 12$.

5. **0** $0 \times 0 = 0$, so $\sqrt{0} = 0$.

6. **20** $20 \times 20 = 400$, so $\sqrt{400} = 20$.

7. **15** $15 \times 15 = 225$, so $\sqrt{225} = 15$.

8. **$\dfrac{1}{2}$** $\dfrac{1}{2} \times \dfrac{1}{2} = \dfrac{1}{4}$, so $\sqrt{\dfrac{1}{4}} = \dfrac{1}{2}$.

Lesson 7.4—ACCUPLACER Challenge (page 171)

1. **(C)** $18 \times 18 = 324$, so $\sqrt{324} = 18$.

2. **(D)** $37 \times 37 = 1{,}369$, so $\sqrt{1369} = 37$.

3. **(B)** $\sqrt{196} = 14$ and $\sqrt{225} = 15$. Therefore, $\sqrt{200}$ would fall between 14 and 15.

4. **(C)** $\sqrt{80}$ falls between 8 and 9. Since $\sqrt{81} = 9$, $\sqrt{80}$ is closer to 9 than 8.

5. **(B)** $\sqrt{5^2} = \sqrt{25} = 5$

Lesson 7.5—Skills Check (page 176)

1. $\mathbf{2\sqrt{5}}$ $\sqrt{20} = \sqrt{4} \times \sqrt{5} = 2 \times \sqrt{5} = 2\sqrt{5}$

2. $\mathbf{2\sqrt{3}}$ $\sqrt{12} = \sqrt{4} \times \sqrt{3} = 2 \times \sqrt{3} = 2\sqrt{3}$

3. $\mathbf{2\sqrt{7}}$ $\sqrt{28} = \sqrt{4} \times \sqrt{7} = 2 \times \sqrt{7} = 2\sqrt{7}$

4. $\mathbf{3\sqrt{6}}$ $\sqrt{54} = \sqrt{9} \times \sqrt{6} = 3 \times \sqrt{6} = 3\sqrt{6}$

5. $\mathbf{5\sqrt{2}}$ $\sqrt{50} = \sqrt{25} \times \sqrt{2} = 5 \times \sqrt{2} = 5\sqrt{2}$

6. $\mathbf{2\sqrt{6}}$ $\sqrt{24} = \sqrt{4} \times \sqrt{6} = 2 \times \sqrt{6} = 2\sqrt{6}$

7. $\mathbf{10\sqrt{2}}$ $\sqrt{200} = \sqrt{100} \times \sqrt{2} = 10 \times \sqrt{2} = 10\sqrt{2}$

8. $\mathbf{3\sqrt{10}}$ $\sqrt{90} = \sqrt{9} \times \sqrt{10} = 3 \times \sqrt{10} = 3\sqrt{10}$

9. $\mathbf{4\sqrt{2}}$ $\sqrt{32} = \sqrt{16} \times \sqrt{2} = 4 \times \sqrt{2} = 4\sqrt{2}$

10. $\mathbf{3\sqrt{5}}$ $\sqrt{45} = \sqrt{9} \times \sqrt{5} = 3 \times \sqrt{5} = 3\sqrt{5}$

11. $\mathbf{3\sqrt{3}}$ $\sqrt{27} = \sqrt{9} \times \sqrt{3} = 3 \times \sqrt{3} = 3\sqrt{3}$

12. $\mathbf{20\sqrt{2}}$ $\sqrt{800} = \sqrt{400} \times \sqrt{2} = 20 \times \sqrt{2} = 20\sqrt{2}$

Lesson 7.5—ACCUPLACER Challenge (page 177)

1. **(A)**
$$5\sqrt{6} + 11\sqrt{6} = (5 + 11)\sqrt{6} = 16\sqrt{6}$$

2. **(C)**
$$\sqrt{90} = \sqrt{9} \times \sqrt{10} = 3 \times \sqrt{10} = 3\sqrt{10}$$
$$\sqrt{40} = \sqrt{4} \times \sqrt{10} = 2 \times \sqrt{10} = 2\sqrt{10}$$
$$3\sqrt{10} - 2\sqrt{10} = \sqrt{10}$$

3. **(B)**
$$\sqrt{5} \times \sqrt{20} = \sqrt{100}$$
$$\sqrt{100} = 10$$

4. **(D)**
$$\sqrt{27} = \sqrt{9} \times \sqrt{3} = 3 \times \sqrt{3} = 3\sqrt{3}$$
$$\sqrt{75} = \sqrt{25} \times \sqrt{3} = 5 \times \sqrt{3} = 5\sqrt{3}$$
$$3\sqrt{3} + 5\sqrt{3} = 8\sqrt{3}$$

5. **(A)**

$$\sqrt{120} \div \sqrt{5} = \sqrt{24}$$

$$\sqrt{24} = \sqrt{4} \times \sqrt{6} = 2 \times \sqrt{6} = 2\sqrt{6}$$

6. **(B)**

$$\sqrt{28} = \sqrt{4} \times \sqrt{7} = 2 \times \sqrt{7} = 2\sqrt{7}$$
$$\sqrt{63} = \sqrt{9} \times \sqrt{7} = 3 \times \sqrt{7} = 3\sqrt{7}$$

$$2\sqrt{7} - 3\sqrt{7} = -\sqrt{7}$$

UNIT TWO
Elementary Algebra

Signed Numbers and Absolute Value

8

Being able to work comfortably with both positive and negative numbers—commonly referred to as **signed numbers**—is crucial to your success in Elementary Algebra. Therefore, the ACCUPLACER is certain to have questions involving signed numbers to help determine your readiness for college algebra coursework.

In this chapter, we will work to master

- determining the absolute value of signed numbers,
- ordering signed numbers,
- adding and subtracting signed numbers, and
- multiplying and dividing signed numbers.

LESSON 8.1—ABSOLUTE VALUE AND ORDERING SIGNED NUMBERS

A good way to start thinking about positive and negative numbers is to consider the following number line.

Positive numbers will always be placed to the right of zero on the number line, which means that they have values that are greater than zero. Negative numbers, like –3, will always be placed to the left of zero, which means that they have values that are less than zero. **A negative number will always have a negative sign in front of it. Positive numbers are usually written with no sign at all. If a number does not have a negative sign in front of it, then the number is positive.**

In discussions of signed numbers, you will often encounter the term **absolute value.** The absolute value of a number is the distance between 0 and that number on the number line. For example, the number 7 is seven units from 0 on the number line, so its absolute value is 7. The same goes for negative numbers. The number –3 is three units from 0 on the number line. Therefore, its absolute value is 3. Absolute value is written with two tall bars around a number or expression. The absolute value of –3 would be written as follows:

$$|-3| = 3$$

To Find the Absolute Value of a Number:

1. Keep positive numbers the same.
2. Change the sign of negative numbers to positive.
3. Drop the absolute value brackets.

➡ Example 1

Determine the absolute value of –8 and the absolute value of 8. Use a number line to help visualize your answer.

Solution

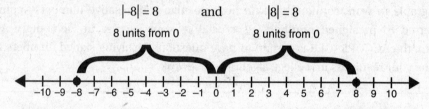

Answer: |–8| = 8 and |8| = 8

➡ Example 2

Mt. Whitney, the highest point of the continental United States, has a maximum elevation of 14,505 feet above sea level. Death Valley, the lowest point of the continental United States, has a minimum elevation of 282 feet below sea level. What is the difference in height between the highest and lowest points of the continental United States?

Solution

It's always a good idea to draw a picture to represent a problem whenever possible. All we really need to consider here is the maximum and minimum heights, not really the mountain or the valley themselves. A simple diagram could look like the following vertical number line.

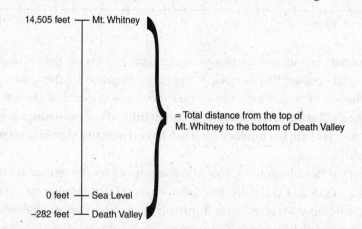

Since Death Valley has an elevation below sea level, this elevation is –282 feet.

We need to find the difference in their heights. Difference is typically thought of as the result of subtraction. The subtraction problem here would be written as 14,505 – (–282) feet. Since we haven't reviewed subtracting signed numbers yet, and because Mt. Whitney and

Death Valley lie on opposite sides of the sea level (0), their distance from one another could also be found by adding the absolute values of their distances from 0. Think about it this way: If Liza is 3 feet immediately to your right, and Ernesto is 4 feet immediately to your left, then Liza and Ernesto are 3 feet + 4 feet, or 7 feet away from one another. This is similar to our situation with Mt. Whitney and Death Valley. Knowing this, we can proceed as follows:

$$|14{,}505| + |-282|$$
$$14{,}505 + 282$$

Answer: The difference in elevation is 14,787 feet.

You can expect to see absolute value questions on all three ACCUPLACER Math tests. You could see questions like those in Examples 1 and 2 or more advanced questions involving algebra. We will cover those higher topics later in Chapter 17, but for now, make sure you feel comfortable with the fundamentals of absolute value.

To Order Signed Numbers:

1. Draw a number line.
2. Place the numbers along the number line.
3. Order the numbers according to where they fall on the number line.

➡ Example 3

Place the following numbers in order from least to greatest.

$$5, -3, 6, -2, 9, 7, -9$$

Solution

First, draw a number line with 0 in the middle. Then place the numbers according to their distance from 0. Negative numbers will be placed further left of 0 as their distance from 0, and hence their absolute value, increases. Positive numbers will be placed further right of 0 as their absolute increases.

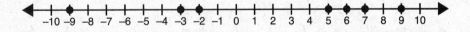

Looking at our number line, we can now write the signed numbers in order from least to greatest.

Answer: The correct order is –9, –3, –2, 5, 6, 7, 9.

➡ Example 4

Place the following numbers in order from least to greatest.

$$-\frac{5}{9}, \frac{2}{9}, \frac{1}{5}, -\frac{3}{5}$$

Solution

All of these numbers are in between –1 and 1, so our number line should look different than before. We need to place fractions with different denominators along the number line so drawing the number line becomes a little difficult. How do we divide 0 to 1 and 0 to –1 on the number line in order to accurately place ninths and fifths? We could start by placing the fifths on the number line.

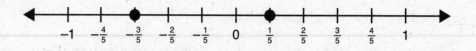

To accurately place the ninths on the number line, we could find an LCD for the four fractions we have. Since we are working with ninths and fifths, the LCD is 45. Do we need to divide the spaces between 0 and 1 and 0 and –1 into 45 pieces? That sounds tedious. Instead, let's convert our fractions into 45ths.

$$-\frac{5}{9} = -\frac{25}{45}$$

$$\frac{2}{9} = \frac{10}{45}$$

$$\frac{1}{5} = \frac{9}{45}$$

$$-\frac{3}{5} = -\frac{27}{45}$$

We now have to put the fractions in order. We don't actually need to place the fractions on the number line perfectly. We just need to know the place of each fraction on the number line in relation to the other fractions. Looking at the positive fractions, we can determine that $\frac{9}{45} < \frac{10}{45}$. Then, looking at the negatives, we can determine that $-\frac{27}{45} < -\frac{25}{45}$. Using this information, we can come up with a general idea of what the number line would look like.

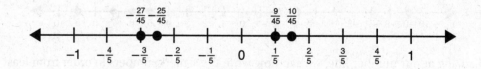

<u>Answer</u>: Using the original fractions, the correct order from least to greatest is:

$$-\frac{3}{5}, \ -\frac{5}{9}, \ \frac{1}{5}, \ \frac{2}{9}$$

LESSON 8.1—SKILLS CHECK

> **Directions:** Put the following numbers in order from least to greatest.

1. $14, -5, 3, 8, -4, -12$

2. $\dfrac{2}{3}, -\dfrac{1}{3}, \dfrac{5}{7}, -\dfrac{2}{7}$

3. $-6.25, 3.375, 0.125, -2.875, -3.375$

4. $-2\dfrac{1}{3}, -4\dfrac{4}{5}, 1\dfrac{9}{10}, -\dfrac{9}{16}, 2\dfrac{1}{2}$

> **Directions:** Determine the absolute value for each number.

5. $|{-12}|$

6. $|25|$

7. $|{-3^5}|$

(Answers are on page 202.)

LESSON 8.1—ACCUPLACER CHALLENGE

> **Directions:** For each of the questions below, choose the best answer from the four choices given.

1. Which of the following choices has the numbers listed in order from least to greatest?

   (A) $-\dfrac{11}{2}, -\dfrac{17}{4}, \dfrac{7}{3}, \dfrac{12}{5}$

   (B) $-\dfrac{11}{2}, -\dfrac{17}{4}, \dfrac{12}{5}, \dfrac{7}{3}$

   (C) $-\dfrac{17}{4}, -\dfrac{11}{2}, \dfrac{12}{5}, \dfrac{7}{3}$

   (D) $-\dfrac{17}{4}, -\dfrac{11}{2}, \dfrac{7}{3}, \dfrac{12}{5}$

2. Which of the following choices has the numbers listed in order from least to greatest?

   (A) $-\dfrac{2}{3}, -\dfrac{3}{8}, -\dfrac{7}{12}, -\dfrac{5}{6}$

   (B) $-\dfrac{3}{8}, -\dfrac{2}{3}, -\dfrac{5}{6}, -\dfrac{7}{12}$

   (C) $-\dfrac{3}{8}, -\dfrac{7}{12}, -\dfrac{2}{3}, -\dfrac{5}{6}$

   (D) $-\dfrac{5}{6}, -\dfrac{2}{3}, -\dfrac{7}{12}, -\dfrac{3}{8}$

3. On the coldest day of the week, the temperature in London was –7°C. On the warmest day of that same week, the temperature in London was 8°C. What was the difference in temperatures between the warmest and coldest days of the week?

    (A) –1°C
    (B) 1°C
    (C) 15°C
    (D) –56°C

4. What number lies exactly halfway between –20 and 4?

    (A) –16
    (B) –8
    (C) 0
    (D) 12

5. The distance from point A to point B is 2.07 units. Given that point B is located at the coordinate 2.01 on a number line, which of the following could be the coordinate of point A on the same number line?

    (A) –0.6
    (B) –0.06
    (C) 0.6
    (D) 4.8

(Answers are on page 202.)

LESSON 8.2—ADDING AND SUBTRACTING SIGNED NUMBERS

When discussing signed numbers, it is helpful to consider the idea of the opposite of a number. The opposite of 4 is –4, and the opposite of –9 is 9. The sum of a number and its opposite is 0.

$$4 + (-4) = 0 \quad \text{and} \quad (-9) + 9 = 0$$

Let's use winning and losing money to help illustrate this point. Think of negative numbers as money that was lost and positive numbers as money that was won. If you won $6 and then lost $6, you would have $0. We would write out this situation like this:

$$6 + (-6) = 0$$

ACCUPLACER TIPS

You probably noticed that, in these examples, there are parentheses around some of the negative numbers. Parentheses often serve two main purposes: to tell us what to do first in a series of operations or to tell us to multiply. In this case, though, the parentheses are only there to separate the + and – signs. This is the way that number sentences involving negative numbers are written in most textbooks and on the ACCUPLACER. When you see parentheses, make sure to think carefully about what purpose they serve!

What if you won $4 and then lost $9? It looks like you lost more than you won! More specifically, you lost $5 in total.

$$4 + (-9) = -5$$

Think about it this way. The $4 that you won is canceled out by the $4 dollars that you lost. You end up $5 further on the "lost" side.

We can use a number line to help further illustrate this. Start at the 4, which represents the $4 that you won. Then, move nine values to the left to illustrate that your winnings are decreasing by $9. As illustrated on the following number line, you will end up at –5, which represents that you have $5 less than when you started.

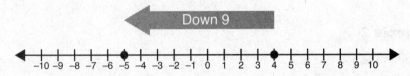

What if you lost $4 and then won $9? This time, you won more than you lost, so altogether you won $5!

$$(-4) + 9 = 5$$

Using a number line for this scenario, we could start with the $4 you lost, or –4, and move 9 places to the right to illustrate that your money is increasing by $9. Here we end at 5, which represents the $5 that you end up with.

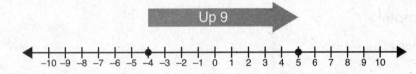

We still haven't covered the most confusing scenario. What happens if you lose $4 and then lose another $9? Intuitively, you probably know that you lost $13 total.

$$-4 + (-9) = -13$$

Don't two negatives equal a positive? Not in this case! That rule only applies when we multiply and divide. When we add two negatives, as we did here, we end up with a negative number. Think about it with money. Losing money twice does not become winning money! Losing money and then losing money again means losing even more money.

There are many ways of thinking about positive and negative numbers. If you don't like using the number line or the analogy of winning and losing money, try memorizing these rules.

To Add Two Numbers with the Same Sign:

1. Ignore the signs and add the two numbers together.
2. Keep their common sign.

To Add Two Numbers with Different Signs:

1. Ignore the signs and subtract the two numbers.
2. Keep the sign of the larger numeral.

➥ Example 1

Find the sum of −5 + (−4).

Solution

The signs of both numbers are negative. Add the 5 and the 4 together. Then, keep the negative sign that they had in common.

$$-5 + (-4) = -9$$

Answer: −9

➥ Example 2

Find the sum of −5 + 4.

Solution

Here, the signs of the numbers are different. Therefore, subtract 5 − 4 to get a difference of 1. Keep the negative sign since −5 has a larger numeral than 4.

$$-5 + 4 = -1$$

Answer: −1

➥ Example 3

Find the sum of 5 + (−4).

Solution

The signs of the numbers are different. Therefore, subtract 5 − 4 to get a difference of 1. Keep the positive sign since 5 has a larger numeral than −4.

$$5 + (-4) = 1$$

Answer: 1

Subtracting signed numbers is a little less intuitive than adding. Instead of trying to think of new subtraction scenarios involving gambling or something else, we are going to stick with what we know, which is adding signed numbers by changing subtraction to addition. We can do this because subtracting a positive number is the same as adding a negative number. For example, −8 − 4 is the same as −8 + (−4). Subtracting a negative number is the same as adding a positive number. For example, 7 − (−9) is the same as 7 + 9. We can use these facts to change our subtraction problems to addition problems.

To Subtract Signed Numbers ("KFC"):

1. Keep the first number and sign.
2. Flip subtraction to addition.
3. Change the second number to its opposite.
4. Follow the rules for adding signed numbers.

➡ Example 4

Solve: $-4 - 7$

Solution

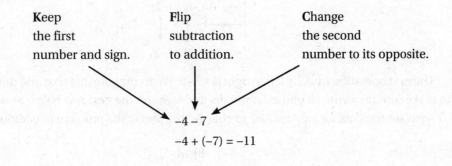

Keep	Flip	Change
the first	subtraction	the second
number and sign.	to addition.	number to its opposite.

$$-4 - 7$$
$$-4 + (-7) = -11$$

Answer: -11

➡ Example 5

At 3:00 P.M., the temperature in a small Nebraska town was –5 degrees Fahrenheit. By 5:00 P.M., the temperature had dropped by 9 degrees. What was the temperature at 5:00 P.M.?

Solution

To solve this problem, first write a mathematical expression. When it comes to temperature, and most measurements, a decrease can be translated to subtraction. If we look at a number line, a drop in temperature would be moving to the left from where we started. Our expression could look like:

$$-5 - 9$$

We can change this subtraction problem to an addition problem by following KFC.

$$-5 - 9 = -5 + (-9)$$

Since both signs are negative, add 5 to 9, and keep the negative sign.

$$-5 + (-9) = -14$$

Answer: The temperature at 5:00 P.M. was –14 degrees Fahrenheit.

To Add and Subtract More Than Two Signed Numbers:

1. Working left to right, add or subtract the first two numbers as directed.
2. Use the result of the first step to add or subtract with the next number as directed.
3. Continue until you have one result.

➡ Example 6

Find the result of –4 + 3 – 9 + (–2).

Solution

Since we have multiple numbers and operations to deal with, we'll add the first two numbers together and work our way through the problem left to right as we go. The first operation in the problem is –4 + 3, so we'll add those together while making sure to rewrite any numbers and signs we haven't dealt with yet. It's a good idea to use parentheses as grouping symbols to help record that you are performing each operation.

$$(-4 + 3) - 9 + (-2)$$
$$-1 \quad - 9 + (-2)$$

The next operation working left to right is –1 – 9. We'll compute that next and then continue to make sure to rewrite all unused numbers and signs on the next row below as we go. Since –1 – 9 is subtraction, we can use KFC to change that part of the problem to addition.

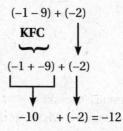

$$(-1 - 9) + (-2)$$
$$\textbf{KFC}$$
$$(-1 + -9) + (-2)$$
$$-10 \quad + (-2) = -12$$

Answer: –12

➡ Example 7

On a game show, a contestant chooses a question with a particular dollar value. If she answers the question correctly, the value of the question is added to her previous total. If she answers the question incorrectly, the value of the question is subtracted from her previous total. Assume a contestant starts with $0. She answers a $200 question correctly and then answers a $400 question correctly. After that, she answers a $1,000 question incorrectly. What is her score?

Solution

First, let's list some important facts.

- If the contestant gets a question correct, we add that dollar amount.
- If the contestant gets a question incorrect, we subtract that dollar amount.
- The contestant started with $0.
- She answered a $200 question correctly, then a $400 question correctly, and then a $1,000 question incorrectly.

Knowing these facts, we can write an expression. The contestant's score equals:

$$\$0 + \$200 + \$400 - \$1,000$$

Next, let's evaluate by working in pairs from left to right.

$$(\$0 + \$200) + \$400 - \$1,000 =$$
$$(\$200 + \$400) - \$1,000 =$$
$$\$600 - \$1,000 =$$

Since we are subtracting a greater number from a lesser number, we can use KFC to make this into an addition problem. The contestant's score equals $\$600 + (-\$1,000)$. The signs of the two numbers are different, so first subtract $1,000 - 600$ to get a difference of 400. Then, keep the negative sign since $-\$1,000$ has a larger numeral than $\$600$.

$$\$600 + (-\$1,000) = -\$400$$

Answer: The contestant's score is $-\$400$.

LESSON 8.2—SKILLS CHECK

Directions: Find the sum or difference.

1. $-14 + 3$

2. $-14 + (-3)$

3. $14 + (-3)$

4. $10 + 2$

5. $-2 + 10$

6. $-10 + (-2)$

7. $2 + (-10)$

8. $4 + (-4)$

9. $-4 + (-4)$

10. $-5 - 9$

11. $5 - 9$

12. $-5 - (-9)$

13. $5 - (-9)$

14. $12 - (-4)$

15. $-12 - 4$

16. $-12 - (-4)$

17. $4 - 12$

18. $5 + (-3) + 9$

19. $-6 + (-2) + (-3) + 4$

20. $-5 - 8 + 2$

(Answers are on page 203.)

LESSON 8.2—ACCUPLACER CHALLENGE

> **Directions:** For each of the questions below, choose the best answer from the four choices given.

1. $3 - 7 + 8 - (-18) =$

    (A) -30
    (B) 6
    (C) 22
    (D) 30

2. Benjamin bought a bicycle for $100 and sold it to a friend for $110. He then bought the bicycle back from his friend for $120, and he later sold it again for $130. What was the financial outcome of these transactions?

    (A) $20 profit
    (B) $30 profit
    (C) $20 loss
    (D) $30 loss

3. In physics, a proton has a charge of 1 and an electron has a charge of -1. What is the total charge of an ion that has 8 protons and 13 electrons?

    (A) -21
    (B) -5
    (C) 5
    (D) 21

4. Walter had $135 in his bank account. He then made a purchase that dropped his bank balance to $-$27. What was the amount of the purchase?

    (A) $27
    (B) $108
    (C) $152
    (D) $162

5. The distance from point C to point D is 3.1 units. Given that point D is located at the coordinate -1.96 on a number line, which of the following could be the coordinate of point C on the same number line?

    (A) -5.06
    (B) 1.26
    (C) 2.04
    (D) 2.86

(Answers are on page 203.)

LESSON 8.3—MULTIPLICATION AND DIVISION OF SIGNED NUMBERS

Multiplying and dividing signed numbers is more straightforward than adding and subtracting signed numbers because the rules stay the same regardless of the size of the numbers.

To Multiply or Divide Two Signed Numbers:

1. When the signs of the two numbers are the same, the result is always positive.
2. When the signs of the two numbers are different, the result is always negative.

That's all there is to it! Therefore, if you are multiplying or dividing two negative numbers, the answer will always be positive. If you are multiplying or dividing a negative number and a positive number, the answer will always be negative. Let's look at some examples.

➡ Example 1

Find the product of –7 and –8.

Solution

According to the rule above, the signs of both numbers are the same, so the result will be positive.

$$-7 \times -8 = 56$$

Answer: 56

➡ Example 2

Find the quotient of –63 and 9.

Solution

The signs of both numbers are different, so the result will be negative.

$$-63 \div 9 = -7$$

Answer: –7

➡ Example 3

Express $\dfrac{-72}{8}$ in simplest terms.

Solution

Remember, fractions are another way of writing division. Therefore, this problem can be thought of as –72 ÷ 8. Since the signs of both numbers are different, the result will be negative.

$$\frac{-72}{8} = -9$$

Answer: –9

To Find the Product or Quotient of More Than Two Signed Numbers:

1. Working left to right, find the product or quotient of the first two signed numbers as directed.
2. Use the result from the first step to find the product or quotient with the next signed number as directed.
3. Repeat working left to right until there is one single result.

➡ Example 4

Which of the following choices will result in a positive product?

(A) $-2 \times 3 \times 4$
(B) $-2 \times (-3) \times 4$
(C) $-2 \times (-3) \times (-4)$
(D) None of the above

Solution

To find which answer choice is correct, separately compute the product of choices A, B, and C. To do that, for each, first find the product of the first two numbers, and then multiply times the third number.

Choice A	Choice B	Choice C
$-2 \times 3 \times 4$	$-2 \times (-3) \times 4$	$-2 \times (-3) \times (-4)$
-6×4	6×4	$6 \times (-4)$
-24	24	-24

Answer: Choice B is the correct answer since it is the only choice that results in a final positive product.

ACCUPLACER TIPS

You can easily predict the sign of a product without actually multiplying at all by counting how many negative numbers are going to be multiplied. This is a great way to eliminate some answer choices on the ACCUPLACER if you aren't completely sure what the correct answer is.

When Multiplying Several Signed Numbers:

- An even number of negative factors will result in a positive product.
- An odd number of negative factors will result in a negative product.

Let's take a look at a few examples.

Number of Negative Factors	Example	The product is...
One negative factor	$(-1) \times 2 \times 3 \times 4 \times 5 = -120$	negative
Two negative factors	$(-1) \times (-2) \times 3 \times 4 \times 5 = 120$	positive
Three negative factors	$(-1) \times (-2) \times (-3) \times 4 \times 5 = -120$	negative
Four negative factors	$(-1) \times (-2) \times (-3) \times (-4) \times 5 = 120$	positive
Five negative factors	$(-1) \times (-2) \times (-3) \times (-4) \times (-5) = -120$	negative

Notice that we didn't concern ourselves with how many positive factors there are. Positive factors do not change the final product's sign, so we don't need to count how many positive factors there are to make our prediction.

Dividing works the same way as multiplication in this regard.

$$-120 \div 5 \div 4 = -6$$
$$-120 \div (-5) \div 4 = 6$$
$$-120 \div (-5) \div (-4) = -6$$

Exponents behave in a similar fashion. Raising a negative number to an even power results in a positive product, whereas raising a negative number to an odd power results in a negative product.

$$(-2)^2 = 4 \qquad (-2)^3 = -8 \qquad (-2)^4 = 16$$
$$(-2)^5 = -32 \qquad (-2)^6 = 64 \qquad (-2)^7 = -128$$

➥ Example 5

Simplify $\dfrac{(-6)(12)(-1)}{9} \times (-5)^3$.

Solution

By simplifying, the question is asking us to make this expression as short, or simple, as we can by performing any operations that we can.

You may notice that there are several parentheses in this problem. Whenever we have a quantity in parentheses immediately next to another quantity in parentheses with nothing in between, we multiply the quantities in the parentheses together.

We can predict whether or not the final result will be positive or negative by counting how many negative factors we have present. In the fraction, we have two negative factors, whereas $(-5)^3$ contains three negative factors. Altogether we have five negative factors, which is an odd number of negative factors, so the final outcome should be negative. Let's make sure though. Remember, this applies only to multiplication and division with signed numbers. Different rules apply for adding and subtracting signed numbers.

First, we will work left to right with the first two quantities in the numerator of the fraction, and we can go ahead and expand $(-5)^3$ at the same time.

$$\frac{(-6)(12)(-1)}{9} \times (-5)^3 =$$

$$\frac{(-72)(-1)}{9} \times (-5)(-5)(-5) =$$

$$\frac{(72)}{9} \times (25)(-5) =$$

$$(8)(-125) = -1,000$$

Answer: $-1,000$

LESSON 8.3—SKILLS CHECK

Directions: Find the product or quotient as directed.

1. 12×-3

2. $(-12)(-3)$

3. $18 \div (-6)$

4. $(-56) \div (-7)$

5. $\dfrac{-34}{-17}$

6. $-\dfrac{15}{3}$

7. $\dfrac{-9}{-36}$

8. $-3 \times 8 \times -6$

9. $(-7)(4)(-3)(-5)$

10. $\dfrac{4-8}{-2}$

11. $2(3-4) - (-3)^2$

12. $16 - 3(8-3)^2 \div 5$

(Answers are on page 204.)

LESSON 8.3—ACCUPLACER CHALLENGE

> **Directions:** For each of the questions below, choose the best answer from the four choices given.

1. $\dfrac{3 - (-15)}{6} =$

   (A) -2

   (B) $\dfrac{1}{3}$

   (C) 2

   (D) 3

2. $\dfrac{-4}{2 - (-14)} =$

   (A) -4

   (B) $-\dfrac{1}{4}$

   (C) $-\dfrac{1}{3}$

   (D) $\dfrac{1}{4}$

3. On a test, a teacher gives 5 points for a correct answer, -2 points for an incorrect answer, and -1 point for leaving a question blank. What is the score for a student who answered 16 questions correctly, 5 questions incorrectly, and left 4 questions blank?

   (A) 2

   (B) 66

   (C) 80

   (D) 94

4. The high temperatures over a 7-day period in Barrow, Alaska were $-4°F$, $-9°F$, $-5°F$, $2°F$, $4°F$, $-1°F$, and $-8°F$. What was the average daily high temperature for this period?

   (A) $-21°F$

   (B) $-5°F$

   (C) $-3°F$

   (D) $4°F$

(Answers are on page 205.)

> To complete a set of practice questions that reviews all of Chapter 8, go to:
> *barronsbooks.com/TP/accuplacer/math29qw/*

Lesson 8.1—Skills Check (page 189)

1. **−12, −5, −4, 3, 8, 14** First, determine the order of the negative values from least to greatest. Then, do the same for the positive values. You could also use a number line like the one below.

2. $-\dfrac{1}{3}, -\dfrac{2}{7}, \dfrac{2}{3}, \dfrac{5}{7}$ Change all fractions to equivalent fractions with a common denominator. Then, order the fractions according to their numerators.

$$-\frac{1}{3} = -\frac{7}{21}, \ -\frac{2}{7} = -\frac{6}{21}, \ \frac{2}{3} = \frac{14}{21}, \ \frac{5}{7} = \frac{15}{21}$$

3. **−6.25, −3.375, −2.875, 0.125, 3.375** Order each number according to the digit in the ones place for each.

4. $-4\dfrac{4}{5}, -2\dfrac{1}{3}, -\dfrac{9}{16}, 1\dfrac{9}{10}, 2\dfrac{1}{2}$ First, determine the order of the negative values from least to greatest. Then, do the same for the positive values.

5. **12** $|{-12}| = 12$

6. **25** $|25| = 25$

7. **243** $|{-3^5}| = |{-243}| = 243$

Lesson 8.1—ACCUPLACER Challenge (pages 189–190)

1. **(A)** Change each of the fractions to equivalent fractions with a common denominator. Then, order the fractions from least to greatest according to their numerators.

$$-\frac{11}{2} = -\frac{330}{60}, \ -\frac{17}{4} = -\frac{255}{60}, \ \frac{7}{3} = \frac{140}{60}, \ \frac{12}{5} = \frac{144}{60}$$

2. **(D)** Change each of the fractions to equivalent fractions with a common denominator. Then, order the fractions from least to greatest according to their numerators.

$$-\frac{5}{6} = -\frac{20}{24}, \ -\frac{2}{3} = -\frac{16}{24}, \ -\frac{7}{12} = -\frac{14}{24}, \ -\frac{3}{8} = -\frac{9}{24}$$

3. **(C)** The difference in the temperatures can be thought of as the distance from −7 to 8 on a number line. This can be found by finding the absolute value of the difference of the temperatures using $|{-7° - 8°}| = |{-15°}| = 15°$ *or* $|8° - (-7°)| = |8° + 7°| = 15°$.

4. **(B)** The distance from −20 to 4 is 24. Exactly halfway between the two numbers would be the middle. The middle is 12 more than −20, which is −20 + 12 = −8. This can also be found by finding the number that is 12 less than 4, which is 4 − 12 = 4 + (−12) = −8.

5. **(B)** Point B is located at 2.01 on a number line. We know that point A is 2.07 units from point B. We don't know whether point A is to the left of point B on the number

line, which would be less, or whether point A is to the right of point B on the number line, which would be more. This means that point A could either be:

$$2.01 + 2.07 = 4.08$$

or

$$2.01 - 2.07 = 2.01 + (-2.07) = -0.06$$

4.08 is not a choice, but –0.06 is, so point A must be at –0.06.

Lesson 8.2—Skills Check (page 195)

1. **–11** $-14 + 3 = -11$
2. **–17** $-14 + (-3) = -17$
3. **11** $14 + (-3) = 11$
4. **12** $10 + 2 = 12$
5. **8** $-2 + 10 = 8$
6. **–12** $-10 + (-2) = -12$
7. **–8** $2 + (-10) = -8$
8. **0** $4 + (-4) = 0$
9. **–8** $-4 + (-4) = -8$
10. **–14** $-5 - 9 = -5 + (-9) = -14$
11. **–4** $5 - 9 = 5 + (-9) = -4$
12. **4** $-5 - (-9) = -5 + 9 = 4$
13. **14** $5 - (-9) = 5 + 9 = 14$
14. **16** $12 - (-4) = 12 + 4 = 16$
15. **–16** $-12 - 4 = -12 + (-4) = -16$
16. **–8** $-12 - (-4) = -12 + 4 = -8$
17. **–8** $4 - 12 = 4 + (-12) = -8$
18. **11** $5 + (-3) + 9 = 2 + 9 = 11$
19. **–7** $-6 + (-2) + (-3) + 4 = -8 + (-3) + 4 = -11 + 4 = -7$
20. **–11** $-5 - 8 + 2 = -5 + (-8) + 2 = -13 + 2 = -11$

Lesson 8.2—ACCUPLACER Challenge (page 196)

1. **(C)** $3 - 7 + 8 - (-18) = 3 + (-7) + 8 + 18 = -4 + 8 + 18 = 4 + 18 = 22$

2. **(A)** Money spent, or lost, can be represented with negative numbers, and money gained, or earned from selling, can be represented using positive numbers. Using this information, we can represent all of the transactions with the following expression:

$$-\$100 + \$110 + (-\$120) + \$130 = \$20$$

Since the outcome is positive, Benjamin earned a $20 profit.

3. **(B)** The total charge of 8 protons (+8), and 13 electrons (–13) can be found by adding.

$$8 + (-13) = -5$$

The total charge is –5.

4. **(D)** Walter's account started at $135 and ended at –$27 after a purchase. The amount of the purchase is equal to the distance from $135 to –$27 on a number line.

$$|\$135 - (-\$27)| = |\$135 + \$27| = \$162$$

Therefore, the purchase cost $162.

5. **(A)** Point D is located at –1.96 on a number line. We know that point C is 3.1 units from point D. We don't know if point C is to the left of point D on the number line, which would be less, or if point C is to the right of point D on the number line, which would be more. This means that point C could be found one of two ways.

$$-1.96 + 3.1 = 1.14$$
$$or$$
$$-1.96 - 3.1 = -5.06$$

1.14 is not a choice, but –5.06 is, so point C must be at –5.06.

Lesson 8.3—Skills Check (page 200)

1. **–36** $12 \times -3 = -36$

2. **36** $(-12)(-3) = 36$

3. **–3** $18 \div (-6) = -3$

4. **8** $(-56) \div (-7) = 8$

5. **2** $\dfrac{-34}{-17} = 2$

6. **–5** $-\dfrac{15}{3} = -5$

7. **$\dfrac{1}{4}$** $\dfrac{-9}{-36} = \dfrac{1}{4}$

8. **144** $-3 \times 8 \times -6 = -24 \times -6 = 144$

9. **–420** $(-7)(4)(-3)(-5) = (-28)(-3)(-5) = (84)(-5) = -420$

10. **2** $\dfrac{4-8}{-2} = \dfrac{-4}{-2} = 2$

11. **–11** $2(3-4) - (-3)^2 = 2(-1) - (9) = -2 - 9 = -11$

12. **1** $16 - 3(8-3)^2 \div 5 =$
$$16 - 3(5)^2 \div 5 =$$
$$16 - 3(25) \div 5 =$$
$$16 - 75 \div 5 =$$
$$16 - 15 = 1$$

Lesson 8.3—ACCUPLACER Challenge (page 201)

1. **(D)** $\dfrac{3-(-15)}{6} = \dfrac{3+15}{6} = \dfrac{18}{6} = 3$

2. **(B)** $\dfrac{-4}{2-(-14)} = \dfrac{-4}{2+14} = \dfrac{-4}{16} = -\dfrac{1}{4}$

3. **(B)** With 16 questions answered correctly, the student receives 16(5) = 80 points. However, the student receives –2(5) = –10 points for 5 questions answered incorrectly, and the student receives –1(4) = –4 points for leaving 4 questions blank. Altogether, the student's score is:

$$80 + (-10) + (-4) = 70 + (-4) = 66 \text{ points}$$

4. **(C)** The average, or mean, temperature for the 7 days can be determined using the following method.

$$\frac{-4+(-9)+(-5)+2+4+(-1)+(-8)}{7} = \frac{-21}{7} = -3°F$$

Introduction to Algebra

9

Algebra involves the mathematical use of symbols, or **variables**, to stand for unknown or changing values. However, algebra is much bigger than this simple definition. In many ways, it helps us do math more efficiently.

- Algebra helps us make generalizations about arithmetic. For example, we can observe that two times any number plus three times that same number is always equal to five times that same number. Instead of writing those words, we can write $2x + 3x = 5x$.
- Algebra helps us to work backwards in order to solve mathematical problems.
- Algebra helps us simplify mathematical procedures.
- Algebra lets us symbolize formulas and algorithms.

Algebra is so useful and important to human progress that most colleges and universities require that their graduates pass at least one college algebra course. As a result, the ACCUPLACER focuses a lot of attention on determining your knowledge and comfort with algebra. In fact, the Elementary Algebra and College-Level Mathematics tests focus almost exclusively on key algebra concepts.

> In this chapter, we will work to master
>
> - evaluating algebraic expressions using substitution and the correct order of operations, and
> - representing words and phrases algebraically.

Before we get to our first objective, let's define some important terms.

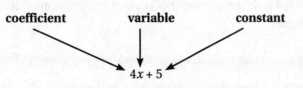

VARIABLE: a symbol, usually a letter, that is used to represent a number that has an unknown value or a value that changes. The word *variable* is related to the words *vary* and *variety*, which both have to do with change and different forms of something.

COEFFICIENT: a number immediately in front of a variable that indicates the number of times we have that variable. For example, $4x$ indicates that we have x four times.

CONSTANT: a number that stands by itself. The value of a constant *does not* change. For example, 5 is a constant. There is no question as to what its value is. It's 5, always.

EXPRESSION: a mathematical statement that can include any combination of variables, coefficients, exponents, operations, and/or constants. An expression doesn't have an equal sign (=).

EQUATION: a mathematical statement with two expressions that are equal. Equations must have an equal sign (=).

EVALUATE: to determine the value of an expression or equation.

LESSON 9.1—EVALUATING ALGEBRAIC EXPRESSIONS

Now that we have reviewed all of those definitions, let's put it all together and evaluate an algebraic expression. We can do this if we are given the value of one or more variables in an expression. If so, we can substitute the given value into the expression and find the expression's total value by performing the operations contained in the expression.

➡ **Example 1**

Evaluate $2a$ when $a = 8$.

Solution

We know that $a = 8$, and the 2 in $2a$ indicates that we have two of these 8s, not just one. To write this, we **substitute** 8 for the variable a, and then multiply.

$$2a \text{ becomes } 2(8) = 16$$

Answer: 16

Notice that we wrote 2 next to 8 inside of parentheses. Whenever we have something inside of parentheses with a quantity immediately outside of the parentheses, we multiply the quantity outside of the parentheses with all of the quantities inside of the parentheses. We could just as easily write 2×8, but using parentheses instead helps avoid confusion between the x variable and the $\times$ for multiplication.

Of course, evaluating algebraic expressions questions are not always going to be that simple. Often, an expression will require you to perform several operations before you arrive at a final answer. What is the proper order of performing these operations? If you're not sure, take a look at Example 2.

➡ **Example 2**

What is the value of the expression ab^2 when $a = 3$ and $b = 5$?

Solution

First, substitute 3 for a and 5 for b.

$$(3)(5)^2$$

What comes next? Do we multiply 3 times 5 and then work with the exponent? Do we work with the exponent and then multiply times 3? In other words, is it 15^2 or $(3)(25)$?

Luckily, there is an established order in which the operations must be performed. That order tells us that we should work with the exponent first, and then multiply.

Answer: $(3)(5)^2 = (3)(5 \times 5) = (3)(25) = 75$

Many students learn a catchy saying to help them remember the order of operations. You may have heard it before. It goes:

Please Excuse My Dear Aunt Sally

The first letters from the words of this mnemonic device form PEMDAS, which in turn help us to remember that we perform operations in the following order:

1. **P**arentheses
2. **E**xponents
3. **M**ultiplication or **D**ivision (whichever appears first, working from left to right)
4. **A**ddition or **S**ubtraction (whichever appears first, working from left to right)

To clarify, PEMDAS says that we first perform any operations we can that are found inside of parentheses, then we work with the exponents, then we perform any multiplication or division, and, finally, we perform any addition or subtraction that is found in the expression.

You probably noticed that, although PEMDAS has six letters, the order of operations only includes four steps. This is because multiplication and division are actually the same step, and addition and subtraction are also the same step. Remember that all multiplication problems can be written as division problems, and vice versa. The same is true of addition and subtraction. All addition problems can be written as subtraction problems, and vice versa.

To Evaluate an Algebraic Expression Using the Order of Operations:

1. Substitute each variable's given value in place of its letter.
2. Perform the operations in the order outlined by PEMDAS.

Let's look at the following examples that include many of the most commonly confused issues on ACCUPLACER questions that involve the order of operations.

➡ Example 3

What is the value of $a + b \times c$ when $a = 4$, $b = 3$, and $c = 5$?

Solution

The first step is to substitute the given values for a, b, and c. Doing so, we rewrite the expression as

$$4 + 3 \times 5$$

Next, we look to PEMDAS. We don't have any parentheses or exponents here, but we do have multiplication, so we need to start there.

$$4 + 3 \times 5 =$$
$$4 + 15 =$$

Finally, we only have addition left to do.

$$4 + 15 = 19$$

Answer: 19

➡ Example 4

What is the value of $p \div q - t \times q$ when $p = 18$, $q = 9$, and $t = 3$?

Solution

Start by substituting the given values for p, q, and t. Then multiply and divide, working left to right. Subtraction will be the last step in this problem.

$$18 \div 9 - 3 \times 9 =$$
$$2 - 27 = -25$$

Answer: -25

Now let's look at a more complicated example—one that involves a lot more steps and a greater degree of precision.

➡ Example 5

What is the value of $\dfrac{\sqrt{kh} + gh}{g^2 - h^2}$ when $k = 50$, $g = 8$, and $h = 2$?

Solution

Start by substituting the given values for k, g, and h.

$$\frac{\sqrt{(50)(2)} + (8)(2)}{8^2 - 2^2}$$

Notice that we now have a fraction and a square root symbol. Neither of those is specifically referenced by any of the letters of PEMDAS. Not to worry. The P in PEMDAS actually refers to all grouping symbols, not just parentheses. That means that we first complete any operations we can that are grouped together by symbols, such as the square root symbol, or those grouped together in the numerator or the denominator of a fraction. Below, we address the square root first and then proceed with PEMDAS in the numerator and the denominator separately.

$$\frac{\sqrt{(50)(2)} + (8)(2)}{8^2 - 2^2} =$$
$$\frac{\sqrt{100} + (8)(2)}{8^2 - 2^2} =$$
$$\frac{10 + (8)(2)}{64 - 4} =$$
$$\frac{10 + 16}{64 - 4} =$$
$$\frac{26}{60} = \frac{13}{30}$$

Answer: $\dfrac{13}{30}$

LESSON 9.1—SKILLS CHECK

Directions: For questions 1–3, evaluate the expression when $a = 2$, $b = 3$, and $c = 4$.

1. $3b - 3c$

2. $6b \div a$

3. $(a + b)^2$

Directions: For questions 4–8, evaluate the expression when $x = 3$, $y = 5$, and $z = -10$.

4. $2z^2$

6. $\dfrac{3}{4}(3z + 9)$

8. $3y^2 + 4z - 2x^2$

5. $y^2 - \dfrac{xz}{8} + 9$

7. $\sqrt{15xy}$

(Answers are on page 219.)

LESSON 9.1—ACCUPLACER CHALLENGE

Directions: For each of the questions below, choose the best answer from the four choices given.

1. What is the value of the expression $4x^2 - 5xy - 3y^2$ when $x = -2$ and $y = 6$?

   (A) -200
   (B) -152
   (C) -64
   (D) -32

2. What is the value of the expression $\dfrac{ab + 1}{a^4 + ab}$ when $a = -3$ and $b = 7$?

   (A) $-\dfrac{1}{3}$

   (B) -3

   (C) $\dfrac{20}{33}$

   (D) $\dfrac{20}{9}$

3. What is the value of $(4mn)^2 \div n^2 - 2m(n + 6)$ when $m = 3$ and $n = -4$?

   (A) 0
   (B) 116
   (C) 132
   (D) 576

(Answers are on page 219.)

LESSON 9.2—TRANSLATING WRITTEN PHRASES INTO ALGEBRA

One of the core uses of algebra is to represent real-world, mathematical situations and relationships symbolically. The ACCUPLACER will test your ability to do just that.

Let's consider the following scenario. At the grocery store, avocados cost $2 each and watermelons cost $5 each. If we wanted to figure out how much it would cost to buy 4 avocados and 6 watermelons, we could just add $2 each time we buy an avocado and $5 each time we buy a watermelon. Remember that we can use multiplication any time we have the same quantity repeatedly. By multiplying the number of avocados that we would like to buy, 4 avocados, times the price of each avocado, $2, and then repeating the process for the number of watermelons, 6 watermelons times $5 each, we can determine how much it would cost to buy all of these fruits.

$$\$2(4) + \$5(6) =$$
$$\$8 + \$30 = \$38$$

Notice that we followed the correct order of operations. We completed all the multiplication first, and then we added.

What does this have to do with algebra? Well, algebra gives us a way of generalizing a mathematical process. In other words, we can use the same process for any number of avocados or watermelons or anything else. When we want to represent a value that can change, we use a variable. Therefore, we can use the same mathematical expression only using variables to represent the number of avocados and watermelons we would like to buy. Unfortunately, we can't change the price of the fruit at the store, only how much of the fruit we buy.

The first step in writing this example as an algebraic expression is to define the variables.

Let a = the number of avocados
Let w = the number of watermelons

To represent the rule for determining the cost of an undetermined number of avocados, or a avocados, at $2 each, and an undetermined number of watermelons, or w watermelons, at $5 each, we write the following:

$$\$2a + \$5w = \text{total cost}$$

Representing real-world situations with algebraic expressions is a valuable skill that will serve you well throughout college algebra as well as in life. On the ACCUPLACER, however, you won't have to directly demonstrate your expertise at this. Instead, you will have to answer questions that ask you to match a written situation with the correct algebraic expression that can be used to represent that situation. To do this, it is helpful to be familiar with many of the common key words that signify mathematical operations or relationships as well as a few test-taking techniques.

To Translate Written Phrases to Algebra:

1. Define the variable(s).
2. Underline key words and phrases.
3. Translate the words and numbers into an algebraic expression that represents them.

➡ Example 1

Leroy goes to the store and buys B bagels and C cookies. Bagels cost 75 cents each, and cookies cost 60 cents each. Which of the following expressions represents the total cost of Leroy's purchase in cents?

(A) $B + C$

(B) $135BC$

(C) $60B + 75C$

(D) $75B + 60C$

Solution

First, let's identify our variables B and C.

$$B = \text{the number of bagels}$$
$$C = \text{the number of cookies}$$

Next, let's identify key phrases and underline key words. "Bagels cost 75 cents <u>each</u>" indicates that it costs 75 cents for one bagel, 75 cents × 2 for two bagels, 75 cents × 3 for three bagels, and so forth. "Cookies cost 60 cents <u>each</u>" indicates that it costs 60 cents for one cookie, 60 cents × 2 for two cookies, 60 cents × 3 for three cookies, and so forth. The word "<u>total</u>" indicates that we want to combine, or add, the cost of all of the bagels and cookies.

We don't know how many bagels or cookies are being bought, but we do know that, regardless of what those amounts are, we can figure out the cost in cents with this expression.

| cost of one bagel | number of bagels | cost of one cookie | number of cookies |

$$75 \times B + 60 \times C$$

<u>Answer</u>: The algebraic expression that represents those steps is choice D.

➡ Example 2

Peter completed p math problems. Chester, his brother, completed 6 more math problems than Peter. Which of the following expressions could be used to represent the total number of math problems that Peter and Chester completed altogether?

(A) $p + 6$
(B) $2p + 6$
(C) $8p$
(D) $12p$

Solution

First, let's identify our variable, p.

$$p = \text{the number of math problems that Peter completed}$$

Next, let's identify key phrases and underline key words. "Chester completed <u>6 more</u> math problems <u>than Peter</u>." This statement indicates that we can take whatever number of math problems that Peter completed and add 6 to it to find how many math problems Chester completed. We only know that Peter completed p math problems. Therefore, Chester completed 6 more than p math problems, or $p + 6$. The word "<u>total</u>" indicates that we need to combine, or add, all of the math problems that Peter and Chester completed.

To represent the total number of math problems that Peter and Chester completed, we add:

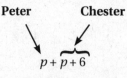

Peter   Chester

$p + p + 6$

None of our answer choices look quite like this. We know that the correct answer can't be choice A, $p + 6$, because that's how many math problems that Chester completed by himself, not the total number of problems completed by both Peter and Chester.

Since $p + p + 6$ is not a choice, one of the choices must offer an equivalent, or equal, expression. Before we move forward, it's important to introduce the concept of like terms. **Like terms** are terms with the same variable(s) to the same power. For example $5x$ and $3x$ are like terms. $4m^2$ and $7m^2$ are also like terms. However, $5x$ and $4m^2$ are not like terms.

When we have like terms, we can combine, or add, them together. We will cover the topic of combining like terms in further detail in Chapter 12. For now, we need to focus on the basics.

In this problem, we have $p + p + 6$. Here, the ps are like terms so they can be combined. To do that, we must recognize that p, by itself, indicates that we have one p or $1p$. To combine p with another p, we simply add $1p + 1p$ to get $2p$. Using our full expression, we have:

$$p + p + 6 = 2p + 6$$

We cannot do any further combining because 6 is a constant. It does not have a p, so it cannot be combined with the ps.

Answer: Therefore, the correct answer is choice B.

➥ Example 3

Kathy sells fish tanks. She earns \$600 a week plus \$50 for every fish tank she sells. Which of the following expressions could be used to calculate the amount of money she earns in one week after selling f fish tanks?

(A) $650f$
(B) $600f + 50$
(C) $50f + 600$
(D) $600(f + 50)$

Solution

First, let's define what our variable, f, represents:

f = the number of fish tanks that Kathy sells

Next, we need to identify key phrases and underline key words. "\$600 a week <u>plus</u> \$50 <u>for every</u> fish tank she sells." We also need to look at the part of the problem where the question identifies the mathematical relation we are supposed to model. This would be, "the amount of money she earns in <u>one week</u> after selling f fish tanks." We know that Kathy gets a flat \$600 a week regardless of the number of fish tanks she sells. Since we are only talking about one week, we are modeling a situation with one \$600. Looking at the answer choices, we can see that choice C is the only choice that is not multiplying the \$600 by some amount. Right there, we could conclude that choice C is the correct answer, but let's check the other facts to be sure.

Kathy earns 600 <u>plus</u> 50 <u>for every fish tank</u>. We could translate this statement as follows:

$$600 + 50 \times f$$
$$600 + 50f$$

Answer: Looking at the answer choices again, choice C is indeed the correct answer. You may notice that choice C is written as $50f + 600$. Remember that addition is commutative, which means that the order doesn't matter. In other words, $600 + 50f = 50f + 600$.

LESSON 9.2—SKILLS CHECK

> **Directions:** For questions 1–6, write an algebraic expression that can represent the age of each person in terms of k, Kate's age, on the line provided. Question 1 has been completed for you.

1. Kate is k years old. Bill is four years older than Kate. Bill's age in terms of k, Kate's age, can be expressed as _____ $k + 4$ _____.

2. Kate is k years old. Nancy is 3 years younger than Kate. Nancy's age in terms of k, Kate's age, can be expressed as _____.

3. Kate is k years old. Maria is twice as old as Kate. Maria's age in terms of k, Kate's age, can be expressed as _____.

4. Kate is k years old. Phyllis is 8 years less than three times Kate's age. Phyllis's age in terms of k, Kate's age, can be expressed as _____.

5. Kate is k years old. Tony is half as old as Kate. Tony's age in terms of k, Kate's age, can be expressed as _____.

6. Kate is k years old. Tony is half as old as Kate. Daquan is four times older than Tony. Daquan's age in terms of k, Kate's age, can be expressed as _____.

(Answers are on page 219.)

LESSON 9.2—ACCUPLACER CHALLENGE

> **Directions:** For each of the questions below, choose the best answer from the four choices given.

1. From a bag of c candies, Ms. Williams eats 8 of the candies and then she shares the rest with her 5 children equally. Which of the following expressions could be used to determine the number of candies that each child receives?

   (A) $\dfrac{c - 5}{8}$

   (B) $\dfrac{8 - 5}{c}$

   (C) $\dfrac{5 - c}{8}$

   (D) $\dfrac{c - 8}{5}$

2. Mateo is 4 years younger than twice his brother's age. If *b* stands for his brother's age, which of the following expressions could be used to represent Mateo's age?

   (A) $2b - 4$
   (B) $2(b - 4)$
   (C) $4 - 2b$
   (D) $2b$

3. In Shalinda's French class, there are *w* women and *m* men. Which of the following expressions could be used to represent the ratio of women to men in Shalinda's French class?

   (A) $\dfrac{m}{w}$

   (B) $\dfrac{w}{m}$

   (C) $m + w$
   (D) $w - m$

4. Tye does *t* pushups. Fred does 10 more pushups than Tye. Which of the following expressions could be used to find how many pushups they both do together in total?

   (A) $t - 10$
   (B) $t + 10$
   (C) $t + 20$
   (D) $2t + 10$

5. Rashida completed *p* math problems on Sunday. Every day for the rest of the week, she completed one more math problem than she did the day before. Which of the following expressions could be used to represent how many math problems Rashida completed that entire week?

   (A) $p + 6$
   (B) $7p + 1$
   (C) $7p + 6$
   (D) $7p + 21$

6. On Saturday, it took Raymond's Newsstand 5 hours to sell out of newspapers. On Thursday, it took Raymond's Newsstand 3 hours to sell the same number of newspapers. If the newsstand's average rate of newspaper sales on Saturday was *n* newspapers per hour, what was their average rate of newspaper sales on Thursday, in terms of *n*?

   (A) $2n$
   (B) $3(5 - n)$

   (C) $\dfrac{3}{5}n$

   (D) $\dfrac{5}{3}n$

7. The image below shows a circle that is inscribed in a square whose sides measure s inches in length. Which of the following expressions could be used to find the area of the shaded region in terms of s?

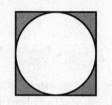

(A) $s^2 - s^2\pi$

(B) $s^2 - \left(\dfrac{s}{2}\right)^2\pi$

(C) $s^2 - 2s\pi$

(D) $s^2 - \left(\dfrac{s}{2}\right)\pi$

8. In Eric's math class, there are m men in a class of s students. Which of the following expressions could be used to represent the ratio of men to women in Eric's math class?

(A) $\dfrac{m}{s}$

(B) $\dfrac{s}{m}$

(C) $\dfrac{m}{s - m}$

(D) $\dfrac{m}{m - s}$

(Answers are on page 220.)

> **To complete a set of practice questions that reviews all of Chapter 9, go to:**
> *barronsbooks.com/TP/accuplacer/math29qw/*

Lesson 9.1—Skills Check (page 211)

1. **–3** $3(3) - 3(4) = 9 - 12 = -3$

2. **9** $6(3) \div 2 = 18 \div 2 = 9$

3. **25** $(2+3)^2 = (5)^2 = 25$

4. **200** $2(-10)^2 = 2(100) = 200$

5. **$37\frac{3}{4}$** $5^2 - \dfrac{(3)(-10)}{8} + 9 = 25 - \dfrac{(-30)}{8} + 9 = 25 + 3\frac{3}{4} + 9 = 37\frac{3}{4}$

6. **$-15\frac{3}{4}$** $\dfrac{3}{4}(3(-10) + 9) = \dfrac{3}{4}(-30 + 9) = \dfrac{3}{4}(-21) = \dfrac{-63}{4} = -15\frac{3}{4}$

7. **15** $\sqrt{15(3)(5)} = \sqrt{225} = 15$

8. **17** $3(5)^2 + 4(-10) - 2(3)^2 = 3(25) - 40 - 2(9) = 75 - 40 - 18 = 17$

Lesson 9.1—ACCUPLACER Challenge (page 211)

1. **(D)** $4(-2)^2 - 5(-2)(6) - 3(6)^2 =$
   $4(4) - (-10)(6) - 3(36) =$
   $16 + 60 - 108 =$
   $76 - 108 = -32$

2. **(A)** $\dfrac{(-3)(7) + 1}{(-3)^4 + (-3)(7)} = \dfrac{-21 + 1}{81 - 21} = \dfrac{-20}{60} = -\dfrac{1}{3}$

3. **(C)** $(4(3)(-4))^2 \div (-4)^2 - 2(3)(-4 + 6) =$
   $(12(-4))^2 \div (-4)^2 - 6(2) =$
   $(-48)^2 \div (-4)^2 - 12 =$
   $2{,}304 \div 16 - 12 =$
   $144 - 12 = 132$

Lesson 9.2—Skills Check (page 216)

1. **$k+4$**   Bill's age = Kate's age + 4 years = $k + 4$

2. **$k-3$**   Nancy's age = Kate's age – 3 years = $k - 3$

3. **$2k$**   Maria's age = Kate's age $\times$ 2 = $k \times 2 = 2k$

4. **$3k-8$**   Phyllis's age = 3 $\times$ Kate's age – 8 years = $3 \times k - 8 = 3k - 8$

5. **$\dfrac{k}{2}$**   Tony's age = $\dfrac{1}{2}$ of Kate's age = $\dfrac{1}{2} \times k = \dfrac{k}{2}$

   or

   Tony's age = Kate's age $\div$ 2 = $\dfrac{k}{2}$

6. **2k**   Daquan's age = Tony's age × 4. We figured out in question 5 that Tony's age is $\frac{k}{2}$, so Daquan = $\frac{k}{2} \times 4 = \frac{k}{2} \times \frac{4}{1} = \frac{4k}{2} = 2k$.

Lesson 9.2—ACCUPLACER Challenge (pages 216–218)

1. **(D)** We need to take 8 candies away from c, the number of candies in the bag, which is $c - 8$. Then, Ms. Williams divides the remainder out among her 5 children equally, so the expression $\frac{c - 8}{5}$ could be used to determine the number of candies that each child receives.

2. **(A)** "Twice his brother's age" translates to $2 \times b$ or $2b$. Since Mateo is "4 years younger than twice his brother's age," we must subtract 4 years from $2b$. The complete expression that could be used to represent Mateo's age is $2b - 4$. We must be careful not to confuse this with choice B, $2(b - 4)$, which would be read as "Mateo is twice four years younger than his brother's age." Another strategy would be to pick an age for Mateo's brother, b. If Mateo's brother is 20 years old, then 4 years younger than twice 20 years old is $2(20) - 4 = 36$ years old. Only choice A will give us the correct age.

3. **(B)** The ratio given is women, w, to men, m. The first part of the ratio given becomes the numerator, and the second part given becomes the denominator. Comparing the women to the men would be expressed as $\frac{w}{m}$.

4. **(D)** Fred does 10 more pushups than Tye. Since t represents how many pushups Tye does, the expression $t + 10$ represents how many pushups Fred does. The expression to find out how many pushups they both do together in total is the two expressions combined, which is $t + t + 10$, which simplifies to $2t + 10$.

5. **(D)** To figure out how many math problems Rashida completed in the week, we need to add the problems that she completed on each of the 7 days together, or Sunday + Monday + Tuesday + Wednesday + Thursday + Friday + Saturday. On the first day of the week, Sunday, Rashida completed p math problems. The next day, she completed one more problem than she did on Sunday, so she completed $p + 1$ on Monday. Tuesday, the next day, Rashida completed one more math problem than she did on Monday, so she completed $p + 1 + 1$, or $p + 2$, math problems. To figure out how many problems she completed altogether, we follow the pattern and write $p + p + 1 + p + 2 + p + 3 + p + 4 + p + 5 + p + 6$ to represent all 7 days. When we simplify by combining like terms, we have a final expression of $7p + 21$.

6. **(D)** On Saturday, Raymond's Newsstand sold n newspapers per hour, which is a rate of $\frac{n \text{ newspapers}}{1 \text{ hour}}$. To represent the entire 5 hours it took to sell the newspapers on Saturday, we use the equivalent fraction $\frac{n \text{ newspapers}}{1 \text{ hour}} = \frac{5n \text{ newspapers}}{5 \text{ hours}}$. In terms of n, the total number of newspapers that Raymond's Newsstand sold on either day can be represented as $5n$. On Thursday, the newsstand sold the same number of newspapers, or $5n$, in 3 hours. That is a rate of $\frac{5n \text{ newspapers}}{3 \text{ hours}}$, or $\frac{5}{3}n$.

7. **(B)** The length of the side of the square is the same as the length of the diameter of the circle, so both measurements can be represented by s. To find the area of a circle, we use the formula $A = \pi r^2$. Knowing that the diameter of the circle can be represented by s, the radius of the circle, r, can be represented by $\frac{s}{2}$. To find the area of the shaded region, we must subtract the area of the circle, $\pi \left(\frac{s}{2} \right)^2$, from the area of the square, s^2. The expression that can be used to find the area of the shaded region in terms of s is $s^2 - \left(\frac{s}{2} \right)^2 \pi$. Choice B simply writes the radius squared, $\left(\frac{s}{2} \right)^2$, before π, which is the same as $s^2 - \pi \left(\frac{s}{2} \right)^2$.

8. **(C)** In this problem, we are only given variables to represent the number of men, m, and the total number of students, s. The number of women would be all of the people in the class who are not men. In other words, the number of women equals the total number of students minus the number of men, or $s - m$. The ratio of men to women could then be expressed as a comparison of m to $s - m$, or $\frac{m}{s - m}$.

Solving Equations and Inequalities of One Variable 10

In the previous chapter, we introduced a lot of vocabulary, and it's worth a brief revisit here. We defined a **variable** as a letter that is used to stand in for an unknown value in a mathematical expression, and an **expression** as a collection of numbers, mathematical symbols, processes, and variables. For example, $12y + 7$ is an expression, and it contains a single variable, y. In this section, we will begin solving **equations** with one variable. An equation differs from an expression in that it contains an equal sign that represents a balance between the expression on the left and the expression on the right. An equation might look like this:

$$5x + 3 = 2x + 12$$

Our task when solving an equation is to undertake a series of steps in order to determine the correct value of the variable, such as x in the example above.

Think of an equation as a balance scale. Say that we have a scale with a 20-gram weight on one side and a 20-gram weight on the other. Since the weights are the same, the scale would be balanced. If we were to add a 5-gram stone to one side of the scale, however, the scale would no longer be balanced. We would need to add a 5-gram stone to the other side too in order to keep the scale balanced. This scale analogy is key to understanding how equations work and the idea of balance represented by the equal sign.

Keeping this in mind, we can perform any number of mathematical operations on an equation—we just have to make sure that what we do to one side, we do to the other. In the end, we will be able to determine a numerical value for the variable in the equation. This is called the **solution**. The solution to an equation is a **value** (or **values**—we'll get to this later) that makes the statement in the equation true. You will be expected to solve all different kinds of equations on the ACCUPLACER. Let's start by looking at some one-step equations and working our way up to the more challenging types of problems you'll see on your exam.

In this chapter, we will work to master

- solving equations with one variable,
- creating equations from word problems, and
- solving inequalities with one variable.

LESSON 10.1—SOLVING ONE-STEP EQUATIONS USING INVERSE OPERATIONS

➤ Example 1

Solve $x + 5 = 15$.

Solution

We need to find a value of x that will make the two sides equal or balanced. To find this value, we need x to be by itself on one side of the equal sign. This equation essentially says, "Some number plus 5 will give us 15." You might be able to see right away that the answer is 10. Let's look at the algebraic process for finding the answer.

Think of the balance scale. If you change something on one side of an equation, you have to do the same thing to the other side. Since we want to get x all by itself, we need to eliminate the "+ 5." To do this, we will perform the **inverse operation** on both sides of the equal sign. Since 5 has been added to x, we'll take 5 away from both sides. This is how we can keep the equation balanced. When you write out this step, it should look like this:

$$\begin{array}{r} x + 5 = 15 \\ \underline{-5 \quad -5} \\ x + 0 = 10 \end{array}$$

Answer: $x = 10$

Notice that after we subtracted 5 from both sides, we drew a line separating the process from the result. It's a good idea to draw a line like this after you add or subtract a value from both sides of an equation—it helps keep your work organized! The solution to this equation is $x = 10$. We know this is true because when we substitute 10 in for x in the original equation, we get $10 + 5 = 15$. This is a true statement, so our answer is correct.

➤ Example 2

Solve the following equation: $b - 8 = 9$

Solution

Our goal is to have b by itself on one side of the equal sign, so we should add 8 to both sides. By adding 8 to the –8, we end up with 0. This step allows us to get b by itself on the left side of the equation. The process looks like this:

$$\begin{array}{r} b - 8 = \ 9 \\ \underline{+8 \quad +8} \\ b + 0 = 17 \\ b = 17 \end{array}$$

By performing the same operation on both sides of the equal sign, we kept the equation balanced and isolated the variable b.

Answer: The solution is $b = 17$.

➡ Example 3

Solve $3y = 45$.

Solution

You can think of this equation as three times a number equals 45, so you might be able to see that the number we are looking for is 15. This time, to get the variable y all by itself algebraically, we need to perform the inverse operation of multiplication, which is division. We can divide both sides of the equation by the **coefficient** of y, which is 3. The process looks like this:

$$\frac{3y}{3} = \frac{45}{3}$$
$$y = 15$$

Answer: Dividing by a coefficient is often the final step in a multistep equation. Here, it's the only step we need to take in order to find the solution, which is 15.

➡ Example 4

Solve $\frac{x}{2} = -4$.

Solution

This equation says that some number, when divided by 2, is equal to –4. To solve it algebraically, we need to eliminate the fraction on the left side of the equal sign. We can think of $\frac{x}{2}$ as $x \div 2$. Since the inverse of division is multiplication, we should multiply both sides of the equation by 2.

$$\left(\frac{2}{1}\right)\left(\frac{x}{2}\right) = -4(2)$$

Notice that $\left(\frac{2}{1}\right)\left(\frac{x}{2}\right) = x$, based on the rules involved in multiplying fractions. The 2s cancel each other out, which leaves us with $\left(\frac{x}{1}\right)$, or x. Multiplying the right side by 2, we get $-4 \times 2 = -8$.

Answer: The solution to this equation is $x = -8$.

➡ Example 5

Solve $\left(\frac{3}{5}\right)x = 12$.

Solution

Some equations, like this one, will have a fractional coefficient of the variable. This doesn't change the goal though. We still need to find a way to eliminate the fraction so that the variable, x, can stand on its own. Think back to the previous example and to working with fractions in general. We know that when we multiply a fraction by its reciprocal, or inverse, the value of the fraction will be equal to 1. This would make the coefficient of x equal to 1, so we would have $1x$, or just x.

$$\left(\frac{5}{3}\right)\left(\frac{3}{5}\right)x = \left(\frac{12}{1}\right)\left(\frac{5}{3}\right)$$

$$x = \frac{60}{3}$$

$$x = 20$$

Answer: By multiplying the fractional coefficient of the variable by its inverse—remember that using inverse operations is the way we solve algebra problems—we are able to cancel it out and arrive at a solution of $x = 20$.

Because solving equations is such a key ACCUPLACER skill, we have provided several practice questions in this unit. To really master this topic, you should try to work through them all, but if you feel like you already have some comfort with this material, skip ahead to the problems that look more challenging.

LESSON 10.1—SKILLS CHECK

Directions: Solve the following one-step equations by using inverse operations.

1.	$c + 12 = 22$	9.	$y - 78 = 22$	17.	$-4m = 8$
2.	$p + 4 = -1$	10.	$g - 9 = -4$	18.	$-5x = -30$
3.	$18 + z = 10$	11.	$k - 19 = 0$	19.	$\frac{x}{9} = 4$
4.	$x + 8 = 20$	12.	$x - 34 = -41$	20.	$\frac{b}{4} = 10$
5.	$r + 1 = -1$	13.	$5x = 30$	21.	$\frac{y}{7} = 6$
6.	$b + 13 = -20$	14.	$7c = 700$	22.	$\frac{2}{3}m = 14$
7.	$x - 11 = 12$	15.	$8r = -32$	23.	$\frac{9}{2}x = 36$
8.	$u - 3 = 17$	16.	$3v = 36$	24.	$\frac{3}{4}n = \frac{16}{27}$

(Answers are on page 240.)

LESSON 10.2—SOLVING MULTISTEP EQUATIONS

On the ACCUPLACER, the linear equations you will be asked to solve will involve several steps, and so you will have to work strategically. There is no one correct order for solving multistep problems, though we will provide a couple of important guidelines here:

- If you can perform the distributive property, do that first to eliminate any parentheses.
- Combine like terms by adding and subtracting first until there is nothing left to add or subtract. This means that you want to get all your variables together and all of your constants together.
- Your last step should always be multiplication or division. Make sure to do all of the adding and subtracting you can *before* you multiply or divide. In other words, get rid of constants before you get rid of coefficients.

➡ Example 1

Solve the following: $3x - 12 = -3$

Solution

Our primary goal is to get the variable by itself. It's always good to keep the end goal in mind as you work through a problem!

According to the guidelines that we outlined, we need to work with like terms first. Notice that we have a -12 and a -3. We can combine these because they're both constants. Since we want to get our constants on one side of the equal sign and our variables on the other, it would make the most sense to add 12 to both sides, like this:

$$
\begin{aligned}
3x - 12 &= -3 \\
+12 \quad &+12 \\
\hline
3x + 0 &= 9 \\
3x &= 9
\end{aligned}
$$

By adding 12 to both sides, we now have an equation that only requires one more step to solve.

$$\frac{3x}{3} = \frac{9}{3}$$
$$x = 3$$

<u>Answer:</u> The solution is $x = 3$.

It's also a good idea to check that the solution works, just to be safe. If we substitute 3 into the original equation in place of x, our answer checks out.

$$
\begin{aligned}
3x - 12 &= -3 \\
3(3) - 12 &= -3 \\
9 - 12 &= -3 \\
-3 &= -3
\end{aligned}
$$

Since the left side of the equation equals the right side of the equation after substituting the solution, 3, for x, we know that we have the correct answer. It is well worth your time on the ACCUPLACER to do this last step. It can save you from selecting wrong answers, and remember, the test is untimed, so you do have the time.

➡ Example 2

Solve $7x - 10 = 5x - 8$.

Solution

In this equation, we have an x on both sides of the equation. We need to combine all of the constant terms and all of the variable terms. We could start with either, but let's try combining the variable here first. To make things easier, let's subtract $5x$ from both sides. Alternatively, we could subtract $7x$ from both sides, but that would result in a negative coefficient of x.

$$
\begin{array}{l}
7x - 10 = 5x - 8 \\
\underline{-5x \qquad -5x} \\
2x - 10 = 0 - 8 \\
2x - 10 = -8
\end{array}
$$

Now we can combine the constants. Since our goal is to have x by itself on the left side of the equation, let's add 10 to both sides.

$$
\begin{array}{l}
2x - 10 = -8 \\
\underline{+10 \quad +10} \\
2x + 0 = 2 \\
2x = 2
\end{array}
$$

The last step is to divide by the coefficient of x.

$$
\frac{2x}{2} = \frac{2}{2}
$$
$$
x = 1
$$

<u>Answer:</u> The solution is $x = 1$. It took several steps to get here, but just take it one small step at a time, and keep reminding yourself that what you do to one side of the equation, you have to do to the other side.

Solving Multistep Equations Involving the Distributive Property

➡ Example 3

Solve $2(3x + 4) = 4(x - 7)$.

Solution

To solve this equation, you need to know how the distributive property works. On the left side of the equation, notice that 2 is multiplied by another quantity within parentheses. There is nothing we can do to simplify $3x + 4$, but we can distribute the 2. To do this on the left side of the equation, multiply 2 by each term inside the parentheses. On the right side of the equation, multiply 4 by each term inside the parentheses.

$$2(3x + 4) = 4(x - 7)$$

$$6x + 8 = 4x - 28$$

Now that we have used the distributive property on both sides, let's rewrite the equation and solve, using steps similar to the ones from the last example.

$$6x + 8 = 4x - 28$$
$$\underline{-4x \qquad -4x}$$
$$2x + 8 = -28$$
$$\underline{\quad -8 \quad -8}$$
$$\frac{2x}{2} = \frac{-36}{2}$$
$$x = -18$$

Answer: The solution is $x = -18$.

Make sure to review this kind of problem carefully—there may be several similar questions on the actual exam that will require you to know and apply the distributive property, as we did here.

LESSON 10.2—SKILLS CHECK

Directions: Solve the following two-step equations using inverse operations.

1. $3x + 10 = 22$

2. $7 = 15 - 2x$

3. $8n - 8 = -40$

4. $40 = 3m + 4$

5. $4n - 2 = -12$

6. $9z + 3 = -6$

7. $9p - 4 = 32$

8. $-6y - 11 = 49$

9. $17 + 8p = 33$

10. $27 = -9 - 2x$

(Answers are on page 241.)

LESSON 10.2—ACCUPLACER CHALLENGE

> **Directions:** Solve for the unknown variable in each of the equations below.

1. If $8t + 7 = 3t + 22$, then $t =$

   (A) -3
   (B) 5
   (C) 3
   (D) -5

2. Solve for x: $4(8 + x) = 50$

   (A) 4.5
   (B) 4.2
   (C) 10.5
   (D) 18

3. What is the solution to the equation
   $16b - 60 = 5(5b - 3)$?

   (A) 1
   (B) 3
   (C) -3
   (D) -5

4. Let $4a - 3 - 2a = -4a - 3$. Find the value of a.

   (A) 0
   (B) -2
   (C) 2
   (D) -1

5. $-2(5c + 6) + 8c = 24$

   (A) 2
   (B) -2
   (C) 18
   (D) -18

6. $\frac{1}{2}(x - 12) = -\frac{1}{2}x + 4$

   (A) -2
   (B) 10
   (C) 16
   (D) 0

(Answers are on page 242.)

LESSON 10.3—SOLVING INEQUALITIES WITH ONE VARIABLE

Solving one-variable inequalities is an extension of solving linear equations. Throughout this chapter, we have followed a series of procedures using inverse operations to isolate a variable. For example, we would solve the linear equation $7x - 9 = 61$ as follows:

$$7x - 9 = 61$$
$$\underline{+9 \quad +9}$$
$$\frac{7x}{7} = \frac{70}{7}$$
$$x = 10$$

In this equation, we find only one solution for x. We use the term **linear equation** for this type of problem because both sides of the equation are equal and because there is only one correct answer. When we solve a **linear inequality**, however, we are actually finding a whole set of possible values for x. The term **inequality** refers to the signs, $<$, $\leq$, $>$, $\geq$, and $\neq$, which we introduced in Chapter 1.

Let's see what would happen if we replaced the equal sign in the previous equation with a "less than" symbol, so that it is now written as $7x - 9 < 61$. Now it is a linear inequality, not an equation. To solve it, though, we would still use the same set of steps that we did before:

$$7x - 9 < 61$$
$$\underline{+9 \quad +9}$$
$$\frac{7x}{7} < \frac{70}{7}$$
$$x < 10$$

In this inequality statement, any value of x that is less than 10 will make the statement true. If we graph this inequality on a number line, it will look like

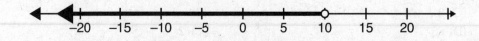

The open circle at 10 means that the number 10 is not included in the set of possible values for x, and the arrow pointing to the left indicates that any number less than 10 could be a possible value for x. In other words, if we substitute the number 3, or the number 0, or the number –25, or any number less than 10 for x in the original inequality statement, it will be a true statement.

➡ Example 1

Solve $6y + 3 \geq 4y - 17$, and plot the solution on a number line.

Solution

Begin by combining all of the constants and all of the variables, and finally, divide by 2, as shown below.

$$6y + 3 \geq 4y - 17$$
$$\underline{-3 \qquad -3}$$
$$6y \geq 4y - 20$$
$$\underline{-4y \quad -4y}$$
$$\frac{2y}{2} \geq \frac{-20}{2}$$
$$y \geq -10$$

<u>Answer</u>: The solution is $y \geq -10$. We read this as "y is greater than or equal to –10." To graph the solution set, we place a closed circle on a number line at –10. The closed circle indicates that –10 *is* included in the solution set. Since the solution is all numbers that are *greater than or equal* to –10, we draw a line to the right. In other words, any number that is –10 or greater will make the inequality true.

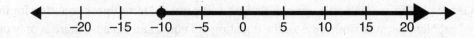

Multiplying or Dividing by a Negative Number with Linear Inequalities

Solving a linear inequality differs most significantly from solving a linear equation when we have to multiply or divide by a negative number. With linear equations, we can multiply or divide by negative numbers without having to change anything. With linear inequalities, **whenever we multiply or divide both sides of the inequality by a negative number, the inequality sign changes direction.**

➡ Example 2

Solve $-5a > 20$, and plot the solution on a number line.

Solution

This is a straightforward, one-step process. To isolate the variable a, we divide both sides of the inequality by -5.

$$\frac{-5a}{-5} > \frac{20}{-5}$$

Here's the big difference: Since we are dividing by a negative number, we need to change the direction of the inequality sign. In this case, it was > (greater than), and it will be flipped to < (less than).

$$\frac{-5a}{-5} > \frac{20}{-5}$$
$$a < -4$$

<u>Answer</u>: Since $a < -4$, the graph of the solution will have an open circle placed at -4, with a line extending to the left, indicating that all numbers less than -4 will make the inequality true.

➡ Example 3

Solve the inequality $\frac{x}{-3} < -2$. Then graph the solution.

Solution

This is another one-step inequality. Here, we would need to multiply both sides of the inequality by -3 to isolate the variable x.

$$\left(\frac{-3}{1}\right)\left(\frac{x}{-3}\right) < -2(-3)$$

Multiplying by –3 changes the direction of the inequality sign. It will go from being < (less than) to > (greater than).

$$\left(\frac{-3}{1}\right)\left(\frac{x}{-3}\right) < -2(-3)$$

$$x > 6$$

Answer: The solution is $x > 6$. To plot this, place an open circle at 6. The open circle indicates that 6 is not included in the set of possible solutions. Finally, draw a line extending from the open circle to the right, as shown below. Any number that is larger than 6 will satisfy the inequality.

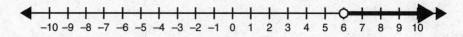

LESSON 10.3—SKILLS CHECK

Directions: Solve each linear inequality below.

1. $b + 12 \geq -7$

2. $-7y \leq 49$

3. $5(q - 12) > 85$

4. $-8n - 4 \geq -12$

5. $\frac{2x}{5} + 10 < 14$

(Answers are on page 242.)

LESSON 10.3—ACCUPLACER CHALLENGE

Directions: For each of the questions below, choose the best answer from the four choices given.

1. $-3(w + 5) > 30 + 2w$

   (A) $w < -9$
   (B) $w > -9$
   (C) $w < 9$
   (D) $w > 9$

2. $15 + \frac{2}{3}x \leq 21$

   (A) $x \leq 4$
   (B) $x \geq 4$
   (C) $x \leq 9$
   (D) $x \geq 9$

3. $2n - 11 \geq \dfrac{5}{2}n - 8$

    (A) $n \geq 1.5$

    (B) $n \leq 3$

    (C) $n \geq 3$

    (D) $n \leq -6$

4. Which of the inequalities below satisfies the following number line?

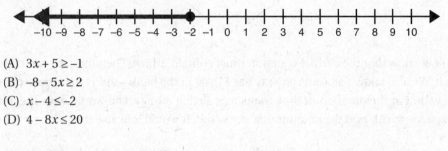

    (A) $3x + 5 \geq -1$

    (B) $-8 - 5x \geq 2$

    (C) $x - 4 \leq -2$

    (D) $4 - 8x \leq 20$

5. Which of the number lines below shows the solution to $20 - 3(4x + 2) \leq 26$?

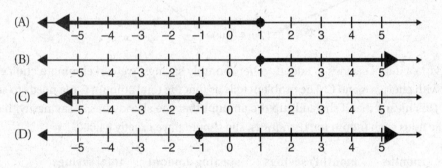

(Answers are on page 243.)

LESSON 10.4—CREATING EQUATIONS FROM WORD PROBLEMS

Up until now in this chapter, we have concentrated on the processes used to solve for unknown variables and how to draw conclusions about possible values of those variables in inequalities. The ACCUPLACER will also ask you to create equations or inequality statements based on verbal prompts, much like the ones that we worked on in the previous chapter. The process for setting up an equation or equations based on a word problem is very similar, only this time there will be one other piece to consider: where to place the equal sign or the inequality sign.

➥ Example 1

Carla is saving up to make a down payment on a new car. She already has $1,600 in her bank account, and she plans to save money each month for one year. Altogether, she needs to have $4,000 for the down payment. Which of the following choices is an equation that Carla could use to determine how much money, m, she should save each month?

(A) $12m + 1,600 = 5,600$

(B) $1,600m + 12 = 4,000$

(C) $12m + 1,600 = 4,000$

(D) $12m - 1,600 = 4,000$

Solution

This problem helps us out a little by telling us that the variable m equals the money that she saves each month. Before we get too hung up on trying to eliminate answer choices, let's think about what we would need to do with the variable. The prompt tells us that Carla plans to save money for one year, or 12 months. It makes sense, then, that she would multiply the amount she saved by the time that she was saving.

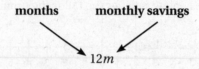

We now know that the correct equation must contain a $12m$. Therefore, we can eliminate choice B. We also know that Carla already has $1,600 in the bank—this is the amount that she started with. For the duration of time that she is saving money, she would have the amount that she started with *plus* the amount that she saved. It would look like this:

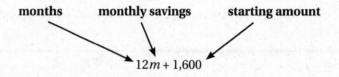

Since Carla's initial savings is added to her monthly savings, we can eliminate choice D. We're left with choices A and C. The problem tells us that the total amount Carla needs to save is $4,000. This means that if she adds up the amount that she already has in savings with the amount she adds each month for 12 months, she should have exactly $4,000.

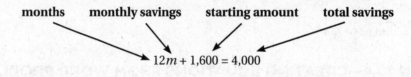

<u>Answer</u>: We can eliminate choice A, which has the wrong total savings. Therefore, choice C is correct.

Now let's look at an example of a similar problem that asks you to create an inequality statement.

➡ Example 2

Nico works at a store where he sells smartphones. He makes a base salary of $350 per week and gets an additional $15 bonus each time he sells a smartphone. Nico has determined that he needs to make at least $470 each week in order to pay his rent and his bills. Which of the following choices is the correct linear inequality Nico could use to determine the number of smartphones, n, that he needs to sell each week?

(A) $350 + 15n \leq 470$
(B) $350 + 15n \geq 470$
(C) $15 + 350n \leq 470$
(D) $15 + 350n \geq 470$

Solution

This problem is similar to the previous example, but it has an important difference: the inequality signs. Notice how, for example, choices A and B are the same except for the direction of the inequality sign. We really have to pay attention to the verbal cues in the problem.

We know that Nico gets $15 each time he sells a smartphone. Therefore, if he sold 2 smartphones, he would get $30. If he sold 3 smartphones, he would get $45, and so on. We know that we need to multiply Nico's $15 bonus times the number of smartphones that he sells. Using the variable n to represent the number of smartphones sold, we have:

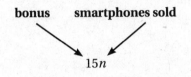

$$15n$$

With just this little bit of information, we can eliminate choices C and D. We also know that Nico gets a $350 salary per week no matter how many smartphones he sells, so his total earnings in a given week would be:

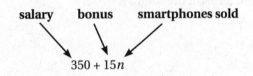

$$350 + 15n$$

We have to choose between choices A and B, and the only difference is the inequality sign. The prompt tells us that Nico needs to make *at least* $470 each week. This means that if he were to make less than $470, he wouldn't be able to pay for everything. Choice A says that Nico's weekly salary could be less than or equal to $470, which isn't true. He needs to make greater than or equal to $470. The inequality should look like this:

$$350 + 15n \geq 470$$

Answer: Choice B is the correct answer.

ACCUPLACER TIPS

There are five different inequality signs, and there are verbal clues that can help you decide which one is appropriate for a given situation. In the previous example, the prompt told us that Nico needed to make at least $470. This means that he could have made exactly $470 and been able to pay his bills, but obviously it would be okay if he made more than that. This is how we knew that the $\geq$ sign was correct. Here are some key words for the inequality signs:

$\geq$: at least
$>$: more than
$\leq$: no more than, at most
$<$: less than
$\neq$: not equal

LESSON 10.4—ACCUPLACER CHALLENGE

> **Directions:** For each of the questions below, choose the best answer from the four choices given.

1. A water tank holds 100 gallons. The tank already has 16 gallons of water in it, and it needs to be filled. Assume that water enters the tank at a rate of 2 gallons per minute. Which of the following equations could be used to find the number of minutes, m, it would take to fill the tank?

   (A) $2m - 16 = 100$
   (B) $16m - 2 = 100$
   (C) $18m = 100$
   (D) $2m + 16 = 100$

2. Roxy has a rectangular garden in her backyard. The length of the garden is 5 feet greater than its width. Which of the following equations could be used to calculate the perimeter, P, in terms of the width, w?

   (A) $w + (w + 5) = P$
   (B) $4w + 10 = P$
   (C) $w(w + 5) = P$
   (D) $5w = P$

3. Megan baked a pan of 40 brownies. She wants to keep exactly 16 brownies for her family, and she wants to give four brownies each to some of her friends. Which of the equations below shows how many friends, n, Megan could give brownies to in order to make sure that she has 16 left for her family?

   (A) $40 - 4n = 16$
   (B) $4n - 16 = 40$
   (C) $40 - 16n = 4$
   (D) $16n + 4 = 40$

4. Walter has $4,050 in his savings account, and he just lost his job. He estimates that he will spend an average of $225 per week. Which of the following inequality statements could be used to find the possible number of weeks, w, that Walter could be unemployed before his savings runs out?

   (A) $225w \geq 4,050$
   (B) $4,050 - w \geq 225$
   (C) $225w \leq 4,050$
   (D) $4,050 - w \leq 225$

5. The town of Greendale got 10 inches of snow last night. The next day, the snow started to melt at a rate of 0.5 inches per hour. Which of the following equations could be used to find the number of hours, t, that it would take for all the snow to melt?

(A) $10t = 0.5$

(B) $0.5(10t) = 0$

(C) $0.5t - 10 = 10$

(D) $10 - 0.5t = 0$

6. Morris is taking a math class, and he wants to make sure that his average is at least a 79.5% so that he will be guaranteed a B minus or better. His class grade is based on four exams, all of which carry equal weight toward his final grade. On the first three exams, he scored 68%, 84%, and 81%. Which of the following inequalities could be used to find the score, x, that Morris would need to get on his last test to make sure that his average in the class is higher than 79%?

(A) $\dfrac{68 + 84 + 81 + 79.5}{4} \leq x$

(B) $\dfrac{68 + 84 + 81 + x}{4} \geq 79.5$

(C) $\dfrac{68 + 84 + 81 + x}{4} \leq 79.5$

(D) $\dfrac{68 + 84 + 81 + 79.5}{4} \geq x$

(Answers are on page 244.)

To complete a set of practice questions that reviews all of Chapter 10, go to:
barronsbooks.com/TP/accuplacer/math29qw/

Lesson 10.1—Skills Check (page 226)

1. $c = 10$

$$c + 12 = 22$$
$$\underline{-12 \quad -12}$$
$$c = 10$$

2. $p = -5$

$$p + 4 = -1$$
$$\underline{-4 \quad -4}$$
$$p = -5$$

3. $z = -8$

$$18 + z = 10$$
$$\underline{-18 \qquad -18}$$
$$z = -8$$

4. $x = 12$

$$x + 8 = 20$$
$$\underline{-8 \quad -8}$$
$$x = 12$$

5. $r = -2$

$$r + 1 = -1$$
$$\underline{-1 \quad -1}$$
$$r = -2$$

6. $b = -33$

$$b + 13 = -20$$
$$\underline{-13 \quad -13}$$
$$b = -33$$

7. $x = 23$

$$x - 11 = 12$$
$$\underline{+11 \quad +11}$$
$$x = 23$$

8. $u = 20$

$$u - 3 = 17$$
$$\underline{+3 \quad +3}$$
$$u = 20$$

9. $y = 100$

$$y - 78 = 22$$
$$\underline{+78 \quad +78}$$
$$y = 100$$

10. $g = 5$

$$g - 9 = -4$$
$$\underline{+9 \quad +9}$$
$$g = 5$$

11. $k = 19$

$$k - 19 = 0$$
$$\underline{+ 19 \quad +19}$$
$$k = 19$$

12. $x = -7$

$$x - 34 = -41$$
$$\underline{+34 \quad +34}$$
$$x = -7$$

13. $x = 6$

$$\frac{5x}{5} = \frac{30}{5}$$
$$x = 6$$

14. $c = 100$

$$\frac{7c}{7} = \frac{700}{7}$$
$$c = 100$$

15. $r = -4$

$$\frac{8r}{8} = \frac{-32}{8}$$
$$r = -4$$

16. $v = 12$

$$\frac{3v}{3} = \frac{36}{3}$$
$$v = 12$$

17. $m = -2$

$$\frac{-4m}{-4} = \frac{8}{-4}$$
$$m = -2$$

18. $x = 6$

$$\frac{-5x}{-5} = \frac{-30}{-5}$$
$$x = 6$$

19. $x = 36$

$$\left(\frac{9}{1}\right)\frac{x}{9} = \frac{4}{1}\left(\frac{9}{1}\right)$$
$$x = 36$$

20. $b = 40$

$$\left(\frac{4}{1}\right)\frac{b}{4} = \frac{10}{1}\left(\frac{4}{1}\right)$$
$$b = 40$$

21. $y = 42$

$$\left(\frac{7}{1}\right)\frac{y}{7} = \frac{6}{1}\left(\frac{7}{1}\right)$$
$$y = 42$$

22. $m = 21$

$$\left(\frac{3}{2}\right)\frac{2}{3}m = \frac{14}{1}\left(\frac{3}{2}\right)$$
$$m = \frac{42}{2}$$
$$m = 21$$

23. $x = 8$

$$\left(\frac{2}{9}\right)\frac{9}{2}x = \frac{36}{1}\left(\frac{2}{9}\right)$$

$$x = \frac{72}{9}$$

$$x = 8$$

24. $n = \dfrac{64}{81}$

$$\left(\frac{4}{3}\right)\frac{3}{4}n = \frac{16}{27}\left(\frac{4}{3}\right)$$

$$n = \frac{64}{81}$$

Lesson 10.2—Skills Check (page 230)

1. $x = 4$

$$3x + 10 = 22$$
$$\underline{-10\ -10}$$
$$\frac{3x}{3} = \frac{12}{3}$$
$$x = 4$$

6. $z = -1$

$$9z + 3 = -6$$
$$\underline{-3\ -3}$$
$$\frac{9z}{9} = \frac{-9}{9}$$
$$z = -1$$

2. $x = 4$

$$7 = 15 - 2x$$
$$\underline{-15\ -15}$$
$$\frac{-8}{-2} = \frac{-2x}{-2}$$
$$4 = x$$

7. $p = 4$

$$9p - 4 = 32$$
$$\underline{+4\ +4}$$
$$\frac{9p}{9} = \frac{36}{9}$$
$$p = 4$$

3. $n = -4$

$$8n - 8 = -40$$
$$\underline{+8\ \ +8}$$
$$\frac{8n}{8} = \frac{-32}{8}$$
$$n = -4$$

8. $y = -10$

$$-6y - 11 = 49$$
$$\underline{+11\ +11}$$
$$\frac{-6y}{-6} = \frac{60}{-6}$$
$$y = -10$$

4. $m = 12$

$$40 = 3m + 4$$
$$\underline{-4-4}$$
$$\frac{36}{3} = \frac{3m}{3}$$
$$12 = m$$

9. $p = 2$

$$17 + 8p = 33$$
$$\underline{-17-17}$$
$$\frac{8p}{8} = \frac{16}{8}$$
$$p = 2$$

5. $n = -2\dfrac{1}{2}$ or -2.5

$$4n - 2 = -12$$
$$\underline{+2\ \ +2}$$
$$\frac{4n}{4} = \frac{-10}{4}$$
$$n = -2\frac{1}{2}$$

10. $x = -18$

$$27 = -9 - 2x$$
$$\underline{+9\ +9}$$
$$\frac{36}{-2} = \frac{-2x}{-2}$$
$$-18 = x$$

Lesson 10.2—ACCUPLACER Challenge (page 231)

1. **(C)**

$$8t + 7 = 3t + 22$$
$$\underline{-7 \qquad -7}$$
$$8t = 3t + 15$$
$$\underline{-3t \quad -3t}$$
$$\frac{5t}{5} = \frac{15}{5}$$
$$t = 3$$

4. **(A)**

$$4a - 3 - 2a = -4a - 3$$
$$2a - 3 = -4a - 3$$
$$\underline{+4a \qquad +4a}$$
$$6a - 3 = -3$$
$$\underline{+3 \quad +3}$$
$$\frac{6a}{6} = \frac{0}{6}$$
$$a = 0$$

2. **(A)**

$$4(8 + x) = 50$$
$$32 + 4x = 50$$
$$\underline{-32 \qquad -32}$$
$$\frac{4x}{4} = \frac{18}{4}$$
$$x = 4.5$$

5. **(D)**

$$-2(5c + 6) + 8c = 24$$
$$-10c - 12 + 8c = 24$$
$$-2c - 12 = 24$$
$$\underline{+12 \quad +12}$$
$$\frac{-2c}{-2} = \frac{36}{-2}$$
$$c = -18$$

3. **(D)**

$$16b - 60 = 5(5b - 3)$$
$$16b - 60 = 25b - 15$$
$$\underline{-16b \qquad -16b}$$
$$-60 = 9b - 15$$
$$\underline{+15 \qquad +15}$$
$$\frac{-45}{9} = \frac{9b}{9}$$
$$-5 = b$$

6. **(B)**

$$\frac{1}{2}(x - 12) = -\frac{1}{2}x + 4$$
$$\frac{1}{2}x - 6 = -\frac{1}{2}x + 4$$
$$\underline{+\frac{1}{2}x \qquad +\frac{1}{2}x}$$
$$x - 6 = 4$$
$$\underline{+6 + 6}$$
$$x = 10$$

Lesson 10.3—Skills Check (page 234)

1. $b \geq -19$

$$b + 12 \geq -7$$
$$\underline{-12 \quad -12}$$
$$b \geq -19$$

3. $q > 29$

$$5(q - 12) > 85$$
$$5q - 60 > 85$$
$$\underline{+60 \quad +60}$$
$$\frac{5q}{5} > \frac{145}{5}$$
$$q > 29$$

2. $y \geq -7$

$$\frac{-7y}{-7} \leq \frac{49}{-7}$$
$$y \geq -7$$

4. $n \leq 1$

$$-8n - 4 \geq -12$$
$$\underline{ + 4 \quad + 4}$$
$$\frac{-8n}{-8} \geq \frac{-8}{-8}$$
$$n \leq 1$$

5. $x < 10$

$$\frac{2x}{5} + 10 < 14$$
$$\underline{\phantom{\frac{2x}{5}} - 10 \quad -10}$$
$$\frac{2x}{5} < 4$$
$$\left(\frac{5}{2}\right)\frac{2x}{5} < \frac{4}{1}\left(\frac{5}{2}\right)$$
$$x < \frac{20}{2}$$
$$x < 10$$

Lesson 10.3—ACCUPLACER Challenge (pages 234–235)

1. **(A)**

$$-3(w + 5) > 30 + 2w$$
$$-3w - 15 > 30 + 2w$$
$$\underline{-2w -2w}$$
$$-5w - 15 > 30$$
$$\underline{ + 15 + 15}$$
$$\frac{-5w}{-5} > \frac{45}{-5}$$
$$w < -9$$

2. **(C)**

$$15 + \frac{2}{3}x \leq 21$$
$$\underline{-15 \phantom{+ \frac{2}{3}x } -15}$$
$$\frac{2}{3}x \leq 6$$
$$\left(\frac{3}{2}\right)\frac{2}{3}x \leq \frac{6}{1}\left(\frac{3}{2}\right)$$
$$x \leq \frac{18}{2}$$
$$x \leq 9$$

3. **(D)**

$$2n - 11 \geq \frac{5}{2}n - 8$$
$$\underline{ + 11 \phantom{\geq \frac{5}{2}n} + 11}$$
$$2n \geq \frac{5}{2}n + 3$$
$$\underline{-\frac{5}{2}n -\frac{5}{2}n}$$
$$-\frac{1}{2}n \geq 3$$
$$\left(-\frac{2}{1}\right)\left(-\frac{1}{2}\right)n \geq (3)(-2)$$
$$n \leq -6$$

4. **(B)** The inequality reads $x \leq -2$. Choice B is the only equation that has this inequality as its solution.

$$-8 - 5x \geq 2$$
$$\underline{+8 +8}$$
$$\frac{-5x}{-5} \geq \frac{10}{-5}$$
$$x \leq -2$$

5. **(D)**

$$20 - 3(4x + 2) \leq 26$$
$$\underline{-20 -20}$$
$$-3(4x + 2) \leq 6$$
$$-12x - 6 \leq 6$$
$$\underline{ + 6 + 6}$$
$$\frac{-12x}{-12} \leq \frac{12}{-12}$$
$$x \geq -1$$

Choice D represents the inequality $x \geq -1$.

Lesson 10.4—ACCUPLACER Challenge (pages 238–239)

1. **(D)** There are already 16 gallons of water in the tank, and it fills at a rate of 2 gallons per minute, m, which would be written as $2m$. Therefore, the amount of water in the tank at any given time would be $2m + 16$. We know that the tank can hold a maximum of 100 gallons, so we would use the equation $2m + 16 = 100$ to find the amount of time it would take to fill the tank.

2. **(B)** All of the answer choices use the variable w for width. Since the length is 5 feet greater than the width, we would write the length as $w + 5$. To find the perimeter of the garden, we would need to find the sum of all four sides.

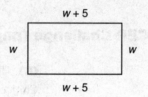

The sum of all the sides is $w + w + (w + 5) + (w + 5)$. After combining the variables, $4w + 10 = P$.

3. **(A)** First, we know that the variable n represents the total number of friends that Megan could give brownies to. If she plans to give each friend 4 brownies, then she would be giving away $4n$ brownies altogether. Since she starts with 40 brownies and gives $4n$ away, the number she has left would be written as $40 - 4n$. Therefore, to find the number of friends she could give brownies away to, we would use the equation $40 - 4n = 16$.

4. **(C)** Walter plans to spend $225 each week. If he spends this same amount over a period of w weeks, then he will spend $225w$. He can't, however, spend more than $4,050 because that's how much is in his savings account. This means that the total he spends, $225w$, must be less than or equal to $4,050, or $225w \le 4{,}050$.

5. **(D)** The snow melts at a rate of 0.5 inches per hour, t, so the total amount melted at a given time would be written as $0.5t$. There were already 10 inches of snow on the ground. As the snow melts at a rate of $0.5t$, these 10 inches will decrease. Therefore, $10 - 0.5t = 0$. The equation is set equal to zero in order to show that after the snow has melted, there will be 0 inches left.

6. **(B)** To find the mean of a set of items, we add all the items in the set and then divide by the total number of items. The variable x represents the score that Morris would need to get on his last test, so he would calculate his final average like this:

$$\frac{68 + 84 + 81 + x}{4}$$

Since he wants to get at least a 79.5% for his average, he would need a total score greater than or equal to 79.5. The correct inequality is $\frac{68 + 84 + 81 + x}{4} \ge 79.5$.

Systems of Linear Equations

11

In the previous chapter, we solved one-variable equations by using inverse operations to isolate the variable. What if we had an equation with two variables—say, an x and a y—like this one:

$$x + y = 12$$

We could solve the equation in terms of x or y, but we wouldn't be able to get a numerical answer for either of the variables because we don't have enough information. There are an infinite number of combinations of x- and y-values that would have a sum of 12! If we had two equations, both of which had the same solutions for x and y, then we could arrive at a definitive solution. For example,

$$x + y = 12$$
$$2x - y = 9$$

For these two equations, there would only be one x and y pair that would satisfy both equations. We call this a **system of linear equations**.

There are several different ways to solve systems of equations like the one above, but, in this chapter, we will focus on the two ways that will serve you best on the ACCUPLACER. We will also look at special cases in which systems of equations have no solutions or infinitely many solutions.

> In this chapter, we will work to master
>
> - solving systems of equations using the addition method,
> - solving systems of equations using the substitution method,
> - identifying systems of equations that have no solutions or an infinite number of solutions,
> - working backwards to solve systems of equations, and
> - choosing either the addition or substitution method based on the problem.

This chapter is going to work a little differently than some of the others. You won't see any multiple-choice questions until the end of the chapter. It's important that you spend some time developing real fluency with solving systems of equations before looking at problems where you are given answers to choose from.

LESSON 11.1—SOLVING SYSTEMS OF EQUATIONS USING THE ADDITION METHOD

The addition method is sometimes called the elimination method. With this method, we have to add the two equations in the system together, with the goal of "eliminating" one of the variables so that we can solve for the other.

➡ Example 1

Solve the following system of equations using the addition method:

$$x + y = 12$$
$$2x - y = 9$$

Solution

Remember that, in the last chapter, we talked about an equation being like a balance scale. We can perform all kinds of operations on the equation itself, as long as we do the same thing to both sides of the equation. This concept also means that if we add the left sides of the two equations together and add the right sides of the two equations together, the single resulting equation will still be balanced.

$$
\begin{array}{r}
x + y = 12 \\
+\ \ 2x - y = \ 9 \\
\hline
3x + 0 = 21
\end{array}
$$

By adding the two equations together, we eliminated the variable y. Now we can solve for x.

$$\frac{3x}{3} = \frac{21}{3}$$

$$x = 7$$

Now that we have established that $x = 7$, we need to substitute that value into one of the two original equations to find the correct y value. Let's plug it into the first equation, since that one seems the most straightforward. Then solve for y.

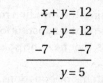

$$
\begin{array}{r}
x + y = 12 \\
7 + y = 12 \\
-7 \quad\quad -7 \\
\hline
y = 5
\end{array}
$$

<u>Answer</u>: The solution to this system of linear equations is $x = 7$, $y = 5$. These x- and y-values satisfy both equations in the system. This solution can also be represented as the ordered pair (7, 5).

In this example, we were able to "eliminate" the variable y by simply adding the two equations together. However, it's not always that simple, as the next example will show.

➥ Example 2

Solve the following system of equations using the addition method:

$$2y - x = 6$$
$$y + 3x = 10$$

Solution

Here, if we try to add the two equations like we did before, we get:

$$2y - x = 6$$
$$\underline{+ \quad y + 3x = 10}$$
$$3y + 2x = 16$$

This isn't helpful. We didn't eliminate either of the two variables, so we can't solve for x or y. We need to change one of the two equations so that either x or y will be eliminated when we add them together. Look at the x terms in both original equations. If we multiply both sides of the top equation by 3, then the equation would still be balanced, and the system's x terms would cancel out when we add them.

$$3(2y - x) = 3(6)$$
$$6y - 3x = 18$$

Now replace the first equation with this new one, and then add.

$$6y - 3x = 18$$
$$\underline{+ \quad y + 3x = 10}$$
$$7y = 28$$

Solve for y.

$$\frac{7y}{7} = \frac{28}{7}$$

$$y = 4$$

Now substitute 4 in place of y in one of the original equations, and solve for x.

$$2y - x = 6$$
$$2(4) - x = 6$$
$$8 - x = 6$$
$$\underline{+ x \quad + x}$$
$$8 = 6 + x$$
$$\underline{-6 \quad -6}$$
$$2 = x$$

Answer: The solution to this system of equations is $x = 2$, $y = 4$, or (2, 4).

➠ Example 3

Solve the following system of equations using the addition method:

$$-4x + 2y = 26$$
$$2x + 5y = 5$$

Solution

Notice that if we multiply the bottom equation by 2, we will be able to add the equations and cancel out the x terms.

$$2(2x + 5y) = 2(5)$$
$$4x + 10y = 10$$

Add the equations.

$$
\begin{array}{r}
-4x + 2y = 26 \\
+\ 4x + 10y = 10 \\
\hline
12y = 36
\end{array}
$$

Solve for y.

$$\frac{12y}{12} = \frac{36}{12}$$
$$y = 3$$

Plug $y = 3$ into either one of the original equations, and solve for x.

$$
\begin{array}{r}
2x + 5y = \quad 5 \\
2x + 5(3) = \quad 5 \\
2x + 15 = \quad 5 \\
-15 \quad -15 \\
\hline
2x = -10
\end{array}
$$

$$\frac{2x}{2} = \frac{-10}{2}$$

$$x = -5$$

<u>Answer</u>: The solution to this system of equations is $x = -5$, $y = 3$, or $(-5, 3)$.

In the last two examples, we only had to multiply one of the two equations in the system in order to solve it. Sometimes, you will need to think strategically and multiply *both* equations by a different constant in order to eliminate one of the variables. Let's look at an example.

➥ Example 4

Solve the following system of linear equations:

$$-2x + 3y = -2$$
$$-11x + 5y = 35$$

Solution

To cancel out one of the variables, we will have to make a choice about what constants we should multiply each of these equations by. Notice that if we multiply the top equation by -11 and the bottom equation by 2, we would create x coefficients that are opposites: $22x$ and $-22y$.

The First Equation

$$-11(-2x + 3y) = -11(-2)$$
$$22x - 33y = 22$$

The Second Equation

$$2(-11x + 5y) = 2(35)$$
$$-22x + 10y = 70$$

Now we can use the addition method with these new equations to cancel out the x-values and solve for y.

$$22x - 33y = 22$$
$$+ \; -22x + 10y = 70$$
$$\overline{\qquad -23y = 92}$$

$$\frac{-23y}{-23} = \frac{92}{-23}$$

$$y = -4$$

Now that we have the y-value, plug it into either one of the original equations.

$$-2x + 3y = -2$$
$$-2x + 3(-4) = -2$$
$$-2x - 12 = -2$$
$$\underline{+12 \quad +12}$$
$$-2x = 10$$
$$\frac{-2x}{-2} = \frac{10}{-2}$$
$$x = -5$$

<u>Answer:</u> The solution to this system of equations is $x = -5$, $y = -4$, or $(-5, -4)$.

LESSON 11.1—SKILLS CHECK

Directions: Solve each of the following systems of equations using the addition method.

1.
$$x + y = 14$$
$$x - y = -2$$

2.
$$6x + y = 17$$
$$-3x - y = 4$$

3.
$$5y - 2x = 6$$
$$-4y + 2x = -4$$

4.
$$7x - 3y = 4$$
$$2x + 3y = 32$$

5.
$$-2x + 6y = -18$$
$$-4x + 3y = 9$$

6.
$$x - 2y = -9$$
$$3x + y = 1$$

7.
$$2x - 2y = 12$$
$$x - 5y = -10$$

8.
$$4y = -5x - 36$$
$$2y = 4x + 60$$

9.
$$-7x + 3y = 6$$
$$-4x + 3y = 42$$

10.
$$2x - 3y = -24$$
$$-3x - 2y = -3$$

11.
$$y + x = -2$$
$$7x - 2y = -5$$

12.
$$2x = 40 + 8y$$
$$4y - 12 = -3x$$

(Answers are on page 259.)

LESSON 11.2—SOLVING SYSTEMS OF EQUATIONS USING THE SUBSTITUTION METHOD

Another way to solve a system of linear equations is to choose one of the two equations, isolate either x or y, and then substitute it into the other equation. Here's how it works.

➡ Example 1

Solve the following system of equations using the substitution method:

$$5x + y = 4$$
$$y = 8x - 9$$

Solution

Notice that this time the second equation is already "solved" in terms of y. Since $y = 8x - 9$, we can plug $8x - 9$ in place of y in the other equation, and then solve for x.

$$5x + y = 4$$
$$5x + (8x - 9) = 4$$
$$13x - 9 = 4$$

Now solve for x.

$$
\begin{array}{r}
13x - 9 = 4 \\
\underline{+9 \quad +9} \\
13x = 13 \\
\dfrac{13x}{13} = \dfrac{13}{13} \\
x = 1
\end{array}
$$

Plug $x = 1$ into either one of the two original equations to find the value for y.

$$
\begin{array}{r}
5x + y = 4 \\
5(1) + y = 4 \\
5 + y = 4 \\
\underline{-5 \quad\quad -5} \\
y = -1
\end{array}
$$

<u>Answer:</u> The solution is $x = 1$, $y = -1$, or the ordered pair $(1, -1)$.

It won't always be the case that one of the equations is already solved in terms of one of the variables. You might have to do that part yourself. The next example will show you how this works.

➡ Example 2

Solve the following system of linear equations using the substitution method:

$$4x + 2y = 58$$
$$x + y = 21$$

Solution

Here, it would be most efficient to solve the second equation, $x + y = 21$, in terms of one of the variables. You can choose to solve for either x or y; as long as you're careful, you will arrive at the correct solution. Let's solve for x.

$$x + y = 21$$
$$\underline{-y \quad -y}$$
$$x = 21 - y$$

Now we can substitute the expression $21 - y$ in place of x in the other equation.

$$4x + 2y = 58$$
$$4(21 - y) + 2y = 58$$
$$84 - 4y + 2y = 58$$
$$84 - 2y = 58$$
$$\underline{-84 \qquad -84}$$
$$-2y = -26$$
$$\frac{-2y}{-2} = \frac{-26}{-2}$$
$$y = 13$$

Now that we know the y-value, we can plug it into either of the two original equations, and solve for x.

$$x + y = 21$$
$$x + 13 = 21$$
$$\underline{-13 \quad -13}$$
$$x = 8$$

<u>Answer</u>: The solution is $x = 8$, $y = 13$, or $(8, 13)$.

➥ Example 3

Solve the following system of equations below:

$$3y = 3x - 12$$
$$5x + 2y = 13$$

Solution

Look at the first equation in this system. All three coefficients are divisible by 3, and if we do divide the entire equation by 3, we will have solved it in terms of y.

$$\frac{3y}{3} = \frac{3x}{3} - \frac{12}{3}$$
$$y = x - 4$$

Now we can substitute the expression $x - 4$ in place of y in the other original equation.

$$5x + 2y = 13$$
$$5x + 2(x - 4) = 13$$
$$5x + 2x - 8 = 13$$
$$7x - 8 = 13$$
$$\underline{+8 \quad +8}$$
$$7x = 21$$
$$\frac{7x}{7} = \frac{21}{7}$$
$$x = 3$$

Substitute 3 in place of x in one of the two original equations.

$$5x + 2y = 13$$
$$5(3) + 2y = 13$$
$$15 + 2y = 13$$
$$\underline{-15 \qquad -15}$$
$$2y = -2$$
$$\frac{2y}{2} = \frac{-2}{2}$$
$$y = -1$$

<u>Answer:</u> The solution is $x = 3$, $y = -1$, or $(3, -1)$.

LESSON 11.2—SKILLS CHECK

Directions: Solve each of the following systems of equations using the substitution method.

1.
$$2x + y = 4$$
$$y = 4x + 10$$

2.
$$x + y = 8$$
$$x = 4y + 3$$

3.
$$y = x$$
$$y = -3x - 8$$

4.
$$y = \frac{2}{3}x - 1$$
$$5x - 6y = 12$$

5.
$$6y = 4x + 12$$
$$2x + y = 10$$

6.
$$\frac{2}{3}x + \frac{1}{3}y = 1$$
$$y = 3x - 17$$

7.
$$7x - 3y = 10$$
$$y = -2x + 14$$

8.
$$x = 3y - 1$$
$$y = \frac{9}{2}x + \frac{9}{2}$$

9.
$$12x + 4y = 24$$
$$3x - 3y = -18$$

10.
$$3x = 6y$$
$$\frac{2}{3}x = y + \frac{1}{3}$$

11.
$$3x + 2y = 10$$
$$y = 2x - 16$$

12.
$$\frac{1}{2}x - \frac{2}{3}y = 1$$
$$y = \frac{3}{2}x + 3$$

(Answers are on page 262.)

LESSON 11.3—SYSTEMS OF EQUATIONS WITH NO SOLUTIONS, SYSTEMS OF EQUATIONS WITH INFINITE SOLUTIONS, AND WORKING BACKWARDS

In the problems we have seen so far, each system of equations has had exactly one (x, y) pair as its solution. There are, however, some systems of equations that have no solutions. Let's look at an example.

➡ Example 1

Solve the following system of equations:

$$3y = 2x + 12$$
$$y - \frac{2}{3}x = 6$$

Solution

Notice that, in the second equation, we can add $\frac{2}{3}x$ to both sides, and then we can use the substitution method. (The addition method would work here as well!)

$$y - \frac{2}{3}x = 6$$
$$\underline{+\frac{2}{3}x + \frac{2}{3}x}$$
$$y = \frac{2}{3}x + 6$$

Now substitute $\frac{2}{3}x + 6$ in place of y in the first equation, and solve for x.

$$3y = 2x + 12$$
$$3\left(\frac{2}{3}x + 6\right) = 2x + 12$$
$$2x + 18 = 2x + 12$$

Something looks odd here. If we try to collect all of the xs on one side of the equation, look what happens.

$$2x + 18 = 2x + 12$$
$$\underline{-2x \qquad -2x}$$
$$18 \neq 12$$

Answer: Obviously, $18 \neq 12$. This means that there is no solution to this system of linear equations. If we were to graph both of these equations, they would be parallel lines that never intersect. We will explore the topic of parallel lines in depth in Chapter 15.

On the opposite end of the spectrum, some systems have infinitely many solutions, as shown in Example 2.

➡ Example 2

How many solutions does the following system of equations have?

$$x - y = \frac{1}{3}$$
$$6x = 2 + 6y$$

Solution

Let's use the substitution method again. We can solve for x in the first equation, and then substitute it into the second one.

$$x - y = \frac{1}{3}$$
$$\underline{+y \quad +y}$$
$$x = \frac{1}{3} + y$$

Now we have the system:

$$6x = 2 + 6y$$
$$6\left(\frac{1}{3} + y\right) = 2 + 6y$$
$$2 + 6y = 2 + 6y$$

The same expression is on both sides of the equal sign! No matter what we plug in for y, the equation would always be true. If we were to go through the process of canceling out terms on both sides, we would end up with $2 = 2$, or $0 = 0$, or $y = y$. All of these statements would be true.

Answer: This system of equations has infinitely many solutions. Any (x, y) pair would satisfy the equation. If we were to graph these two equations, they would actually be the same line.

A system of linear equations can either have no solution, one solution, or infinitely many solutions. If you see a question on the ACCUPLACER that asks how many solutions exist for a given system of equations, "two solutions" will *never* be the correct answer.

ACCUPLACER TIPS

We saved this tip for now because we wanted to make sure that you got plenty of practice using the addition and substitution methods first. This tip, however, is an important one, and it could really help you on the ACCUPLACER. Since the ACCUPLACER consists of multiple-choice questions, you can always plug in the answer choices for x and y in the system and see which (x, y) pair satisfies both equations. This is a great way to get the correct answer if the other methods aren't working. Let's practice this method.

Which of the following choices is the solution to the system of equations below?

$$4x - 2y = 8$$
$$2x + y = 20$$

(A) $(-6, -8)$
(B) $(6, -8)$
(C) $(6, 8)$
(D) $(-6, 8)$

Let's start by trying the x- and y-values from choice A. We can plug them into either equation, but remember: In order for these values to be correct, they have to satisfy both equations! Let's plug $x = -6$, $y = -8$ into the second equation.

$$2x + y = 20$$
$$2(-6) + (-8) = 20$$
$$-12 + (-8) = 20$$
$$-20 \neq 20$$

Since -20 is not equal to 20, choice A can't be correct. The only answer that works is choice C. When we plug $x = 6$, $y = 8$ into both equations, we get true statements.

$4x - 2y = 8$	$2x + y = 20$
$4(6) - 2(8) = 8$	$2(6) + 8 = 20$
$24 - 16 = 8$	$12 + 8 = 20$
$8 = 8$	$20 = 20$

Working backwards is often a great strategy for solving multiple-choice problems like the ones on the ACCUPLACER. Try it on some of the problems in this chapter.

LESSON 11.3—ACCUPLACER CHALLENGE

Directions: For each of the questions below, choose the best answer from the four choices given.

1. How many solutions (x, y) exist for the following system of linear equations?

$$x + y = 12$$
$$2x + 2y = 20$$

(A) 0
(B) 1
(C) 2
(D) More than 2

2. How many solutions (x, y) exist for the following system of linear equations?

$$3x + 6y = 12$$
$$y = -\frac{1}{2}x + 2$$

(A) 0
(B) 1
(C) 2
(D) More than 2

3. How many solutions (x, y) exist for the following system of linear equations?

$$3x - 2y = 6$$
$$x - 3y = -12$$

(A) 0
(B) 1
(C) 2
(D) More than 2

4. How many solutions (x, y) exist for the following system of linear equations?

$$y = 2x - 10$$
$$2x - y = 4$$

(A) 0
(B) 1
(C) 2
(D) More than 2

(Answers are on page 266.)

To complete a set of practice questions that reviews all of Chapter 11, go to:
barronsbooks.com/TP/accuplacer/math29qw/

Lesson 11.1—Skills Check (page 250)

1. **$x = 6, y = 8$** No multiplication is necessary. Add the two equations to eliminate the y-terms.

$$
\begin{aligned}
x + y &= 14 \\
+\, x - y &= -2 \\
\hline
\frac{2x}{2} &= \frac{12}{2} \\
x &= 6
\end{aligned}
$$

Plug $x = 6$ into either of the original equations, and solve for y.

$$
\begin{aligned}
x + y &= 14 \\
6 + y &= 14 \\
-6 \quad\;\; &-6 \\
\hline
y &= 8
\end{aligned}
$$

2. **$x = 7, y = -25$** No multiplication is necessary. Add the equations to eliminate the y-terms.

$$
\begin{aligned}
6x + y &= 17 \\
+\, -3x - y &= 4 \\
\hline
\frac{3x}{3} &= \frac{21}{3} \\
x &= 7
\end{aligned}
$$

Plug $x = 7$ into either of the original equations, and solve for y.

$$
\begin{aligned}
6x + y &= 17 \\
6(7) + y &= 17 \\
42 + y &= 17 \\
-42 \quad\;\; &-42 \\
\hline
y &= -25
\end{aligned}
$$

3. **$x = 2, y = 2$** Add the two equations to eliminate the x-terms.

$$
\begin{aligned}
5y - 2x &= 6 \\
+\, -4y + 2x &= -4 \\
\hline
y &= 2
\end{aligned}
$$

Plug $y = 2$ into either of the original equations, and solve for x.

$$
\begin{aligned}
5y - 2x &= 6 \\
5(2) - 2x &= 6 \\
10 - 2x &= 6 \\
-10 \quad\;\; &-10 \\
\hline
\frac{-2x}{-2} &= \frac{-4}{-2} \\
x &= 2
\end{aligned}
$$

4. **$x = 4, y = 8$** Add the equations to cancel out the y-terms.

$$
\begin{aligned}
7x - 3y &= 4 \\
+\, 2x + 3y &= 32 \\
\hline
\frac{9x}{9} &= \frac{36}{9} \\
x &= 4
\end{aligned}
$$

Plug $x = 4$ into either of the original equations, and solve for y.

$$
\begin{aligned}
2x + 3y &= 32 \\
2(4) + 3y &= 32 \\
8 + 3y &= 32 \\
-8 \quad\;\; &-8 \\
\hline
\frac{3y}{3} &= \frac{24}{3} \\
y &= 8
\end{aligned}
$$

5. **$x = -6, y = -5$** Multiply both sides of the first equation by -2 so that the x-terms will be eliminated when the equations are added.

$$
\begin{aligned}
-2(-2x + 6y) &= -2(-18) \\
4x - 12y &= 36
\end{aligned}
$$

Add the equations.

$$4x - 12y = 36$$
$$+ \; -4x + 3y = \; 9$$
$$\frac{-9y}{-9} = \frac{45}{-9}$$
$$y = -5$$

Plug $y = -5$ into either of the original equations, and solve for x.

$$-2x + 6y = -18$$
$$-2x + 6(-5) = -18$$
$$-2x - 30 = -18$$
$$+30 \quad +30$$
$$\frac{-2x}{-2} = \frac{12}{-2}$$
$$x = -6$$

6. $x = -1, y = 4$ Multiply both sides of the second equation by 2 so that the y-terms cancel out when the equations are added.

$$2(3x + y) = 2(1)$$
$$6x + 2y = 2$$

Add the equations.

$$x - 2y = -9$$
$$+ \; 6x + 2y = \; 2$$
$$\frac{7x}{7} = \frac{-7}{7}$$
$$x = -1$$

Plug $x = -1$ into either of the original equations, and solve for y.

$$x - 2y = -9$$
$$-1 - 2y = -9$$
$$+1 \qquad +1$$
$$\frac{-2y}{-2} = \frac{-8}{-2}$$
$$y = 4$$

7. $x = 10, y = 4$ Multiply both sides of the second equation by -2 so that the x-terms will be eliminated when the equations are added.

$$-2(x - 5y) = -2(-10)$$
$$-2x + 10y = 20$$

Add the equations.

$$2x - 2y = 12$$
$$+ \; -2x + 10y = 20$$
$$\frac{8y}{8} = \frac{32}{8}$$
$$y = 4$$

Plug $y = 4$ into either of the original equations, and solve for x.

$$2x - 2y = 12$$
$$2x - 2(4) = 12$$
$$2x - 8 = 12$$
$$+8 \quad +8$$
$$\frac{2x}{2} = \frac{20}{2}$$
$$x = 10$$

8. $x = -12, y = 6$ Multiply both sides of the second equation by -2 so that the y-terms will be eliminated when the equations are added.

$$-2(2y) = -2(4x + 60)$$
$$-4y = -8x - 120$$

Add the equations.

$$4y = -5x - 36$$
$$+ \; -4y = -8x - 120$$
$$0 = -13x - 156$$
$$+156 \qquad +156$$
$$\frac{156}{-13} = \frac{-13x}{-13}$$
$$-12 = x$$

Plug $x = -12$ into either of the original equations, and solve for y.

$$4y = -5x - 36$$
$$4y = -5(-12) - 36$$
$$4y = 60 - 36$$
$$\frac{4y}{4} = \frac{24}{4}$$
$$y = 6$$

9. **$x = 12$, $y = 30$** Multiply both sides of either equation by -1 so that the y-terms will cancel out when the equations are added. Let's multiply both sides of the first equation by -1.

$$-1(-7x + 3y) = -1(6)$$
$$7x - 3y = -6$$

Add the equations.

$$\begin{array}{r} 7x - 3y = -6 \\ +\ -4x + 3y = 42 \\ \hline \frac{3x}{3} = \frac{36}{3} \\ x = 12 \end{array}$$

Plug $x = 12$ into either of the original equations, and solve for y.

$$\begin{array}{r} -7x + 3y = \ 6 \\ -7(12) + 3y = \ 6 \\ -84 + 3y = \ 6 \\ +84 \qquad +84 \\ \hline \frac{3y}{3} = \frac{90}{3} \\ y = 30 \end{array}$$

10. **$x = -3$, $y = 6$** We need to multiply both equations by a constant in order to use the addition method. Here, we multiply the top equation by 3 and the bottom equation by 2, so that the x-terms will be eliminated.

$$3(2x - 3y) = 3(-24)$$
$$6x - 9y = -72$$

$$2(-3x - 2y) = 2(-3)$$
$$-6x - 4y = -6$$

Add the new equations.

$$\begin{array}{r} 6x - 9y = -72 \\ +\ -6x - 4y = \ -6 \\ \hline \frac{-13y}{-13} = \frac{-78}{-13} \\ y = 6 \end{array}$$

Plug $y = 6$ into either of the original equations, and solve for x.

$$\begin{array}{r} 2x - 3y = -24 \\ 2x - 3(6) = -24 \\ 2x - 18 = -24 \\ +18 \qquad +18 \\ \hline \frac{2x}{2} = \frac{-6}{2} \\ x = -3 \end{array}$$

11. **$x = -1$, $y = -1$** First, rearrange the order of the terms in the first equation to make it easier to use the addition method.

$$x + y = -2$$
$$7x - 2y = -5$$

Next, multiply the top equation by 2 so that the y-terms will be eliminated when the equations are added.

$$2(x + y) = 2(-2)$$
$$2x + 2y = -4$$

Add the equations.

$$2x + 2y = -4$$
$$+\ 7x - 2y = -5$$
$$\frac{9x}{9} = \frac{-9}{9}$$
$$x = -1$$

Plug $x = -1$ into either of the original equations, and solve for y.

$$x + y = -2$$
$$-1 + y = -2$$
$$\underline{+1\qquad +1}$$
$$y = -1$$

12. **$x = 8, y = -3$** First, rearrange the order of the terms in the first equation to make it easier to use the addition method.

$$8y + 40 = \ 2x$$
$$4y - 12 = -3x$$

Next, multiply the second equation by -2 so that the y-terms are eliminated when the equations are added.

$$-2(4y - 12) = -2(-3x)$$
$$-8y + 24 = 6x$$

Add the equations.

$$8y + 40 = 2x$$
$$+\ -8y + 24 = 6x$$
$$\frac{64}{8} = \frac{8x}{8}$$
$$8 = x$$

Plug $x = 8$ into either of the original equations, and solve for y.

$$8y + 40 = 2x$$
$$8y + 40 = 2(8)$$
$$8y + 40 = 16$$
$$\underline{-40\ -40}$$
$$\frac{8y}{8} = \frac{-24}{8}$$
$$y = -3$$

Lesson 11.2—Skills Check (page 254)

1. **$x = -1, y = 6$** The second equation is already solved in terms of y. Substitute $4x + 10$ in place of y in the first equation, and solve for x.

$$2x + y = \ 4$$
$$2x + (4x + 10) = \ 4$$
$$6x + 10 = \ 4$$
$$\underline{-10\ -10}$$
$$\frac{6x}{6} = \frac{-6}{6}$$
$$x = -1$$

Now substitute $x = -1$ into either of the two original equations.

$$y = 4x + 10$$
$$y = 4(-1) + 10$$
$$y = -4 + 10$$
$$y = 6$$

2. **$x = 7, y = 1$** The second equation is already solved in terms of x. Substitute $4y + 3$ in place of x in the first equation, and solve for y.

$$x + y = 8$$
$$(4y + 3) + y = 8$$
$$5y + 3 = 8$$
$$\underline{-3\ -3}$$
$$\frac{5y}{5} = \frac{5}{5}$$
$$y = 1$$

Plug $y = 1$ into either of the two original equations, and solve for x.

$$x + y = 8$$
$$x + 1 = 8$$
$$\underline{-1\ -1}$$
$$x = 7$$

3. **$x = -2$, $y = -2$** Since $y = x$, we can replace the y in the second equation with x, and then solve for x.

$$y = -3x - 8$$
$$x = -3x - 8$$
$$\underline{+3x \quad\quad +3x}$$
$$\frac{4x}{4} = \frac{-8}{4}$$
$$x = -2$$

Since $y = x$, then y is also equal to -2.

4. **$x = 6$, $y = 3$** The first equation is already solved in terms of y, so we can plug $\frac{2}{3}x - 1$ in place of y in the second equation, and then solve for x.

$$5x - 6y = 12$$
$$5x - 6\left(\frac{2}{3}x - 1\right) = 12$$
$$5x - 4x + 6 = 12$$
$$x + 6 = 12$$
$$\underline{\quad\quad -6 \quad -6}$$
$$x = 6$$

Plug $x = 6$ into either of the two original equations, and solve for y.

$$y = \frac{2}{3}x - 1$$
$$y = \left(\frac{2}{3}\right)6 - 1$$
$$y = 4 - 1$$
$$y = 3$$

5. **$x = 3$, $y = 4$** Solve the second equation in terms of y.

$$2x + y = 10$$
$$\underline{-2x \quad\quad -2x}$$
$$y = 10 - 2x$$

Now substitute $10 - 2x$ in place of y in the first equation. Solve for x.

$$6y = 4x + 12$$
$$6(10 - 2x) = 4x + 12$$
$$60 - 12x = 4x + 12$$
$$\underline{-12 \quad\quad\quad -12}$$
$$48 - 12x = \quad 4x$$
$$\underline{+12x \quad +12x}$$
$$\frac{48}{16} = \frac{16x}{16}$$
$$3 = x$$

Now plug $x = 3$ into either of the two original equations, and solve for y.

$$2x + y = 10$$
$$2(3) + y = 10$$
$$6 + y = 10$$
$$\underline{-6 \quad\quad -6}$$
$$y = 4$$

6. **$x = 4$, $y = -5$** First, multiply both sides of the first equation by 3 to eliminate the fractions. It isn't necessary to do this first, but it makes the numbers much easier to work with.

$$3\left(\frac{2}{3}x + \frac{1}{3}y\right) = 3(1)$$
$$2x + y = 3$$

The second equation is already solved in terms of y. Plug $3x - 17$ in place of y in the first equation, and solve for x.

$$2x + y = \quad 3$$
$$2x + (3x - 17) = \quad 3$$
$$5x - 17 = \quad 3$$
$$\underline{+17 \quad +17}$$
$$\frac{5x}{5} = \frac{20}{5}$$
$$x = 4$$

Plug $x = 4$ into either of the two original equations, and solve for y.

$$y = 3x - 17$$
$$y = 3(4) - 17$$
$$y = 12 - 17$$
$$y = -5$$

7. **$x = 4, y = 6$** The second equation is already written in terms of y. Plug $-2x + 14$ in place of y in the first equation, and solve for x.

$$7x - 3y = 10$$
$$7x - 3(-2x + 14) = 10$$
$$7x + 6x - 42 = 10$$
$$13x - 42 = 10$$
$$\underline{+42 \quad +42}$$
$$\frac{13x}{13} = \frac{52}{13}$$
$$x = 4$$

Plug $x = 4$ into either of the two original equations, and solve for y.

$$y = -2x + 14$$
$$y = -2(4) + 14$$
$$y = -8 + 14$$
$$y = 6$$

8. **$x = -1, y = 0$** To eliminate the fractions in the second equation, multiply the entire equation by 2.

$$2(y) = 2\left(\frac{9}{2}x + \frac{9}{2}\right)$$
$$2y = 9x + 9$$

Since the first equation is already written in terms of x, plug $3y - 1$ in place of x in the equation $2y = 9x + 9$, and solve for y.

$$2y = 9x + 9$$
$$2y = 9(3y - 1) + 9$$
$$2y = 27y - 9 + 9$$
$$2y = 27y$$
$$\underline{-2y \quad -2y}$$
$$0 = 25y$$
$$0 = y$$

Plug $y = 0$ into either of the two original equations, and solve for x.

$$x = 3y - 1$$
$$x = 3(0) - 1$$
$$x = 0 - 1$$
$$x = -1$$

9. **$x = 0, y = 6$** If we divide both sides of the second equation by 3, we can solve for x easily.

$$\frac{3x - 3y}{3} = \frac{-18}{3}$$
$$x - y = -6$$
$$\underline{+y \quad +y}$$
$$x = y - 6$$

Now plug $y - 6$ in place of x in the first equation, and solve for y.

$$12x + 4y = 24$$
$$12(y - 6) + 4y = 24$$
$$12y - 72 + 4y = 24$$
$$16y - 72 = 24$$
$$\underline{+72 \quad +72}$$
$$\frac{16y}{16} = \frac{96}{16}$$
$$y = 6$$

Plug $y = 6$ into either of the two original equations, and solve for x.

$$12x + 4y = 24$$
$$12x + 4(6) = 24$$
$$12x + 24 = 24$$
$$\underline{-24 \quad -24}$$
$$12x = 0$$
$$x = 0$$

10. **$x = 2, y = 1$** Divide both sides of the first equation by 3 in order to solve for x.

$$\frac{3x}{3} = \frac{6y}{3}$$
$$x = 2y$$

Plug $2y$ in place of x in the second equation, and solve for y.

$$\frac{2}{3}x = y + \frac{1}{3}$$
$$\frac{2}{3}(2y) = y + \frac{1}{3}$$
$$\frac{4}{3}y = y + \frac{1}{3}$$
$$\underline{-y \quad -y}$$
$$\frac{\frac{1}{3}y}{\frac{1}{3}} = \frac{\frac{1}{3}}{\frac{1}{3}}$$
$$y = 1$$

Plug $y = 1$ into either of the two original equations, and solve for x.

$$3x = 6y$$
$$3x = 6(1)$$
$$3x = 6$$
$$x = 2$$

11. **$x = 6, y = -4$** The second equation is already written in terms of y. Plug $2x - 16$ in place of y in the first equation, and solve for x.

$$3x + 2y = 10$$
$$3x + 2(2x - 16) = 10$$
$$3x + 4x - 32 = 10$$
$$7x - 32 = 10$$
$$\underline{+32 \quad +32}$$
$$\frac{7x}{7} = \frac{42}{7}$$
$$x = 6$$

Plug $x = 6$ into either of the two original equations, and solve for y.

$$y = 2x - 16$$
$$y = 2(6) - 16$$
$$y = 12 - 16$$
$$y = -4$$

12. **$x = -6, y = -6$** Multiply both sides of the first equation by 6 to eliminate the fractions.

$$6\left(\frac{1}{2}x - \frac{2}{3}y\right) = 6(1)$$
$$3x - 4y = 6$$

The second equation is already solved in terms of y. Plug $\frac{3}{2}x + 3$ in place of y in the modified version of the first equation, and solve for x.

$$3x - 4y = 6$$
$$3x - 4\left(\frac{3}{2}x + 3\right) = 6$$
$$3x - 6x - 12 = 6$$
$$-3x - 12 = 6$$
$$\underline{+12 \quad +12}$$
$$\frac{-3x}{-3} = \frac{18}{-3}$$
$$x = -6$$

Now plug $x = -6$ into either of the two original equations, and solve for y.

$$y = \frac{3}{2}x + 3$$
$$y = \left(\frac{3}{2}\right)(-6) + 3$$
$$y = -9 + 3$$
$$y = -6$$

Lesson 11.3—ACCUPLACER Challenge (page 258)

1. **(A)** To solve this system using the addition method, multiply both sides of the first equation by –2.

$$-2(x + y) = -2(12)$$
$$-2x - 2y = -24$$

Now add this to the second equation.

$$\begin{array}{r} -2x - 2y = -24 \\ + \quad 2x + 2y = 20 \\ \hline 0 \neq -4 \end{array}$$

This statement is untrue. There are no solutions to this system.

2. **(D)** We can solve this system by substitution. Substitute the second equation into the first one.

$$3x + 6y = 12$$
$$3x + 6\left(-\frac{1}{2}x + 2\right) = 12$$
$$3x - 3x + 12 = 12$$
$$12 = 12$$

This statement is true. This system has infinitely many solutions.

3. **(B)** We can solve this system using the substitution method. Solve the second equation for x.

$$\begin{array}{r} x - 3y = -12 \\ +3y \quad +3y \\ \hline x = 3y - 12 \end{array}$$

Substitute $3y - 12$ in place of x in the first equation, and solve for y.

$$\begin{array}{r} 3x - 2y = 6 \\ 3(3y - 12) - 2y = 6 \\ 9y - 36 - 2y = 6 \\ 7y - 36 = 6 \\ +36 \quad +36 \\ \hline \dfrac{7y}{7} = \dfrac{42}{7} \\ y = 6 \end{array}$$

We don't need to solve any further. Since there is one unique value of y, there will only be one unique value of x. This system has exactly one solution.

4. **(A)** We can solve this system by substitution. Plug $2x - 10$ in place of y in the second equation, and solve for x.

$$2x - y = 4$$
$$2x - (2x - 10) = 4$$
$$2x - 2x + 10 = 4$$
$$10 \neq 4$$

This statement is untrue, so this system has no real solutions.

Working with Polynomials

12

In order to proceed with a discussion of polynomials, we must first agree on the following important mathematical facts:

2 crocodiles + 3 crocodiles = 5 crocodiles

2 zebras + 3 zebras = 5 zebras

Not equal symbol

2 crocodiles + 3 zebras ≠ 5 crocozebras

These facts may seem painfully obvious, ridiculous even, but there is an important mathematical concept here: We can only combine things that are the same type of thing. This notion extends to working with polynomials where we can only combine like terms. We will define what that means in Lesson 12.1.

In this chapter, we will work to master

- combining like terms by adding and subtracting polynomials,
- multiplying and dividing monomials, and
- multiplying by a polynomial using the distributive property.

LESSON 12.1—COMBINING LIKE TERMS BY ADDING AND SUBTRACTING POLYNOMIALS

Lesson 12.1A—Combining Like Terms

Let's review the concept of like terms.

TERM: a number, a variable, or any product of numbers and variables. For example, 3, $3x$, $3x^5$, and $3x^5y^3$ are each separate terms.

Like terms are terms with the same variables raised to the same powers.

Like Terms	Not Like Terms	
5 and 9	5 and $9a$	(different variable)
$4a$ and $-7a$	$4a$ and $-7b$	(different variable)
$-3a^2$ and $2a^2$	$-3a^2$ and $2a^5$	(different exponent)
a^2b and $6a^2b$	a^2b and $6a^2c^3$	(different variable and different exponent)
$4a^2bc^3$ and $6a^2bc^3$	$4a^2bc^3$ and $6a^2b^5d^3$	(different variable and different exponent)

You may have noticed that we said that like terms have the same variables, and we listed 5 and 9 as like terms. Neither has a variable, so they are constants. Constants are always like terms.

When we have like terms, they can be combined. Whether adding or subtracting, this means that like terms can be combined to make one term. For example, 5 zebras can be added to 10 zebras to make 15 zebras. On the other hand, 5 zebras cannot be added to 10 crocodiles to make 15 zebras, 15 crocodiles, or 15 crocozebras. None of those would be true.

To Combine Like Terms:

1. Add or subtract the coefficients.
2. Keep the variables and the exponents.

➡ Example 1

$3x + 5x = ?$

Solution

$$3x + 5x = 8x$$

Answer: $8x$

➡ Example 2

$5n^2 - 2n^2 = ?$

Solution

$$5n^2 - 2n^2 = 3n^2$$

Answer: $3n^2$ (Notice how, in this example, the exponent doesn't change.)

➡ Example 3

$10p^5 - p^5 = ?$

Solution

Notice that, in this example, there is no coefficient in front of p^5. When there is no visible coefficient, the coefficient is considered to be 1 since $1p^5 = p^5$.

$$10p^5 - p^5 = 10p^5 - 1p^5 = 9p^5$$

Answer: $9p^5$

➡ Example 4

$-\dfrac{1}{2}m + 7m = ?$

Solution

$$-\dfrac{1}{2}m + 7m = 6\dfrac{1}{2}m$$

Answer: $6\dfrac{1}{2}m$

➡ Example 5

$-0.25j^4 - 0.03j^4 = ?$

Solution

$$-0.25j^4 - 0.03j^4 = -0.28j^4$$

Answer: $-0.28j^4$

LESSON 12.1A—SKILLS CHECK

Directions: Find the sum or difference as directed.

1. $7x + 3x$

2. $6n^2 + 15n^2$

3. $p^4 - 8p^4$

4. $\dfrac{1}{2}m + \dfrac{2}{3}m$

5. $-1.25j^8 - 2.47j^8$

(Answers are on page 285.)

Lesson 12.1B—Adding Polynomials

If an expression is composed of one term, as in $5m^3$, we call it a **monomial**. If two or more like terms combine into a single term, then their result is also called a monomial, as in $5m^3 + 2m^3 = 7m^3$ (here $7m^3$ is the monomial).

If, on the other hand, an expression contains terms that are not like terms, such as in the expression $4x + 3n$, then the expression cannot be combined into a single term. When this happens, the terms remain separated by either an addition sign or a subtraction sign. This is what we call a **polynomial**, which we can loosely define as two or more unlike terms separated by addition or subtraction signs. We say loosely because this definition should also state that we cannot have division by a variable or negative exponents or some other things that aren't all that important to us at this point.

You should familiarize yourself with the following definitions:

MONOMIAL: one term, as in 3, $3x$, $3x^5$, or $3x^5y^3$

BINOMIAL: two unlike terms separated by an addition sign or a subtraction sign, as in $3x^2 + 5x$, or $2m^{43} - 8p^{43}$

TRINOMIAL: three unlike terms separated by an addition sign or a subtraction sign, as in $3x^2 + 5x - 4$, or $2m^{43} - 8p^{42} + 17$

POLYNOMIAL: two or more unlike terms separated by an addition sign or a subtraction sign. Binomials, trinomials, and expressions with four or more terms can all be referred to as polynomials. The prefix *poly-* just means more than one, not a specific number of terms.

STANDARD FORM: terms in a polynomial are ordered according to their exponents in descending order, as in $3p^4 + 5p^3 - 2p^2 + 9p - 16$

We cannot combine the terms in a polynomial, but we often come across like terms that can be combined when we add a polynomial to another polynomial as in the example:

$$(3m^3 + 2m^2 - 6) + (5m^3 + 7m^2 - 4)$$
$$8m^3 + 9m^2 - 10$$

To Add Polynomials:

1. Remove parentheses, if there are any.
2. Identify like terms.
3. Rearrange the expression so that like terms are next to one another, making sure to keep the sign found immediately to the left of each term with that term.
4. Combine like terms.
5. Ensure that the terms are arranged according to standard form.

➡ Example 1

Find the sum of $(4x^3 + 6x^2 - 3) + (5x^3 - 2x^2 + 7)$.

Solution

Start by removing the parentheses. They do not change the problem here.

$$(4x^3 + 6x^2 - 3) + (5x^3 - 2x^2 + 7)$$
$$4x^3 + 6x^2 - 3 + 5x^3 - 2x^2 + 7$$

Next, identify like terms. You can keep track of which terms are like terms by drawing distinct shapes around each set of like terms.

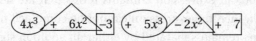

Now rearrange the terms so that like terms are next to one another. Make sure that the sign to the left of each term stays with that term as you move them. In other words, the minus sign stays with $-2x^2$ and -3, whereas the plus sign stays with $+5x^3$, $+6x^2$, and $+7$. In front of $4x^3$, there was no sign, so it is considered positive, or +.

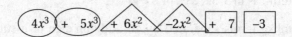

Combine like terms.

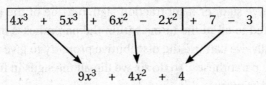

$$4x^3 + 5x^3 + 6x^2 - 2x^2 + 7 - 3$$

$$9x^3 + 4x^2 + 4$$

Answer: The sum, in standard form, is $9x^3 + 4x^2 + 4$.

➡ Example 2

Find the sum of $(-3m^5 + 7m^3 + 8m^2 - m + 10) + (-7m^4 + 6m^3 + 4m - 3)$.

Solution

First, remove the parentheses. Pay special attention to what happens to the term $-7m^4$ once the parentheses are removed. $+ -7m^4$ becomes $-7m^4$ because adding a negative is the same as subtracting a positive.

$$-3m^5 + 7m^3 + 8m^2 - m + 10 - 7m^4 + 6m^3 + 4m - 3$$

Identify like terms, and rearrange the order of the terms so that like terms are next to one another. Notice that not all the terms have a like term to combine with.

$$-3m^5 + 7m^3 + 6m^3 + 8m^2 - m + 4m + 10 - 3 - 7m^4$$

Combine like terms.

$$-3m^5 + 7m^3 + 6m^3 + 8m^2 - m + 4m + 10 - 3 - 7m^4$$

$$-3m^5 + 13m^3 + 8m^2 + 3m + 7 - 7m^4$$

Rearrange the terms in standard form so that their exponents are in descending order.

Answer: The sum, in standard form, is $-3m^5 - 7m^4 + 13m^3 + 8m^2 + 3m + 7$.

LESSON 12.1B—SKILLS CHECK

> **Directions:** Find the sum of the polynomials below.

1. $(3g^4 + 4g) + (9g^4 + 7g)$

2. $(7y^3 - 3y^2 + 4y) + (8y^4 + 3y^2)$

3. $(7n^2 - 3n + 4n) + (8n^2 + 3n^2 + 4n)$

4. $(5a^3 - 2b^3 + 2ab) + (3a^3 + 24b^2 - 2ab)$

5. $(12x^2y - 4xy^2) + (10xy^2 - 4x^2y)$

(Answers are on page 285.)

Lesson 12.1C—Subtracting Polynomials

When there is a negative sign, or a subtraction sign, outside of the parentheses, then that sign must be distributed to all of the terms within the parentheses once those parentheses are removed. Essentially, we can use the distributive property to give the negative sign to all the terms inside of the parentheses. To do so, we flip all the signs in front of each term to its opposite. Here's what we mean.

$$-(x^3 - 2x^2 + 5x - 3) = -x^3 + 2x^2 - 5x + 3$$

Above, we took away the parentheses and the negative sign outside of the parentheses, and we flipped all of the signs inside the parentheses to their opposites. Forgetting to flip all of the signs is the most common error that students make when subtracting polynomials. As a result, the ACCUPLACER will likely try and see if you will make this mistake by including a problem that involves subtracting polynomials.

To Subtract Polynomials:

1. Remove the parentheses, and flip all the signs of the second polynomial to their opposite.
2. Identify like terms.
3. Rearrange the expression so that the like terms are next to one another, making sure to keep the sign that is found immediately to the left of each term with that term.
4. Combine like terms.
5. Ensure that the terms are arranged according to standard form.

➥ Example 1

Solve: $(8x^3 + 3x^2 + 5x - 9) - (x^3 - 2x^2 + 5x - 3)$

Solution

First, remove the parentheses, and then flip all the signs of the second polynomial. Our new expression becomes:

$$8x^3 + 3x^2 + 5x - 9 - x^3 + 2x^2 - 5x + 3$$

Next, identify like terms, and rearrange the expression so that like terms are next to one another. Remember to keep the sign to the left of each term together with its corresponding term if you move it.

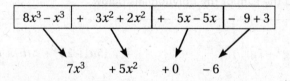

Arrange the terms in standard form.

Answer: $7x^3 + 5x^2 - 6$

➡ Example 2

Subtract $(n^2 + 3)$ from $(5n^2 - 20)$.

Solution

Here, we must first write out the subtraction problem mathematically. Be careful! The wording is tricky since $(n^2 + 3)$ is being subtracted *from* $(5n^2 - 20)$, which is written like this:

$$(5n^2 - 20) - (n^2 + 3)$$

Next, remove the parentheses, and flip all the signs of the second polynomial.

$$5n^2 - 20 - n^2 - 3$$

Next, identify like terms, and rearrange the expression so that the like terms are next to one another. Remember to keep the sign to the left of the term with the term if you move it.

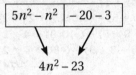

Answer: $4n^2 - 23$

LESSON 12.1C—SKILLS CHECK

Directions: Find the difference of the polynomials below.

1. $(17n^4 + 2n^3) - (10n^4 + n^3)$

2. $(6w^2 - 3w + 1) - (w^2 + w - 9)$

3. $(-7z^3 + 3z - 1) - (-6z^2 + z + 4)$

4. Subtract $x^2 + 6x - 2$ from $3x^2 - 5x + 3$.

5. $5(ab^4 - a^3b + 10) - 2(-ab^3 + a^3b + a^2b)$

(Answers are on page 286.)

ACCUPLACER TIPS

For multiple-choice, algebraic equivalence problems, you can always try substitution as an alternative method to finding the correct answer choice. If two algebraic expressions are equivalent, then they will always be equivalent no matter what value we give to the variable. Let's make sense of this with an example.

$(17n^3 + 2n^2 + 5) - (-10n^3 + n^2 + 2) =$

(A) $7n^3 + 2n^2 + 7$

(B) $7n^3 + n^2 + 3$

(C) $27n^3 + n^2 + 3$

(D) $27n^3 + 3n^2 + 7$

Imagine you get to the test, and you just aren't sure which is the correct answer. We can determine equivalence by substituting any number we like for the variable n. We will demonstrate by replacing all the ns with the number 2.

$$(17n^3 + 2n^2 + 5) - (-10n^3 + n^2 + 2) =$$
$$(17(2^3) + 2(2^2) + 5) - (-10(2^3) + 2^2 + 2) =$$
$$(17(8) + 8 + 5) - (-10(8) + 4 + 2) =$$
$$(136 + 8 + 5) - (-80 + 4 + 2) =$$
$$149 - (-74) = 223$$

Next, evaluate each of the answer choices for when $n = 2$.

(A) $7n^3 + 2n^2 + 7 = 7(2^3) + 2(2^2) + 7 = 7(8) + 2(4) + 7 = 56 + 8 + 7 = 71$

(B) $7n^3 + n^2 + 3 = 7(2^3) + 2^2 + 3 = 7(8) + 4 + 3 = 56 + 4 + 3 = 63$

(C) $27n^3 + n^2 + 3 = 27(2^3) + 2^2 + 3 = 27(8) + 4 + 3 = 216 + 4 + 3 = \mathbf{223}$

(D) $27n^3 + 3n^2 + 7 = 27(2^3) + 3(2^2) + 7 = 27(8) + 3(4) + 7 = 216 + 12 + 7 = 235$

After evaluating each answer choice for $n = 2$, only choice C had the same outcome of 223. Therefore, choice C is the correct answer.

Yes, this can be a lot of work, but you have the time. Even if you do think you know the correct answer, you can also use this method to check to see if your answer choice is the correct one. The simplest number to substitute is 0. You can try that, but be careful to work all of the answer choices out because substituting 0 can sometimes make more than one choice appear equivalent. That's okay. If that happens, just repeat the process by substituting another number, like 2, into the original expression and the answer choices that appear equivalent. Choose the expression that is consistently equivalent.

LESSON 12.1—ACCUPLACER CHALLENGE

> **Directions:** For each of the questions below, choose the best answer from the four choices given.

1. $(2x^3 + 4x^2 - 5) + (2x^3 + 5x - 6) =$

   (A) $4x^3 + 4x^2 + 5x - 11$

   (B) $4x^6 + 4x^2 + 5x - 11$

   (C) $4x^6 + 9x^3 - 11$

   (D) $4x^3 + 4x^2 + 5x + 11$

2. $(17n^4 + 2n^3 + 5) - (-10n^4 + n^3 + 2) =$

   (A) $7n^4 + 2n^3 + 7$

   (B) $7n^4 + n^3 + 3$

   (C) $27n^4 + n^3 + 3$

   (D) $27n^4 + n^3 + 7$

3. $(4p^5 + 2p^3) - (5p^4 + 2p^3) =$

   (A) $9p^9 + 4p^6$

   (B) $4p^5 - 5p^4 + 4p^3$

   (C) $4p^5 - 5p^4$

   (D) $-p$

4. $(g^6 - 3g^4 + 19g^2 - 8) + (4g^5 - 3g^4 - 19g^2 + 2g - 4) =$

   (A) $5g^{11} - 6g^8 + 2g - 12$

   (B) $g^6 + 4g^5 - 6g^4 + 2g - 12$

   (C) $g^6 - 4g^5 - 6g^4 + 2g - 12$

   (D) $g^6 - 6g^4 - 12$

5. $(a^2b + 2ab^2 - ac) - (4ab^2 - ab + a^2b) =$

   (A) $2a^2b - 2ab^2 - ac - ab$

   (B) $-2ab^2 - ac - ab$

   (C) $-2ab^2 - ac + ab$

   (D) $2a^2b + 6ab^2 - ac + ab$

(Answers are on page 288.)

LESSON 12.2—MULTIPLYING AND DIVIDING MONOMIALS

Lesson 12.2A—Multiplying Monomials

To understand multiplying monomials, it helps to first expand the monomials and then multiply them. Here is what we mean by expanding.

$$2p^3 \text{ in expanded form} = 2 \times p \times p \times p$$

To expand $2p^3$, we looked to the exponent, 3, which is attached to the variable p, and wrote p that many times with multiplication ($\times$) symbols in between. It's important to note that we only wrote 2 one time. The exponent here only applies to the p, not to the 2. If we had wanted the exponent to apply to both the coefficient, 2, and the variable, p, we would have written $(2p)^3$, which is different.

$$(2p)^3 \text{ in expanded form} = 2p \times 2p \times 2p = 2 \times 2 \times 2 \times p \times p \times p$$

Now, let's look at an example of multiplying monomials by first expanding and then multiplying.

➡ Example 1

$2p^3 \times 6p^5 =$

Solution

To expand these monomials, we are going to look to the exponents attached to the variable, p, and write p that many times with $\times$ symbols in between.

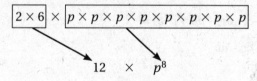

We can now reorganize our factors and multiply like this.

$$\boxed{2 \times 6} \times \boxed{p \times p \times p \times p \times p \times p \times p \times p}$$
$$12 \quad \times \quad p^8$$

<u>Answer:</u> $12p^8$

You may have noticed that we multiplied terms that are not like terms, which is completely allowed. You can multiply unlike terms; you cannot add or subtract unlike terms.

Having found out that $2p^3 \times 6p^5 = 12p^8$, we can now write a rule.

To Multiply Monomials:

   1. Multiply the coefficients.

   2. Keep the variables, and add the exponents of like bases.

➡ Example 2

Simplify $(4m^5n^2)(8m^2n)$.

Solution

To simplify means to rewrite an expression in a simpler, often shorter, way. Here, we can do that by multiplying the monomials. To do so, we first multiply 4×8, and then rewrite the exponents as addition problems. It's important to remember that a variable without a visible exponent is always considered to have an exponent of 1. In this case $n = n^1$.

$$(4m^5n^2)(8m^2n) = (4 \times 8)m^{5+2}n^{2+1}$$

Answer: $32m^7n^3$

LESSON 12.2A—SKILLS CHECK

> **Directions:** Simplify each expression.

1. $5(3k)$

2. $8m^2 \times 3m^4$

3. $5x^3(4x)$

4. $-4p^5(4p^5)$

5. $\frac{2}{3}n^4(2n^5p^2)$

(Answers are on page 289.)

LESSON 12.2B—DIVIDING MONOMIALS

Division is the inverse of multiplication. As such, the steps for dividing monomials are quite similar to multiplying them, only we do things in reverse.

To Divide Monomials:

   1. Divide the coefficients.

   2. Keep the variables, and subtract the exponents of like bases.

➡ Example 1

$24x^5 \div 6x^3 =$

Solution

First divide the coefficients, and then subtract the exponents of like bases.

$$(24 \div 6)x^{5-3}$$

Answer: $4x^2$

➡ **Example 2**

$$\frac{42m^5p^6}{6m^2p} =$$

Solution

First, it's important to remember that fractions are another very common way of writing division. We could also rewrite the problem like this:

$$42m^5p^6 \div 6m^2p$$

This example is similar to the last one except that we have more than one variable. We only subtract exponents of like bases. Notice that we subtract the exponents originally found in the denominator from the exponents originally found in the numerator.

$$(42 \div 6)m^{(5-2)}p^{(6-1)}$$

Answer: $7m^3p^5$

LESSON 12.2B—SKILLS CHECK

Directions: Simplify each expression.

1. $15m^6 \div 3m^4$

2. $\dfrac{12n^9}{3n^4}$

3. $\dfrac{-28x^{12}}{2x^5}$

4. $18b^3 \div 4b$

5. $\dfrac{-225a^{23}b^3}{-15a^{23}b^2}$

(Answers are on page 290.)

LESSON 12.2—ACCUPLACER CHALLENGE

Directions: For each of the questions below, choose the best answer from the four choices given.

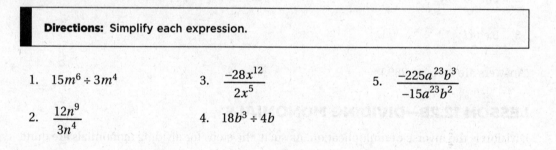

1. $5x^4 \bullet 5x^4 \bullet 5x^4 =$

   (A) $15x^4$

   (B) $125x^4$

   (C) $125x^{12}$

   (D) $125x^{64}$

2. $3p^4 \times 2p^4 + 10p^6 \times 3p =$

   (A) $6p^8 + 30p^7$

   (B) $6p^{16} + 30p^6$

   (C) $36p^{15}$

   (D) $36p^{23}$

3. $\dfrac{24m^5n^8}{6mn^2} =$

   (A) $\dfrac{1}{4}m^5n^6$

   (B) $4m^4n^6$

   (C) $4m^5n^6$

   (D) $30m^6n^{10}$

4. $0.25b^3 =$

   (A) $0.5b^2 \times 5b$

   (B) $12b^5 \div 3b^2$

   (C) $3b^5 \div 12b^2$

   (D) $0.20b^2 \times 0.05b$

5. $24x^2y \div 6x^2y =$

   (A) 0

   (B) 4

   (C) 18

   (D) $4x^2y$

(Answers are on page 291.)

LESSON 12.3—MULTIPLYING BY A POLYNOMIAL USING THE DISTRIBUTIVE PROPERTY

Lesson 12.3A—Multiplying a Monomial by a Polynomial Using the Distributive Property

As we saw previously, the distributive property states the following:

$$a(b + c + d) = ab + ac + ad$$

This means that when we multiply the monomial a, by a polynomial such as $b + c + d$, we must distribute, or give out, the a to all the terms of the polynomial. Said another way, we must multiply all of the terms inside the parentheses by the term found immediately outside the parentheses.

To Multiply a Monomial by a Polynomial:

  1. Separately multiply the monomial times each term of the polynomial.

➥ Example 1

Simplify the expression $4x^2(3x^4 + 5x^3 - 10x^2 - 12)$.

Solution

Let's first distribute.

$$4x^2(3x^4 + 5x^3 - 10x^2 - 12)$$

$$4x^2(3x^4) + 4x^2(5x^3) - 4x^2(10x^2) - 4x^2(12)$$

Then multiply each term separately, remembering to multiply the coefficients, keep the variables, and add the exponents.

$$12x^{2+4} + 20x^{2+3} - 40x^{2+2} - 48x^2$$

<u>Answer:</u> $12x^6 + 20x^5 - 40x^4 - 48x^2$

➥ Example 2

Simplify the expression $2g^3 - g^2(2g^2 - g)$.

Solution

Visually, this problem is a little tricky. We must first recognize that the first term, $2g^3$, is not being multiplied by any of the terms. We also must remember that the minus sign in front of $-g^2$ belongs to this term. When we distribute, we must also distribute this minus, which has the same effect as the negative sign, along with g^2.

Let's first distribute.

$$2g^3 - g^2(2g^2 - g)$$
$$2g^3 - g^2(2g^2) - g^2(-g)$$
$$2g^3 - 2g^{2+2} - (-g^{2+1})$$
$$2g^3 - 2g^4 + g^3$$

We are almost done, but we ended up with like terms, $2g^3$ and g^3, that can be combined. After combining those terms, we have a final, simplified binomial.

Answer: $-2g^4 + 3g^3$

LESSON 12.3A—SKILLS CHECK

> **Directions:** Simplify each expression.

1. $3x^4(5x^2 + 2x)$

2. $2y(y^6 - 4y^3 + 2)$

3. $-4x^3y^2(5x^2y - 3y)$

4. $\dfrac{m}{2}\left(6m^3 + 8m - 6\right)$

5. $5p^2 - 2p^2(4p^2 - p + 2)$

(Answers are on page 292.)

Lesson 12.3B—Multiplying a Polynomial by a Polynomial Using the Distributive Property

The key idea behind multiplying polynomials is not unlike the idea behind combinations. Here's what we mean. If you have two pairs of pants and two shirts, then how many one pants–one shirt outfits can we make? Take a look at the diagram below.

We could have four possible outfits, as shown below:

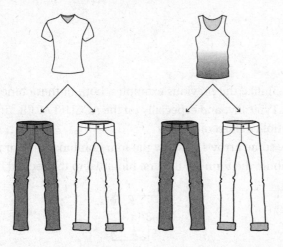

Now let's look at the following example: $(a + b)(c + d)$. Since the two binomials are next to each other without anything in between, we know that we must multiply the two binomials. To use our shirts and pants analogy, we know that if a is multiplied by c and d, and b is also multiplied by c and d, there are four possible pairs.

$$(a + b)(c + d)$$

$$ac + ad + bc + bd$$

What if we were to multiply a binomial times a trinomial, as in $(e + f)(g + h + i)$? Well, since there are two terms times three terms, then there should be six possible pairs. We can pair e with g, h, and i, and we can pair f with g, h, and i.

$$(e + f)(g + h + i)$$

$$eg + eh + ei + fg + fh + fi$$

You can probably guess what would happen if we multiplied a trinomial (three terms) times a four-term polynomial—you would end up with twelve terms. That is exactly what would happen if you had three shirts and four pairs of pants—you could make exactly twelve shirt–pants outfits.

To Multiply Polynomials:

1. Separately multiply the first term of the first polynomial by all the terms of the second polynomial.
2. Next, separately multiply the second term of the first polynomial by all the terms of the second polynomial.
3. Continue to distribute the terms of the first polynomial to all the terms of the second polynomial until you have distributed all the terms of the first polynomial.
4. Combine like terms.

➥ Example 1

Simplify $(x + 3)(x + 2)$.

Solution

You may notice that, unlike the previous examples, both of these binomials contain a variable plus a constant. Typically, and especially on the ACCUPLACER, this is how you will see binomials being multiplied most often.

First, we will draw some arrows to show the four combinations, or factor pairs, that will result when we distribute the terms of the first binomial to the second binomial.

$$(x + 3)(x + 2)$$

$$(x)(x) + (x)(2) + (3)(x) + (3)(2)$$

$$x^2 + 2x + 3x + 6$$

When we had two shirts and two pants, we could make four distinct outfits. The same thing happens here. When we multiplied two binomials, we ended up with the four factor pairs: $x^2 + 2x + 3x + 6$. However, the middle terms are like terms, so we can combine them. In the end, the simplified form of this expression is $x^2 + 5x + 6$.

$$x^2 + 2x + 3x + 6$$

Answer: $x^2 + 5x + 6$

➥ Example 2

Simplify $(x - 4)(x + 7)$.

Solution

$$(x - 4)(x + 7)$$

$$(x)(x) + (7)(x) + (-4)(x) + (-4)(7)$$

$$x^2 + 7x - 4x - 28$$

Answer: $x^2 + 3x - 28$

➡ Example 3

Simplify $(2m - 4)(3m^2 - 5m + 6)$.

Solution

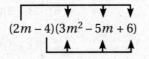

$$(2m)(3m^2) + (2m)(-5m) + (2m)(6) + (-4)(3m^2) + (-4)(-5m) + (-4)(6)$$

$$6m^3 - 10m^2 + 12m - 12m^2 + 20m - 24$$

Notice that by multiplying two terms times three terms, we initially end up with six distinct terms. From there, we combine like terms as shown below.

$$6m^3 - 10m^2 - 12m^2 + 12m + 20m - 24$$

Answer: $6m^3 - 22m^2 + 32m - 24$

➡ Example 4

Simplify $(x + 6)(x - 6)$.

Solution

Pay special attention to this "special case" example; these problems are favorites of test writers, ACCUPLACER and otherwise.

$$(x + 6)(x - 6)$$

$$(x)(x) + (-6)(x) + (6)(x) + (6)(-6)$$

$$x^2 - 6x + 6x - 36$$

Here's where it gets interesting. Notice that the middle terms, $-6x$ and $6x$, are opposites. Whenever we combine opposites, we are left with 0 because they cancel each other out. Therefore, our final polynomial here is only two terms long!

Answer: $x^2 - 36$

We are left with the **difference of squares binomial**, since the terms x^2 and 36 are each perfect squares separated by a subtraction sign (difference).

LESSON 12.3B—SKILLS CHECK

Directions: Simplify each expression.

1. $(x + 4)(x + 2)$

2. $(x + 1)(x - 2)$

3. $(a - 3)(a + 5)$

4. $(m - 1)(m - 9)$

5. $(n + 2)(4n^2 - 3n + 5)$

6. $(x + 3)(x - 3)$

7. $(x + 4)^2$

8. $(2a - 3)(a + 4)$

9. $(5b + 6c)(3b + 2c)$

10. $(8d - g)(8d + g)$

(Answers are on page 293.)

LESSON 12.3—ACCUPLACER CHALLENGE

Directions: For each of the questions below, choose the best answer from the four choices given.

1. $4(p^2 - 2p - 1) - 3(p^2 + 2p + 5) =$

   (A) $p^2 - 14p - 19$
   (B) $p^2 + 6$
   (C) $p^2 + 4p - 4$
   (D) $p^2 + 4p - 14$

2. $(x + 3)(x - 4) =$

   (A) $x^2 + x - 1$
   (B) $x^2 - x + 12$
   (C) $x^2 + x - 12$
   (D) $x^2 - x - 12$

3. $(p + 8)(p - 8) =$

   (A) $9p - 64$
   (B) $p^2 - 64$
   (C) $p^2 + 64$
   (D) $p^2 - 16 - 64$

4. $(m - 4)(2m^2 + 3m - 4) =$

   (A) $8m^2 - 11m + 16$
   (B) $2m^3 + 5m^2 + 16m + 16$
   (C) $2m^3 - 5m^2 + 16m + 16$
   (D) $2m^3 - 5m^2 - 16m + 16$

5. $(5x - 3y)^2 =$

   (A) $25x^2 - 9y^2$
   (B) $25x^2 + 9y^2$
   (C) $25x^2 + 9y^2 + 30xy$
   (D) $25x^2 + 9y^2 - 30xy$

(Answers are on page 295.)

To complete a set of practice questions that reviews all of Chapter 12, go to:
barronsbooks.com/TP/accuplacer/math29qw/

Lesson 12.1A—Skills Check (page 269)

1. **$10x$**

$$7x + 3x$$

Add the coefficients. Keep the common variable.

$$10x$$

2. **$21n^2$**

$$6n^2 + 15n^2$$

Add the coefficients. Keep the common variable and exponent.

$$21n^2$$

3. **$-7p^4$**

$$p^4 - 8p^4$$

Add the coefficients. Keep the common variable and exponent.

$$-7p^4$$

4. **$\frac{7}{6}m$ or $1\frac{1}{6}m$**

$$\frac{1}{2}m + \frac{2}{3}m$$

Raise the fractions to equivalents with a common denominator, and then add the coefficients. Keep the common variable.

$$\frac{1}{2}m + \frac{2}{3}m = \frac{3}{6}m + \frac{4}{6}m = \frac{7}{6}m \text{ or } 1\frac{1}{6}m$$

5. **$-3.72j^8$**

$$-1.25j^8 - 2.47j^8$$

Subtract the coefficients. Keep the common variable and exponent.

$$-3.72j^8$$

Lesson 12.1B—Skills Check (page 271)

1. **$12g^4 + 11g$**

$$(3g^4 + 4g) + (9g^4 + 7g)$$

Remove the parentheses, and reorganize the terms according to like terms.

$$3g^4 + 9g^4 + 4g + 7g$$

Combine like terms.

$$12g^4 + 11g$$

2. **$8y^4 + 7y^3 + 4y$**

$$(7y^3 - 3y^2 + 4y) + (8y^4 + 3y^2)$$

Remove the parentheses, and reorganize the terms according to like terms.

$$8y^4 + 7y^3 - 3y^2 + 3y^2 + 4y$$

Combine like terms.

$$8y^4 + 7y^3 + 4y$$

3. $18n^2 + 5n$

$$(7n^2 - 3n + 4n) + (8n^2 + 3n^2 + 4n)$$

Remove the parentheses, and reorganize the terms according to like terms.

$$7n^2 + 8n^2 + 3n^2 - 3n + 4n + 4n$$

Combine like terms.

$$18n^2 + 5n$$

4. $8a^3 - 2b^3 + 24b^2$

$$(5a^3 - 2b^3 + 2ab) + (3a^3 + 24b^2 - 2ab)$$

Remove the parentheses, and reorganize the terms according to like terms.

$$5a^3 + 3a^3 - 2b^3 + 24b^2 + 2ab - 2ab$$

Combine like terms.

$$8a^3 - 2b^3 + 24b^2$$

5. $8x^2y + 6xy^2$

$$(12x^2y - 4xy^2) + (10xy^2 - 4x^2y)$$

Remove the parentheses, and reorganize the terms according to like terms.

$$12x^2y - 4x^2y - 4xy^2 + 10xy^2$$

Combine like terms.

$$8x^2y + 6xy^2$$

Lesson 12.1C—Skills Check (page 273)

1. $7n^4 + n^3$

$$(17n^4 + 2n^3) - (10n^4 + n^3)$$

Remove the parentheses, and flip the signs of all the terms in the second polynomial to their opposite sign.

$$17n^4 + 2n^3 - 10n^4 - n^3$$

Reorganize the terms according to like terms.

$$17n^4 - 10n^4 + 2n^3 - n^3$$

Combine like terms.

$$7n^4 + n^3$$

2. $5w^2 - 4w + 10$

$$(6w^2 - 3w + 1) - (w^2 + w - 9)$$

Remove the parentheses, and flip the signs of all the terms in the second polynomial to their opposite sign.

$$6w^2 - 3w + 1 - w^2 - w + 9$$

Reorganize the terms according to like terms.

$$6w^2 - w^2 - 3w - w + 1 + 9$$

Combine like terms.

$$5w^2 - 4w + 10$$

3. $-7z^3 + 6z^2 + 2z - 5$

$$(-7z^3 + 3z - 1) - (-6z^2 + z + 4)$$

Remove the parentheses, and flip the signs of all the terms in the second polynomial to their opposite sign.

$$-7z^3 + 3z - 1 + 6z^2 - z - 4$$

Reorganize the terms according to like terms.

$$-7z^3 + 6z^2 + 3z - z - 1 - 4$$

Combine like terms.

$$-7z^3 + 6z^2 + 2z - 5$$

4. $2x^2 - 11x + 5$

$$(3x^2 - 5x + 3) - (x^2 + 6x - 2)$$

Remove the parentheses, and flip the signs of all the terms in the second polynomial to their opposite sign.

$$3x^2 - 5x + 3 - x^2 - 6x + 2$$

Reorganize the terms according to like terms.

$$3x^2 - x^2 - 5x - 6x + 3 + 2$$

Combine like terms.

$$2x^2 - 11x + 5$$

5. $5ab^4 - 7a^3b + 2ab^3 - 2a^2b + 50$

$$5(ab^4 - a^3b + 10) - 2(-ab^3 + a^3b + a^2b)$$

First, distribute the numbers outside of the parentheses. Multiply the numbers outside of each set of parentheses by all the terms inside each set of parentheses.

$$5(ab^4) - 5(a^3b) + 5(10) - 2(-ab^3) - 2(a^3b) - 2(a^2b)$$

$$5ab^4 - 5a^3b + 50 + 2ab^3 - 2a^3b - 2a^2b$$

Reorganize the terms according to like terms.

$$5ab^4 - 5a^3b - 2a^3b + 2ab^3 - 2a^2b + 50$$

Combine like terms.

$$5ab^4 - 7a^3b + 2ab^3 - 2a^2b + 50$$

Lesson 12.1—ACCUPLACER Challenge (page 275)

1. **(A)**

$$(2x^3 + 4x^2 - 5) + (2x^3 + 5x - 6)$$

Remove the parentheses.

$$2x^3 + 4x^2 - 5 + 2x^3 + 5x - 6$$

Rearrange the terms according to like terms.

$$2x^3 + 2x^3 + 4x^2 + 5x - 5 - 6$$

Combine like terms.

$$4x^3 + 4x^2 + 5x - 11$$

2. **(C)**

$$(17n^4 + 2n^3 + 5) - (-10n^4 + n^3 + 2)$$

Remove the parentheses, and flip all the signs of the second polynomial to their opposite.

$$17n^4 + 2n^3 + 5 + 10n^4 - n^3 - 2$$

Rearrange the terms according to like terms.

$$17n^4 + 10n^4 + 2n^3 - n^3 + 5 - 2$$

Combine like terms.

$$27n^4 + n^3 + 3$$

3. **(C)**

$$(4p^5 + 2p^3) - (5p^4 + 2p^3)$$

Remove the parentheses, and flip all the signs of the second polynomial to their opposite.

$$4p^5 + 2p^3 - 5p^4 - 2p^3$$

Rearrange the terms according to like terms.

$$4p^5 - 5p^4 + 2p^3 - 2p^3$$

Combine like terms.

$$4p^5 - 5p^4$$

4. **(B)**

$$(g^6 - 3g^4 + 19g^2 - 8) + (4g^5 - 3g^4 - 19g^2 + 2g - 4)$$

Remove the parentheses.

$$g^6 - 3g^4 + 19g^2 - 8 + 4g^5 - 3g^4 - 19g^2 + 2g - 4$$

Rearrange the terms according to like terms.

$$g^6 + 4g^5 - 3g^4 - 3g^4 + 19g^2 - 19g^2 + 2g - 8 - 4$$

Combine like terms.

$$g^6 + 4g^5 - 6g^4 + 2g - 12$$

5. **(C)**

$$(a^2b + 2ab^2 - ac) - (4ab^2 - ab + a^2b)$$

Remove the parentheses, and flip all the signs of the second polynomial to their opposite.

$$a^2b + 2ab^2 - ac - 4ab^2 + ab - a^2b$$

Rearrange the terms according to like terms.

$$a^2b - a^2b + 2ab^2 - 4ab^2 - ac + ab$$

Combine like terms.

$$-2ab^2 - ac + ab$$

Lesson 12.2A—Skills Check (page 277)

1. **15k**

$$5(3k)$$

Multiply the coefficients, and keep the variable.

$$15k$$

2. **24m^6**

$$8m^2 \times 3m^4$$

Multiply the coefficients, keep the variable, and add the exponents.

$$(8)(3)m^{2+4}$$

$$24m^6$$

3. **20x^4**

$$5x^3(4x)$$

Multiply the coefficients, keep the variable, and add the exponents.

$$(5)(4)x^{3+1}$$

$$20x^4$$

4. **−16p^{10}**

$$-4p^5(4p^5)$$

Multiply the coefficients, keep the variable, and add the exponents.

$$(-4)(4)p^{5+5}$$

$$-16p^{10}$$

5. **$\frac{4}{3}n^9p^2$**

$$\frac{2}{3}n^4\left(2n^5p^2\right)$$

Multiply the coefficients, keep the variables, and add the exponents of like bases.

$$\left(\frac{2}{3}\right)\left(\frac{2}{1}\right)n^{4+5}p^2$$

$$\frac{4}{3}n^9p^2$$

Lesson 12.2B—Skills Check (page 278)

1. $5m^2$

$$15m^6 \div 3m^4$$

Divide the coefficients, and subtract the exponents.

$$(15 \div 3)m^{(6-4)}$$

$$5m^2$$

2. $4n^5$

$$\frac{12n^9}{3n^4}$$

Divide the coefficients, and subtract the exponents.

$$(12 \div 3)n^{(9-4)}$$

$$4n^5$$

3. $-14x^7$

$$\frac{-28x^{12}}{2x^5}$$

Divide the coefficients, and subtract the exponents.

$$(-28 \div 2)x^{(12-5)}$$

$$-14x^7$$

4. $4\frac{1}{2}b^2 \text{ or } 4.5b^2$

$$18b^3 \div 4b$$

Divide the coefficients, and subtract the exponents.

$$(18 \div 4)b^{(3-1)}$$

$$4\frac{1}{2}b^2 \text{ or } 4.5b^2$$

5. $15b$

$$\frac{-225a^{23}b^3}{-15a^{23}b^2}$$

Divide the coefficients, and subtract the exponents.

$$(-225 \div -15)a^{(23-23)}b^{(3-2)}$$

$$15a^0b^1$$

Anything to the zero power is 1, so $a^0 = 1$.

$$15(1)b$$

$$15b$$

Lesson 12.2—ACCUPLACER Challenge (pages 278-279)

1. **(C)**

$$5x^4 \bullet 5x^4 \bullet 5x^4$$

Multiply the coefficients, and add the exponents.

$$(5 \bullet 5 \bullet 5)x^{(4+4+4)}$$

$$125x^{12}$$

2. **(A)**

$$3p^4 \times 2p^4 + 10p^6 \times 3p$$

Multiply the coefficients, and add the exponents.

$$(3 \times 2)p^{(4+4)} + (10 \times 3)p^{(6+1)}$$

$$6p^8 + 30p^7$$

3. **(B)**

$$\frac{24m^5n^8}{6mn^2}$$

Divide the coefficients, and subtract the exponents of like bases.

$$(24 \div 6)m^{(5-1)}n^{(8-2)}$$

$$4m^4n^6$$

4. **(C)**

$$0.25b^3$$

Simplify each choice to find an equivalent.

(A) $0.5b^2 \times 5b = (0.5 \times 5)b^{(2+1)} = 2.5b^3$
(B) $12b^5 \div 3b^2 = (12 \div 3)b^{(5-2)} = 4b^3$
(C) $3b^5 \div 12b^2 = (3 \div 12)b^{(5-2)} = 0.25b^3$
(D) $0.20b^2 \times 0.05b = (0.20 \times 0.05)b^{(2+1)} = 0.01b^3$

Only choice C produces a result equivalent to $0.25b^3$.

5. **(B)**

$$24x^2y \div 6x^2y$$

Divide the coefficients, and subtract the exponents of like bases.

$$(24 \div 6)x^{(2-2)}y^{(1-1)}$$

$$4x^0y^0$$

$$4(1)(1)$$

$$4$$

Lesson 12.3A—Skills Check (page 280)

1. $15x^6 + 6x^5$

$$3x^4(5x^2 + 2x)$$

Distribute.

$$3x^4(5x^2) + 3x^4(2x)$$
$$15x^6 + 6x^5$$

2. $2y^7 - 8y^4 + 4y$

$$2y(y^6 - 4y^3 + 2)$$

Distribute.

$$2y(y^6) + 2y(-4y^3) + 2y(2)$$
$$2y^7 - 8y^4 + 4y$$

3. $-20x^5y^3 + 12x^3y^3$

$$-4x^3y^2(5x^2y - 3y)$$

Distribute.

$$-4x^3y^2(5x^2y) - 4x^3y^2(-3y)$$
$$-20x^5y^3 + 12x^3y^3$$

4. $3m^4 + 4m^2 - 3m$

$$\frac{m}{2}\left(6m^3 + 8m - 6\right)$$

Distribute.

$$\frac{m}{2}(6m^3) + \frac{m}{2}(8m) + \frac{m}{2}(-6)$$
$$\frac{m}{2}\left(\frac{6m^3}{1}\right) + \frac{m}{2}\left(\frac{8m}{1}\right) + \frac{m}{2}\left(\frac{-6}{1}\right)$$
$$\frac{6m^4}{2} + \frac{8m^2}{2} - \frac{6m}{2}$$
$$3m^4 + 4m^2 - 3m$$

5. $-8p^4 + 2p^3 + p^2$

$$5p^2 - 2p^2(4p^2 - p + 2)$$

Distribute, but be careful to recognize that the $5p^2$ does not get distributed here because it is not immediately outside of the parentheses.

$$5p^2 - 2p^2(4p^2) - 2p^2(-p) - 2p^2(2)$$
$$5p^2 - 8p^4 + 2p^3 - 4p^2$$

Reorganize in standard form, and combine like terms.

$$-8p^4 + 2p^3 + 5p^2 - 4p^2$$
$$-8p^4 + 2p^3 + p^2$$

Lesson 12.3B—Skills Check (page 284)

1. $x^2 + 6x + 8$

$$(x + 4)(x + 2)$$

Distribute.

$$x(x) + x(2) + 4(x) + 4(2)$$
$$x^2 + 2x + 4x + 8$$

Combine like terms.

$$x^2 + 6x + 8$$

2. $x^2 - x - 2$

$$(x + 1)(x - 2)$$

Distribute.

$$x(x) + x(-2) + 1(x) + 1(-2)$$
$$x^2 - 2x + x - 2$$

Combine like terms.

$$x^2 - x - 2$$

3. $a^2 + 2a - 15$

$$(a - 3)(a + 5)$$

Distribute.

$$a(a) + a(5) - 3(a) - 3(5)$$
$$a^2 + 5a - 3a - 15$$

Combine like terms.

$$a^2 + 2a - 15$$

4. $m^2 - 10m + 9$

$$(m - 1)(m - 9)$$

Distribute.

$$m(m) + m(-9) - 1(m) - 1(-9)$$
$$m^2 - 9m - m + 9$$

Combine like terms.

$$m^2 - 10m + 9$$

5. $4n^3 + 5n^2 - n + 10$

$$(n + 2)(4n^2 - 3n + 5)$$

Distribute.

$$n(4n^2) + n(-3n) + n(5) + 2(4n^2) + 2(-3n) + 2(5)$$
$$4n^3 - 3n^2 + 5n + 8n^2 - 6n + 10$$

Combine like terms.

$$4n^3 - 3n^2 + 8n^2 + 5n - 6n + 10$$
$$4n^3 + 5n^2 - n + 10$$

6. $x^2 - 9$

$$(x + 3)(x - 3)$$

Distribute.

$$x(x) + x(-3) + 3(x) + 3(-3)$$
$$x^2 - 3x + 3x - 9$$

Combine like terms.

$$x^2 - 9$$

7. $x^2 + 8x + 16$

$$(x + 4)^2$$

Expand.

$$(x + 4)(x + 4)$$

Distribute.

$$x(x) + x(4) + 4(x) + 4(4)$$
$$x^2 + 4x + 4x + 16$$

Combine like terms.

$$x^2 + 8x + 16$$

8. $2a^2 + 5a - 12$

$$(2a - 3)(a + 4)$$

Distribute.

$$2a(a) + 2a(4) - 3(a) - 3(4)$$
$$2a^2 + 8a - 3a - 12$$

Combine like terms.

$$2a^2 + 5a - 12$$

9. $15b^2 + 12c^2 + 28bc$

$$(5b + 6c)(3b + 2c)$$

Distribute.

$$5b(3b) + 5b(2c) + 6c(3b) + 6c(2c)$$
$$15b^2 + 10bc + 18bc + 12c^2$$

Combine like terms, and place in standard form.

$$15b^2 + 12c^2 + 28bc$$

10. $64d^2 - g^2$

$$(8d - g)(8d + g)$$

Distribute.

$$8d(8d) + 8d(g) - g(8d) - g(g)$$
$$64d^2 + 8dg - 8dg - g^2$$

Combine like terms.

$$64d^2 - g^2$$

Lesson 12.3—ACCUPLACER Challenge (page 284)

1. **(A)**

$$4(p^2 - 2p - 1) - 3(p^2 + 2p + 5)$$

Distribute.

$$4(p^2) + 4(-2p) + 4(-1) - 3(p^2) - 3(2p) - 3(5)$$
$$4p^2 - 8p - 4 - 3p^2 - 6p - 15$$

Rearrange according to like terms.

$$4p^2 - 3p^2 - 8p - 6p - 4 - 15$$

Combine like terms.

$$p^2 - 14p - 19$$

2. **(D)**

$$(x + 3)(x - 4)$$

Distribute.

$$x(x) + x(-4) + 3(x) + 3(-4)$$
$$x^2 - 4x + 3x - 12$$

Combine like terms.

$$x^2 - x - 12$$

3. **(B)**

$$(p + 8)(p - 8)$$

Distribute.

$$p(p) + p(-8) + 8(p) + 8(-8)$$
$$p^2 - 8p + 8p - 64$$

Combine like terms.

$$p^2 - 64$$

4. **(D)**

$$(m - 4)(2m^2 + 3m - 4)$$

Distribute.

$$m(2m^2) + m(3m) + m(-4) - 4(2m^2) - 4(3m) - 4(-4)$$
$$2m^3 + 3m^2 - 4m - 8m^2 - 12m + 16$$

Rearrange according to like terms.

$$2m^3 + 3m^2 - 8m^2 - 4m - 12m + 16$$

Combine like terms.

$$2m^3 - 5m^2 - 16m + 16$$

5. **(D)**

$$(5x - 3y)^2$$

Expand.

$$(5x - 3y)(5x - 3y)$$

Distribute.

$$5x(5x) + 5x(-3y) - 3y(5x) - 3y(-3y)$$
$$25x^2 - 15xy - 15xy + 9y^2$$

Combine like terms.

$$25x^2 - 30xy + 9y^2$$

Rearrange in standard form.

$$25x^2 + 9y^2 - 30xy$$

Factoring Polynomials 13

Generally, a factor is something that contributes to a certain outcome. For example, if someone lives a long life, then it could be said that good genes or a good diet were factors. In mathematics, a factor is defined very similarly. For example, $3 \times 4 = 12$. The factors 3 and 4 contribute to the outcome of 12 by multiplication. In mathematics, a **factor** is a number or a quantity that multiplies with another factor to produce a given number or expression.

On the ACCUPLACER, you will have to turn products back into their factors, but you will have to do much more than simply factor numbers.

In this chapter, we will work to master

- factoring a monomial from a polynomial,
- factoring a quadratic expression in the form $ax^2 + bx + c$ when $a = 1$, and
- factoring a quadratic expression in the form $ax^2 + bx + c$ when $a \neq 1$.

LESSON 13.1—FACTORING A MONOMIAL FROM A POLYNOMIAL

Before we factor any polynomials, let's remember the fundamentals of the distributive property. For example, $2(x + 3)$ tells us to multiply both the x and the 3 inside the parentheses by the 2 found outside the parentheses. That would give us $2x + 6$. When we factor a polynomial, we start with the polynomial, such as $2x + 6$, and turn it into $2(x + 3)$. In other words, we "undistribute."

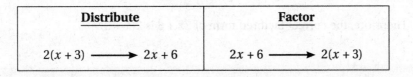

Distribute	Factor
$2(x + 3) \longrightarrow 2x + 6$	$2x + 6 \longrightarrow 2(x + 3)$

To factor a polynomial, we divide each term by the greatest common factor of all the terms. We don't eliminate the GCF; we simply separate it from the other terms. As a result, the GCF will be found outside of the parentheses. Let's look at this more closely.

To Factor a Monomial from a Polynomial:

1. Determine the GCF of all of the terms.
2. Put the polynomial in parentheses.
3. Using fraction lines, write the GCF under each term of the polynomial, and also write the GCF outside of the parentheses.
4. Divide each term inside the parentheses by the GCF.
5. Check your work by distributing the term outside of the parentheses to each term inside of the parentheses using multiplication. The result should be the original polynomial from the beginning.

➡ Example 1

Factor $2x + 6$.

Solution

The only factor that $2x$ and 6 share in common is 2, so the GCF here is 2.

Next, after placing parentheses around the polynomial, we write each term over 2 and put a 2, the GCF, outside of the parentheses.

$$2\left(\frac{2x}{2} + \frac{6}{2}\right)$$

Then, divide each term inside the parentheses by the GCF.

$$2\left(\frac{\overset{1}{\cancel{2}x}}{\underset{1}{\cancel{2}}} + \frac{\overset{3}{\cancel{6}}}{\underset{1}{\cancel{2}}}\right)$$

$$2(x + 3)$$

Finally, check your work using the distributive property.

$$2(x + 3) = 2(x) + 2(3) = 2x + 6$$

<u>Answer:</u> Therefore, the correct factored form of $2x + 6$ is $2(x + 3)$.

➥ Example 2

Factor $6p^5 - 15p^3 + 21p^2$.

Solution

First, identify the GCF for each term. In this case, all the terms have the same variable, p, to a power. Therefore, the GCF will include p to a power. To find the GCF for all the terms, we separately list the factors of the coefficients and variable raised to a power for each term.

$$\text{Factors of } 6p^5: 1, 2, \textcircled{3}, 6, p, \textcircled{p^2}, p^3, p^4, p^5$$

$$\text{Factors of } 15p^3: 1, \textcircled{3}, 5, 15, p, \textcircled{p^2}, p^3$$

$$\text{Factors of } 21p^2: 1, \textcircled{3}, 7, 21, p, \textcircled{p^2}$$

All three lists include 3 and p^2, so the GCF is $3p^2$.

Having found the GCF, place parentheses around the polynomial, write the GCF outside of the polynomial, and then divide each term inside the parentheses by the GCF.

$$6p^5 - 15p^3 + 21p^2 =$$

$$3p^2 \left(\frac{6p^5}{3p^2} - \frac{15p^3}{3p^2} + \frac{21p^2}{3p^2} \right) = 3p^2 \left(2p^3 - 5p + 7 \right)$$

To check to see if we have properly factored the polynomial, we use the distributive property.

$$3p^2 \left(2p^3 - 5p + 7 \right) =$$

$$\left(3p^2 \right)\left(2p^3 \right) - \left(3p^2 \right)(5p) + \left(3p^2 \right)(7) = 6p^5 - 15p^3 + 21p^2$$

Answer: Therefore, $3p^2(2p^3 - 5p + 7)$ is the correct factored form of $6p^5 - 15p^3 + 21p^2$.

LESSON 13.1—SKILLS CHECK

Directions: Completely factor each polynomial.

1. $15x^3 + 35x^2 + 5x$

2. $18m^6 + 9m^4 - 6m^2$

3. $-14h^2 - 7h$

4. $36n^{27} + 24n^{15} + 6n^7$

5. $7k^3 - 35k^2 + 70k$

(Answers are on page 310.)

LESSON 13.1—ACCUPLACER CHALLENGE

Directions: For each of the questions below, choose the best answer from the four choices given.

1. $15x^2 + 25x =$

    (A) $40x^3$
    (B) $15x(x + 10)$
    (C) $5x(3x + 5)$
    (D) $5x(3x^2 + 5x)$

2. If $2x$ is one factor of $2x^2 - 8x$, what is the other factor?

    (A) $2x$
    (B) $x^2 - 8x$
    (C) $x - 4$
    (D) $x + 4$

3. If one factor of $52x^4y^2 - 39x^2y^3$ is $13x^2y^2$, what is the other factor?

    (A) $4x^2 - 3y$
    (B) $4x^2 - 3y^2$
    (C) $4x^2y - 3xy^2$
    (D) $4x^2y - 3xy^3$

4. The area of a rectangle is expressed as $4x^3 - 8x^2 + 32x$. What is the length of the rectangle if the width is $4x$?

    (A) $4x^3 - 8x^2 + 36x$
    (B) $x^2 - 2x + 8$
    (C) $x^3 - 2x^2 + 8x$
    (D) $16x^4 - 32x^3 + 128x^2$

5. The area of a rectangle is expressed as $(x^2 + 3x)$ square feet. Which of the following could be used to represent the perimeter, in feet, of the rectangle given that one side of the rectangle is x feet long?

    (A) $x + 3$
    (B) $2x + 3$
    (C) $4x + 6$
    (D) $10x$

(Answers are on page 311.)

LESSON 13.2—FACTORING A QUADRATIC EXPRESSION IN THE FORM $ax^2 + bx + c$ WHEN $a = 1$

An expression that has a variable squared as the highest power in the polynomial is called a **quadratic expression**. For example, the expressions x^2, $b^2 - 4$, and $x^2 + 5x + 6$ are all examples of quadratic expressions. College algebra classes place high emphasis on factoring quadratic expressions such as $x^2 + 5x + 6$. In turn, the ACCUPLACER will be sure to test your ability to do just that.

Looking at $x^2 + 5x + 6$, you may first notice that there is no factor that is common to all three terms. Therefore, it would only be natural to conclude that this polynomial cannot be factored. However, we must remember what happens when we multiply two binomials, such as $(x + 3)(x + 2)$, together.

$$(x)(x) + (x)(2) + (3)(x) + (3)(2)$$

$$x^2 + 2x + 3x + 6$$

When we multiply a binomial (two terms) times another binomial (two terms), we end up with four terms. Most often, as in the example above, the two middle terms can be combined. Here we end up with three terms.

$$x^2 + 5x + 6$$

Since $(x + 3)$ and $(x + 2)$ multiply to produce $x^2 + 5x + 6$, these binomials are considered to be factors of $x^2 + 5x + 6$. But, how do we get there?

It helps to first define that the standard form for a quadratic expression is written in the form $ax^2 + bx + c$. In the case of $x^2 + 5x + 6$, there is no coefficient written in front of x so $a = 1$, $b = 5$, and $c = 6$.

We have already determined that $(x + 3)(x + 2) = x^2 + 5x + 6$. Now, let's examine the equation. We have the numbers 3 and 2 on the left side, and the numbers 5 and 6 on the right side. It is no coincidence that $3 \times 2 = 6$ and $3 + 2 = 5$. Our whole number factors, 3 and 2, multiply to c and add to b. Knowing this fact, we can now work on factoring some quadratic expressions.

To Factor a Quadratic Expression in the Form $ax^2 + bx + c$ when $a = 1$:

1. Identify a, b, and c.
2. List the factor pairs of c.
3. Choose the factor pair of c that adds to b.
4. Complete the parentheses $(x + \underline{})(x + \underline{})$ with the two factors that multiply to c and add to b.
5. Check your work by multiplying both binomials by each other.

➡ Example 1

Factor $x^2 + 7x + 12$.

Solution

First, identify a, b, and c.

$$a = 1$$
$$b = 7$$
$$c = 12$$

List the factor pairs of c, 12 in this case.

1 and 12   2 and 6   3 and 4   –1 and –12   –2 and –6   –3 and –4

The factor pair of c, 12, that adds to b, 7, is 3 and 4.

$$3 + 4 = 7$$
$$3 \times 4 = 12$$

Use the factors 3 and 4 to complete two binomials with x as the first term in each.

$$(x + 3)(x + 4)$$

Check by multiplying the two binomials together.

$$(x + 3)(x + 4)$$
$$x(x) + x(4) + 3(x) + 3(4)$$
$$x^2 + 4x + 3x + 12$$
$$x^2 + 7x + 12$$

<u>Answer:</u> Therefore, $(x + 4)$ and $(x + 3)$ are factors of $x^2 + 7x + 12$.

➡ Example 2

Factor $m^2 + 2m - 8$.

Solution

First, identify a, b, and c.

$$a = 1$$
$$b = 2$$
$$c = -8$$

Notice that c here is considered negative because the minus sign is in front of 8 in the original expression.

List the factor pairs of c, –8 in this case.

1 and –8   –1 and 8   –2 and 4   2 and –4

The factor pair of c, –8, that adds to b, 2, is –2 and 4.

$$-2 + 4 = 2$$
$$-2 \times 4 = -8$$

Write the factors as binomials. Here the lead term in each binomial is m instead of x because $m \times m = m^2$.

$$(m - 2)(m + 4)$$

Check by multiplying the two binomials together.

$$(m - 2)(m + 4)$$
$$m^2 + 2m - 8$$

Answer: Therefore $(m - 2)$ and $(m + 4)$ are factors of $m^2 + 2m - 8$.

➡ Example 3

Factor the trinomial $p^2 - 8p + 15$.

Solution

First, identify a, b, and c.

$$a = 1$$
$$b = -8$$
$$c = 15$$

List the factor pairs of c, 15 in this case.

$$1 \text{ and } 15 \quad 3 \text{ and } 5 \quad -1 \text{ and } -15 \quad -3 \text{ and } -5$$

The factor pair of c, 15, that adds to b, -8, is -3 and -5.

$$-3 + -5 = -8$$
$$-3 \times -5 = 15$$

Use the factors -3 and -5 to complete two binomials with p as the first term in each.

$$(p - 5)(p - 3)$$

Check by multiplying the two binomials together.

$$(p - 5)(p - 3)$$
$$p^2 - 8p + 15$$

Answer: Therefore, $(p - 5)$ and $(p - 3)$ are factors of $p^2 - 8p + 15$.

➡ Example 4

Factor $x^2 - 16$.

Solution

We pointed out in Chapter 12 that problems like these are considered a special case called **difference of squares binomials**. Examples such as these frequently find their way onto ACCUPLACER questions so it is important to recognize this special case. x^2 is referred to as a **square** because $(x)(x) = x^2$, and 16 is a **square** because $(4)(4) = 4^2 = 16$. Any difference of two squares can fit the form $(x + a)(x - a) = x^2 - a^2$.

Knowing this we can conclude that:

$$(x + 4)(x - 4) = x^2 - 16$$

However, to make sure of this, we can still follow the same steps we have been using to factor $x^2 - 16$.

To do so, first identify a, b, and c.

$$a = 1$$
$$b = 0$$
$$c = -16$$

Next, list the factor pairs of c, -16 in this case.

$$-1 \text{ and } 16 \quad 1 \text{ and } -16 \quad 4 \text{ and } -4$$

The factor pair of c, -16, that adds to b, 0, is 4 and -4.

$$-4 + 4 = 0$$
$$-4 \times 4 = -16$$

Use the factors 4 and -4 to complete two binomials with x as the first term in each.

$$(x + 4)(x - 4)$$

Check by multiplying the two binomials together.

$$(x + 4)(x - 4)$$
$$x^2 - 16$$

<u>Answer</u>: Therefore, $(x + 4)$ and $(x - 4)$ are factors of $x^2 - 16$.

LESSON 13.2—SKILLS CHECK

Directions: Factor each quadratic expression into two binomials.

1. $k^2 + 9k + 8$

2. $x^2 + 12x + 35$

3. $x^2 - 5x + 6$

4. $r^2 - 10r - 11$

5. $x^2 + 2x - 15$

6. $y^2 + 13y - 48$

7. $x^2 - 9$

8. $x^2 - 64$

9. $p^2 + 8p + 16$

10. $m^2 - 14m + 49$

(Answers are on page 312.)

LESSON 13.2—ACCUPLACER CHALLENGE

> **Directions:** For each of the questions below, choose the best answer from the four choices given.

1. $p^2 + 5p - 24 =$

   (A) $(p-8)(p+3)$
   (B) $(p+12)(p-2)$
   (C) $(p+5)(p-24)$
   (D) $(p+8)(p-3)$

2. $(x-2)$ is a factor of

   (A) $x^2 + 2x - 8$
   (B) $x^2 + x - 2$
   (C) $x^2 - x - 6$
   (D) $x^2 - 2x - 24$

3. Given that a is a whole number, $(x+a)(x-a)$ could be the factorization of all of the following EXCEPT

   (A) $(x^2 - 9)$
   (B) $(x^2 - 8)$
   (C) $(x^2 - 16)$
   (D) $(x^2 - 144)$

4. Which of the following choices is a factor of both $x^2 + x - 2$ and $x^2 - 3x + 2$?

   (A) $x - 1$
   (B) $x + 2$
   (C) $x - 2$
   (D) $x + 1$

5. If $(x-5)$ is a factor of $x^2 - 11x + p$, then the value of p is

   (A) -6
   (B) 6
   (C) 10
   (D) 30

(Answers are on page 314.)

LESSON 13.3—FACTORING A QUADRATIC EXPRESSION IN THE FORM $ax^2 + bx + c$ WHEN $a \neq 1$

Earlier, when you worked on factoring quadratic expressions when $a = 1$, you probably asked, what about factoring quadratic expressions when a does not equal ($\neq$) 1? This is a relatively involved process. Unfortunately, questions that require you to factor quadratic expressions with a coefficient other than 1 can appear on the ACCUPLACER. Most likely, they will appear on the College-Level Mathematics test, but you could find a question requiring this skill on the Elementary Algebra test as well.

To Factor a Quadratic Expression in the Form $ax^2 + bx + c$ when $a \neq 1$:

1. Identify a, b, c, and ac.
2. Identify the factor pair of ac that adds to b.
3. Rewrite the original polynomial by separating the b term into two addends using the factors of ac you chose as coefficients on the common variables.
4. Regroup the first two terms into a binomial in parentheses plus the second two terms written as another binomial in parentheses.
5. Factor out the GCF from each binomial.
6. Factor out the common binomial and regroup the terms as a binomial times a binomial.
7. Check by multiplying the two binomials together.

➡ Example 1

Factor $6x^2 + 17x + 5$.

Solution

STEP 1 Identify a, b, c, and ac.

$$a = 6, b = 17, c = 5, \text{ and } ac = (6)(5) = 30$$

STEP 2 Identify the factor pair of ac that adds to b.

Factor pairs of 30: 1 and 30, −1 and −30, 2 and 15, −2 and −15, 3 and 10, −3 and −10

The pair that adds to b is 2 and 15 since $2 + 15 = 17$, which is b, and $2 \times 15 = 30$, which is ac.

STEP 3 Rewrite the original polynomial by separating the b term into $2x$ and $15x$ (2 and 15 are the factors that we found in Step 2).

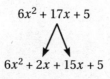

$$6x^2 + 17x + 5$$

$$6x^2 + 2x + 15x + 5$$

STEP 4 Regroup the first two terms into a binomial in parentheses plus the second two terms written as another binomial in parentheses.

$$(6x^2 + 2x) + (15x + 5)$$

STEP 5 Factor out the GCF from each binomial.

$$2x\left(\frac{6x^2}{2x}+\frac{2x}{2x}\right)+5\left(\frac{15x}{5}+\frac{5}{5}\right)=2x\left(\frac{\overset{3x}{\cancel{6x^2}}}{\underset{1}{\cancel{2x}}}+\frac{\overset{1}{\cancel{2x}}}{\underset{1}{\cancel{2x}}}\right)+5\left(\frac{\overset{3x}{\cancel{15x}}}{\underset{1}{\cancel{5}}}+\frac{\overset{1}{\cancel{5}}}{\underset{1}{\cancel{5}}}\right)$$

$$2x(3x+1)+5(3x+1)$$

STEP 6 Factor out the common binomial $(3x + 1)$, and regroup the terms.

$$(3x+1)\left(\frac{2x\overset{1}{\cancel{(3x+1)}}}{\underset{1}{\cancel{(3x+1)}}}+\frac{5\overset{1}{\cancel{(3x+1)}}}{\underset{1}{\cancel{(3x+1)}}}\right)$$

$$(3x + 1)(2x + 5)$$

STEP 7 Check by multiplying the two binomial factors.

Answer: $(3x + 1)(2x + 5) = 6x^2 + 17x + 5$

➡ Example 2

Factor $3x^2 - 10x - 8$.

Solution

STEP 1 Identify a, b, c, and ac.

$$a = 3, \ b = -10, \ c = -8, \text{ and } ac = (3)(-8) = -24$$

STEP 2 Identify the factor pair of ac that adds to b.

Factor pairs of –24: –1 and 24, 1 and –24, –2 and 12, 2
and –12, –3 and 8, 3 and –8, –4 and 6, 4 and –6

The pair that adds to b is 2 and –12 since $2 + (-12) = -10$, which is b.

STEP 3 Rewrite the original polynomial by separating the b term into $-12x$ and $2x$ (–12 and 2 are the factors that we found in Step 2).

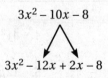

$$3x^2 - 10x - 8$$

$$3x^2 - 12x + 2x - 8$$

STEP 4 Regroup the first two terms into a binomial in parentheses plus the second two terms written as another binomial in parentheses.

$$(3x^2 - 12x) + (2x - 8)$$

STEP 5 Factor out the GCF from each binomial.

$$3x\left(\frac{3x^2}{3x} - \frac{12x}{3x}\right) + 2\left(\frac{2x}{2} - \frac{8}{2}\right) = 3x\left(\frac{\overset{1x}{\cancel{3x^2}}}{\underset{1}{\cancel{3x}}} - \frac{\overset{4}{\cancel{12x}}}{\underset{1}{\cancel{3x}}}\right) + 2\left(\frac{\overset{1x}{\cancel{2x}}}{\underset{1}{\cancel{2}}} - \frac{\overset{4}{\cancel{8}}}{\underset{1}{\cancel{2}}}\right)$$

$$3x(x-4) + 2(x-4)$$

STEP 6 Factor out the common binomial $(x-4)$, and regroup the terms.

$$(x-4)\left(\frac{3x(x-4)}{(x-4)} + \frac{2(x-4)}{(x-4)}\right) = (x-4)\left(\frac{3x\overset{1}{\cancel{(x-4)}}}{\underset{1}{\cancel{(x-4)}}} + \frac{2\overset{1}{\cancel{(x-4)}}}{\underset{1}{\cancel{(x-4)}}}\right)$$

$$(x-4)(3x+2)$$

STEP 7 Check by multiplying the two binomial factors.

Answer: $(x-4)(3x+2) = 3x^2 - 10x - 8$

ACCUPLACER TIPS

Factoring trinomials where $a \neq 1$ can be difficult and sometimes frustrating. Your math has to be extremely precise, and sometimes, when you sit down to take a test like the ACCUPLACER, you freeze up. Remember that working backwards from the answer choices is always a great way to solve problems like these. Let's say that you are tasked with factoring the trinomial $12x^2 - 11x - 15$. Only one of the answer choices will contain factors that, when you multiply them together, bring you back to the original trinomial. If you get stuck factoring, regroup, and try working backwards!

LESSON 13.3—SKILLS CHECK

Directions: Completely factor each polynomial.

1. $2x^2 + 13x + 6$

2. $2m^2 + 11m - 6$

3. $5x^2 + 13x - 6$

4. $4y^2 - 9$

5. $12p^2 - 5p - 2$

(Answers are on page 315.)

LESSON 13.3—ACCUPLACER CHALLENGE

Directions: For each of the questions below, choose the best answer from the four choices given.

1. $2x^2 + 9x + 10 =$

   (A) $(2x + 5)(x + 2)$
   (B) $(2x + 5)(x - 2)$
   (C) $(2x - 5)(x + 2)$
   (D) $(2x - 5)(x - 2)$

2. $p(3p + 4) - 2(3p + 4) =$

   (A) $(3p + 4)(p - 2)$
   (B) $(3p + 4)(p + 2)$
   (C) $(3p - 4)(p - 2)$
   (D) $(3p - 4)(p + 2)$

3. All of the polynomials below can be factored into two binomials EXCEPT

   (A) $2x^2 + 7x + 3$
   (B) $2x^2 + 5x + 3$
   (C) $2x^2 - x - 3$
   (D) $2x^2 - x + 3$

4. Factored completely, the expression $6x^4 + 10x^3 - 4x^2$ is equivalent to

   (A) $2(3x^2 - x)(x^2 + 2x)$
   (B) $x^2(6x - 2)(x + 2)$
   (C) $2x^2(3x + 1)(x - 2)$
   (D) $2x^2(3x - 1)(x + 2)$

5. $3x^2 - 75 =$

   (A) $(3x - 25)(3x + 25)$
   (B) $3(x - 5)(x + 5)$
   (C) $3x(x - 25)$
   (D) $(x + 15)(x - 15)$

(Answers are on page 317.)

To complete a set of practice questions that reviews all of Chapter 13, go to:
barronsbooks.com/TP/accuplacer/math29qw/

Lesson 13.1—Skills Check (page 299)

1. $5x(3x^2 + 7x + 1)$

$$15x^3 + 35x^2 + 5x$$

Place the GCF, $5x$, outside of the parentheses, and then write all the terms from the polynomial over $5x$ inside of the parentheses.

$$5x\left(\frac{15x^3}{5x} + \frac{35x^2}{5x} + \frac{5x}{5x}\right)$$

Reduce all the terms inside the parentheses.

$$5x(3x^2 + 7x + 1)$$

Check using the distributive property.

$$5x(3x^2 + 7x + 1) = 15x^3 + 35x^2 + 5x$$

2. $3m^2(6m^4 + 3m^2 - 2)$

$$18m^6 + 9m^4 - 6m^2$$

Place the GCF, $3m^2$, outside of the parentheses, and then write all the terms from the polynomial over $3m^2$ inside of the parentheses.

$$3m^2\left(\frac{18m^6}{3m^2} + \frac{9m^4}{3m^2} - \frac{6m^2}{3m^2}\right)$$

Reduce all the terms inside the parentheses.

$$3m^2(6m^4 + 3m^2 - 2)$$

Check using the distributive property.

$$3m^2(6m^4 + 3m^2 - 2) = 18m^6 + 9m^4 - 6m^2$$

3. $-7h(2h + 1)$

$$-14h^2 - 7h$$

Place the GCF, $-7h$, outside of the parentheses, and then write all the terms from the polynomial over $-7h$ inside of the parentheses.

$$-7h\left(\frac{-14h^2}{-7h} - \frac{7h}{-7h}\right)$$

Reduce all the terms inside the parentheses.

$$-7h(2h + 1)$$

Check using the distributive property.

$$-7h(2h + 1) = -14h^2 - 7h$$

4. $6n^7(6n^{20} + 4n^8 + 1)$

$$36n^{27} + 24n^{15} + 6n^7$$

Place the GCF, $6n^7$, outside of the parentheses, and then write all the terms from the polynomial over $6n^7$ inside of the parentheses.

$$6n^7\left(\frac{36n^{27}}{6n^7} + \frac{24n^{15}}{6n^7} + \frac{6n^7}{6n^7}\right)$$

Reduce all the terms inside the parentheses.

$$6n^7(6n^{20} + 4n^8 + 1)$$

Check using the distributive property.

$$6n^7(6n^{20} + 4n^8 + 1) = 36n^{27} + 24n^{15} + 6n^7$$

5. $7k(k^2 - 5k + 10)$

$$7k^3 - 35k^2 + 70k$$

Place the GCF, $7k$, outside of the parentheses, and then write all the terms from the polynomial over $7k$ inside of the parentheses.

$$7k\left(\frac{7k^3}{7k} - \frac{35k^2}{7k} + \frac{70k}{7k}\right)$$

Reduce all the terms inside the parentheses.

$$7k(k^2 - 5k + 10)$$

Check using the distributive property.

$$7k(k^2 - 5k + 10) = 7k^3 - 35k^2 + 70k$$

Lesson 13.1—ACCUPLACER Challenge (page 300)

1. **(C)**

$$15x^2 + 25x$$

The GCF of both terms is $5x$. Write the GCF outside of the parentheses. Inside of the parentheses, write each term over the GCF.

$$5x\left(\frac{15x^2}{5x} + \frac{25x}{5x}\right)$$

Divide the terms inside the parentheses by the GCF.

$$5x(3x + 5)$$

2. **(C)** If $2x$ is a factor of $2x^2 - 8x$, then we need to divide the polynomial by $2x$ to determine what the other factor is.

$$\frac{2x^2}{2x} - \frac{8x}{2x} = x - 4$$

3. **(A)** If $13x^2y^2$ is a factor of $52x^4y^2 - 39x^2y^3$, then we must divide the polynomial by $13x^2y^2$ to determine what the other factor is.

$$\frac{52x^4y^2}{13x^2y^2} - \frac{39x^2y^3}{13x^2y^2} = 4x^2 - 3y$$

4. **(B)** Knowing the area of a rectangle and the width of the rectangle, we can find the length of the rectangle using the following formula:

$$\frac{\text{Area}}{\text{Width}} = \text{Length}$$

$$\frac{4x^3 - 8x^2 + 32x}{4x} =$$

$$\frac{4x^3}{4x} - \frac{8x^2}{4x} + \frac{32x}{4x} = x^2 - 2x + 8$$

5. **(C)** First, find the length of the other side of the rectangle using $\dfrac{\text{Area}}{\text{Width}} = \text{Length}$.

$$\frac{x^2 + 3x}{x} =$$

$$\frac{x^2}{x} + \frac{3x}{x} = x + 3$$

To find the perimeter, add the length of the two sides that each measure x feet to the two sides that each measure $(x + 3)$ feet.

$$\text{Perimeter} = 2(x + 3) + 2(x)$$

$$\text{Perimeter} = 2x + 6 + 2x$$

$$\text{Perimeter} = (4x + 6) \text{ feet}$$

Lesson 13.2—Skills Check (page 304)

1. $(k + 1)(k + 8)$

$$k^2 + 9k + 8$$

After listing the factor pairs of 8, we find that 1 and 8 are the factors that multiply to 8 and also add to 9.

$$(k + 1)(k + 8)$$

2. $(x + 5)(x + 7)$

$$x^2 + 12x + 35$$

After listing the factor pairs of 35, we find that 5 and 7 are the factors that multiply to 35 and also add to 12.

$$(x + 5)(x + 7)$$

3. $(x-3)(x-2)$

$$x^2 - 5x + 6$$

After listing the factor pairs of 6, we find that –3 and –2 are the factors that multiply to 6 and also add to –5.

$$(x-3)(x-2)$$

4. $(r+1)(r-11)$

$$r^2 - 10r - 11$$

After listing the factor pairs of –11, we find that 1 and –11 are the factors that multiply to –11 and also add to –10.

$$(r+1)(r-11)$$

5. $(x+5)(x-3)$

$$x^2 + 2x - 15$$

After listing the factor pairs of –15, we find that 5 and –3 are the factors that multiply to –15 and also add to 2.

$$(x+5)(x-3)$$

6. $(y+16)(y-3)$

$$y^2 + 13y - 48$$

After listing the factor pairs of –48, we find that 16 and –3 are the factors that multiply to –48 and also add to 13.

$$(y+16)(y-3)$$

7. $(x+3)(x-3)$

$$x^2 - 9$$

After listing the factor pairs of –9, we find that 3 and –3 are the factors that multiply to –9 and also add to 0.

$$(x+3)(x-3)$$

8. $(x+8)(x-8)$

$$x^2 - 64$$

After listing the factor pairs of –64, we find that 8 and –8 are the factors that multiply to –64 and also add to 0.

$$(x+8)(x-8)$$

9. $(p+4)(p+4)$ *or* $(p+4)^2$

$$p^2 + 8p + 16$$

After listing the factor pairs of 16, we find that 4 and 4 are the factors that multiply to 16 and also add to 8.

$$(p+4)(p+4)$$

Since both factors are the same, this can also be written as the square of $(p + 4)$.

$$(p + 4)^2$$

10. $(m - 7)(m - 7)$ or $(m - 7)^2$

$$m^2 - 14m + 49$$

After listing the factor pairs of 49, we find that –7 and –7 are the factors that multiply to 49 and also add to –14.

$$(m - 7)(m - 7)$$

Since both factors are the same, this can also be written as the square of $(m - 7)$.

$$(m - 7)^2$$

Lesson 13.2—ACCUPLACER Challenge (page 305)

1. **(D)**

$$p^2 + 5p - 24$$

The factor pair that multiples to –24 and adds to 5 is 8 and –3.

$$8 + {-3} = 5$$

$$8 \times {-3} = -24$$

Completing the binomials and checking by multiplication gives us:

$$(p + 8)(p - 3) = p^2 + 5p - 24$$

2. **(A)** To determine which polynomial $(x - 2)$ is a factor of, we need to separately factor each of the answer choices.

Factors of choice A: $\qquad x^2 + 2x - 8 = (x + 4)(x - 2)$

Factors of choice B: $\qquad x^2 + x - 2 = (x + 2)(x - 1)$

Factors of choice C: $\qquad x^2 - x - 6 = (x - 3)(x + 2)$

Factors of choice D: $\qquad x^2 - 2x - 24 = (x - 6)(x + 4)$

The only expression that contains the factor $(x - 2)$ is choice A.

3. **(B)** Multiplying $(x + a)(x - a)$ produces the expression $x^2 - a^2$. Since a is a whole number, then a^2 must be the square of a whole number. All of the answer choices contain the difference of two squares except choice B because there is no whole number that could fit $a^2 = 8$.

4. **(A)** To find the factor that both $x^2 + x - 2$ and $x^2 - 3x + 2$ share in common, we can separately factor both polynomials.

$$x^2 + x - 2 = (x + 2)(x - 1)$$

$$x^2 - 3x + 2 = (x - 2)(x - 1)$$

The only factor that both polynomials share in common is $(x - 1)$.

5. **(D)** Looking at the polynomial $x^2 - 11x + p$, we can determine that $a = 1$, $b = -11$, and c (i.e., p) is a positive number.

We know that one of the factors is $(x - 5)$, and we know that the second factor must fit $(x - n)$. The missing term, n, must have a minus sign in front of it because it must multiply with -5 to give us a positive p. Only a negative times a negative will produce a positive p. In other words, $-5 \times -n = p$.

We also know that $-5 + -n = -11$ because the middle term in the polynomial is $-11x$. That is, b is -11. This means that $-n = -6$ because $-5 + -6 = -11$. Knowing that, we can complete our factors.

$$(x - 5)(x - 6)$$

Multiplying them together, we get our full polynomial.

$$(x - 5)(x - 6) = x^2 - 11x + 30$$

Going back to the original problem of finding p in $x^2 - 11x + p$, we can say that $p = 30$.

Lesson 13.3—Skills Check (page 308)

1. **$(2x + 1)(x + 6)$**

$$2x^2 + 13x + 6$$

Identify a, b, c, and ac.

$$a = 2, b = 13, c = 6, \text{ and } ac = (2)(6) = 12$$

Identify the two factors of ac, 12, that add to b, 13.

$$1 + 12 = 13$$
$$1 \times 12 = 12$$

Rewrite the polynomial by separating the b term into $1x$ and $12x$.

$$2x^2 + x + 12x + 6$$

Regroup the terms into binomials.

$$(2x^2 + x) + (12x + 6)$$

Separately factor out the GCF of each binomial.

$$x(2x + 1) + 6(2x + 1)$$

Factor out the common binomial $(2x + 1)$.

$$(2x + 1)(x + 6)$$

2. **$(2m - 1)(m + 6)$**

$$2m^2 + 11m - 6$$

Identify a, b, c, and ac.

$$a = 2, b = 11, c = -6, \text{ and } ac = (2)(-6) = -12$$

Identify the two factors of ac, -12, that add to b, 11.

$$-1 + 12 = 11$$

$$-1 \times 12 = -12$$

Rewrite the polynomial by separating the b term into $-m$ and $12m$.

$$2m^2 - m + 12m - 6$$

Regroup the terms into binomials.

$$(2m^2 - m) + (12m - 6)$$

Separately factor out the GCF of each binomial.

$$m(2m - 1) + 6(2m - 1)$$

Factor out the common binomial.

$$(2m - 1)(m + 6)$$

3. **$(5x - 2)(x + 3)$**

$$5x^2 + 13x - 6$$

Identify a, b, c, and ac.

$$a = 5,\ b = 13,\ c = -6,\ \text{and}\ ac = (5)(-6) = -30$$

Identify the two factors of ac, -30, that add to b, 13.

$$15 + (-2) = 13$$

$$15 \times (-2) = -30$$

Rewrite the polynomial by separating the b term into $15x$ and $-2x$.

$$5x^2 + 15x - 2x - 6$$

Regroup the terms into binomials.

$$(5x^2 + 15x) + (-2x - 6)$$

Separately factor out the GCF of each binomial.

$$5x(x + 3) - 2(x + 3)$$

Factor out the common binomial $(x + 3)$.

$$(5x - 2)(x + 3)$$

4. **$(2y + 3)(2y - 3)$**

$$4y^2 - 9$$

The polynomial above fits the form of the special case $a^2 - b^2 = (a + b)(a - b)$, a difference of two squares. Knowing that, we can factor the binomial like so.

$$4y^2 - 9 = (2y + 3)(2y - 3)$$

5. $(4p + 1)(3p - 2)$

$$12p^2 - 5p - 2$$

Identify a, b, c, and ac.

$$a = 12, b = -5, c = -2, \text{ and } ac = (12)(-2) = -24$$

Identify the two factors of ac, –24, that add to b, –5.

$$3 + -8 = -5$$

$$3 \times (-8) = -24$$

Rewrite the polynomial by separating the b term into $3p$ and $-8p$.

$$12p^2 + 3p + -8p - 2$$

Regroup the terms into binomials.

$$(12p^2 + 3p) + (-8p - 2)$$

Separately factor out the GCF of each binomial.

$$3p(4p + 1) + -2(4p + 1)$$

Factor out the common binomial $(4p + 1)$.

$$(4p + 1)(3p - 2)$$

Lesson 13.3—ACCUPLACER Challenge (page 309)

1. **(A)**

$$2x^2 + 9x + 10$$

Identify a, b, c, and ac.

$$a = 2, b = 9, c = 10, \text{ and } ac = (2)(10) = 20$$

Identify the two factors of ac, 20, that add to b, 9.

$$4 + 5 = 9$$

$$4 \times 5 = 20$$

Rewrite the polynomial by separating the b term into $4x$ and $5x$.

$$2x^2 + 4x + 5x + 10$$

Regroup the terms into binomials.

$$(2x^2 + 4x) + (5x + 10)$$

Separately factor out the GCF of each binomial.

$$2x(x + 2) + 5(x + 2)$$

Factor out the common binomial $(x + 2)$.

$$(2x + 5)(x + 2)$$

2. **(A)**

$$p(3p + 4) - 2(3p + 4)$$

The easiest way to address this problem is to notice that $(3p + 4)$ is a common factor. That means we can factor out this common factor to find an equivalent expression. To do so, we divide out $(3p + 4)$.

$$(3p+4)\left(\frac{p\overset{1}{\cancel{(3p+4)}}}{\underset{1}{\cancel{(3p+4)}}} - \frac{2\overset{1}{\cancel{(3p+4)}}}{\underset{1}{\cancel{(3p+4)}}} \right)$$

$$(3p + 4)(p - 2)$$

3. **(D)** All of the polynomials listed in the answer choices can be factored into two binomials *except* for $2x^2 - x + 3$. This polynomial cannot be factored because there are no two numbers that can multiply to the product of ac and also add to b. That is, $ac = (2)(3) = 6$, and $b = -1$. The factor pairs of 6 are 1 and 6, -1 and -6, 2 and 3, and -2 and -3. The sum of 1 and 6 is 7, the sum of -1 and -6 is -7, the sum of 2 and 3 is 5, and the sum of -2 and -3 is -5. There is no way to find two whole numbers that multiply to 6 and add to -1, so $2x^2 - x + 3$ cannot be factored.

4. **(D)** The first step in factoring $6x^4 + 10x^3 - 4x^2$ is to factor out the common factor, $2x^2$, that all three terms share in common.

$$2x^2\left(\frac{6x^4}{2x^2} + \frac{10x^3}{2x^2} - \frac{4x^2}{2x^2} \right)$$

$$2x^2\left(3x^2 + 5x - 2 \right)$$

Next, separately factor the trinomial $3x^2 + 5x - 2$ into the two binomials $(3x - 1)(x + 2)$. The complete factorization is $2x^2(3x - 1)(x + 2)$.

5. **(B)** Both $3x^2$ and -75 are products of the factor 3, so 3 can be factored out of the binomial first.

$$3\left(\frac{3x^2}{3} - \frac{75}{3} \right)$$

$$3\left(x^2 - 25 \right)$$

After factoring out 3, we are left with the difference of two squares $(x^2 - 25)$, which can be factored to $(x - 5)(x + 5)$. The complete factorization is then $3(x - 5)(x + 5)$.

Rational Expressions

14

Previously, we defined a **ratio** as a comparison of two quantities that is usually expressed as a fraction. In this chapter, we will now look at **rational expressions**, which is a comparison of two expressions expressed as a fraction. For example, $\dfrac{2x^2 + 4x}{2x}$ is a rational expression.

When we use fractions, we often express them in simplified form by reducing the numerator and the denominator by the greatest common factor (GCF). We do the same for rational expressions. Consider the following: If $\dfrac{2x}{2x} = 1$, then what is $\dfrac{x+1}{x+1}$ equal to? Furthermore, what is $\dfrac{2(x+1)}{x+1}$ equal to? This chapter will provide you with the tools to answer these types of questions and more.

> In this chapter, we will work to master
>
> - simplifying rational expressions, and
> - adding and subtracting rational expressions.

LESSON 14.1—SIMPLIFYING RATIONAL EXPRESSIONS

It's important to go back to the fundamentals of reducing fractions in order to proceed with simplifying rational expressions.

Fractions represent division and, generally, any quantity divided by itself equals 1. Here are some examples.

$$\frac{3}{3} = 1$$

$$\frac{-297}{-297} = 1$$

$$\frac{x}{x} = 1$$

$$\frac{x+1}{x+1} = 1$$

Knowing that, when we find a common factor in the numerator and the denominator of a rational expression, we can divide those common factors by one another and thus turn them into 1. Effectively, the common factors cancel out one another.

For example,

$$\frac{6(x+1)}{x+1} = \frac{6\overset{1}{\cancel{(x+1)}}}{\underset{1}{\cancel{x+1}}} = \frac{6(1)}{1} = 6$$

Of course, it often won't be quite so easy. In most problems, you will have to factor a polynomial in order to uncover a factor common to both the numerator and the denominator.

To Simplify Rational Expressions:

1. Separately factor the numerator and the denominator, if possible.
2. Turn common factors, found in both the numerator and the denominator, to 1 by division.
3. Further simplify, if possible.

➡ Example 1

Simplify $\frac{4(y+3)}{(y+3)}$.

Solution

In this example, we already have $(y+3)$ common to both the numerator and the denominator without having to factor anything, so we can start by dividing those by one another.

$$\frac{4\overset{1}{\cancel{(y+3)}}}{\underset{1}{\cancel{(y+3)}}} = \frac{4(1)}{1} = 4$$

Answer: In the end, we are left with 4, so $\frac{4(y+3)}{(y+3)} = 4$.

➡ Example 2

Simplify $\frac{5x+15}{x+3}$.

Solution

In this example, we need to start by factoring out the 5 in the numerator.

$$\frac{5x+15}{x+3} = \frac{5(x+3)}{x+3}$$

Then, we find that both the numerator and the denominator contain the common factor $(x+3)$, so we can divide those.

$$\frac{5\overset{1}{\cancel{(x+3)}}}{\underset{1}{\cancel{x+3}}} = \frac{5(1)}{1} = 5$$

Answer: In the end, we are left with 5, so $\frac{5x+15}{x+3} = 5$.

➡ Example 3

Simplify $\dfrac{3x^6 - 9x^4}{3x^6 - 6x^4}$.

Solution

It's important to note here that we cannot start out by dividing $3x^6$ by $3x^6$. That is a common mistake. We can only divide by factors that are shared by all the terms, and $3x^6$ is not a factor of $-9x^4$. Instead, we need to start out by factoring the numerator and the denominator.

$$\frac{3x^6 - 9x^4}{3x^6 - 6x^4} = \frac{3x^4(x^2 - 3)}{3x^4(x^2 - 2)}$$

After factoring the numerator and the denominator, we can cancel out the common factor, $3x^4$.

$$\frac{\overset{1}{\cancel{3x^4}}(x^2 - 3)}{\underset{1}{\cancel{3x^4}}(x^2 - 2)} = \frac{x^2 - 3}{x^2 - 2}$$

Answer: In the end, we are left with $\dfrac{x^2 - 3}{x^2 - 2}$.

➡ Example 4

If $y > -3$, then $\dfrac{y^2 + 5y + 6}{y^2 + 7y + 12} =$

(A) $\dfrac{1}{2}$

(B) $\dfrac{y + 2}{y + 4}$

(C) $\dfrac{y + 1}{y + 2}$

(D) $\dfrac{11y^3}{20y^3}$

Solution

First, it's important to note that for one reason or another, problems such as this example are commonly used in college algebra classes so there is a good chance you will find a question similar to this on the Elementary Algebra test or the College-Level Mathematics test.

Second, it's also important to note that rational expression questions often start out with, or include, statements like "If $y > -3$" or "If $x < 2$" or "If $a \neq 5$." When simplifying expressions, especially rational expressions, we sometimes encounter situations where the simplified expression is equivalent in almost every situation, but not every situation. To avoid these inconsistencies, there are restrictions on the values that variables can be. That is what statements like "If $y > -3$" are for. These inconsistencies are not of particular importance at this point. What is important is that those statements really don't affect how you simplify these rational expressions on the ACCUPLACER. In fact, you can pretty much ignore these statements for all of the problems you see in this chapter.

In this example, at first glance, it does not appear that the numerator and the denominator will have common factors. To be sure, we need to factor the numerator and the denominator first.

$$\frac{y^2 + 5y + 6}{y^2 + 7y + 12} = \frac{(y+3)(y+2)}{(y+3)(y+4)}$$

After factoring, we can see that both the numerator and the denominator include the common factor $(y + 3)$. Therefore, we proceed by dividing the common factor by itself.

$$\frac{\overset{1}{\cancel{(y+3)}}(y+2)}{\underset{1}{\cancel{(y+3)}}(y+4)} = \frac{(y+2)}{(y+4)}$$

<u>Answer:</u> The simplified rational expression is $\frac{(y+2)}{(y+4)}$, choice B.

LESSON 14.1—SKILLS CHECK

Directions: Simplify the following rational expressions.

1. $\dfrac{3(x+3)}{x+3}$

2. $\dfrac{12x - 8}{4x + 8}$

3. $\dfrac{25p^2 - 5p^2}{5p^2}$

4. $\dfrac{30m^4 - 24m^3}{6m}$

5. $\dfrac{4k^5}{16k^2 - 12k}$

6. $\dfrac{8p^8 - 32p^6}{2p^3}$

7. $\dfrac{h+3}{h^2 - 9}$

8. $\dfrac{x^2 + 2x - 3}{x - 1}$

9. $\dfrac{2x^2 + 11x + 12}{x + 4}$

10. $\dfrac{2x^2 + 13x + 6}{2x^2 + 9x - 18}$

(Answers are on page 328.)

LESSON 14.1—ACCUPLACER CHALLENGE

> **Directions:** For each of the questions below, choose the best answer from the four choices given.

1. $\dfrac{12p^8 + 8p^4}{4p^4} =$

   (A) $5p^3$
   (B) $5p^5$
   (C) $3p^4 + 2$
   (D) $3p^4 + 2p$

2. $\dfrac{12m^5 - 4m^4}{2m^3} =$

   (A) $\dfrac{4}{m^2}$
   (B) $6m^2 - 2m$
   (C) $2m(3m + 1)$
   (D) $12m^5 - 2m$

3. If $x > 3$, then $\dfrac{x^2 + 3x - 10}{x^2 + 2x - 15} =$

   (A) $\dfrac{3x - 10}{2x - 15}$
   (B) $\dfrac{x + 2}{x + 3}$
   (C) $\dfrac{x - 2}{x - 3}$
   (D) $\dfrac{7}{13}$

4. If $p > 4$, then $\dfrac{p^2 + 2p - 8}{p^2 - 16} =$

   (A) $\dfrac{p - 1}{p - 2}$
   (B) $\dfrac{p - 2}{p - 4}$
   (C) p
   (D) $p + 2p - 2$

5. If $x \neq 3$, then $\dfrac{\left(x^2 - 9\right) + (x + 3)}{x^2 + 6x + 9} =$

   (A) $x - 2$
   (B) $x - 3$
   (C) $\dfrac{x + 3}{6x}$
   (D) $\dfrac{x - 2}{x + 3}$

(Answers are on page 329.)

LESSON 14.2—ADDING AND SUBTRACTING RATIONAL EXPRESSIONS

Lesson 14.2A—Adding and Subtracting Rational Expressions with Like Denominators

To add and subtract rational expressions, we have to follow all the rules we learned about working with fractions as well as all the rules we learned about adding and subtracting polynomials. Most importantly, we can only combine fractions with the same denominators, and we can only combine terms with the same variables to the same powers.

To Combine Rational Expressions with Like Denominators:

1. Add or subtract the numerators, and keep the common denominator.
2. Reduce when possible.

➡ Example 1

If $a \neq 0$, then $\dfrac{4}{a} + \dfrac{3}{a} = ?$

Solution

In this example, we have a like denominator: a. Therefore, we add the 4 and the 3 in the numerator, and keep a as the common denominator.

$$\frac{4}{a} + \frac{3}{a} = \frac{4+3}{a} = \frac{7}{a}$$

Answer: $\dfrac{7}{a}$

➡ Example 2

If $b \neq 0$, then $\dfrac{5}{2b} + \dfrac{1}{2b} = ?$

Solution

Here we add $5 + 1$ in the numerator, and we keep $2b$ as our common denominator. Then we have to reduce because 6 in the numerator and $2b$ in the denominator contain the common factor of 2. Therefore, we divide both the numerator and the denominator by 2.

$$\frac{5}{2b} + \frac{1}{2b} = \frac{5+1}{2b} = \frac{6}{2b} = \frac{\overset{3}{\cancel{6}}}{\underset{1}{\cancel{2}b}} = \frac{3}{b}$$

Answer: $\dfrac{3}{b}$

LESSON 14.2A—SKILLS CHECK

> **Directions:** Simplify the following rational expressions.

1. $\dfrac{b}{2} + \dfrac{b}{2}$

2. $\dfrac{2}{m} + \dfrac{2}{m}$

3. $\dfrac{11y}{3x} - \dfrac{2y}{3x}$

4. $\dfrac{3p+1}{2p} - \dfrac{10p-7}{2p}$

5. $\dfrac{5x+9}{x+2} + \dfrac{x+4}{x+2} - \dfrac{x+3}{x+2}$

(Answers are on page 331.)

Lesson 14.2B—Adding and Subtracting Rational Expressions with Unlike Denominators

Just as when we work with any other fraction, we can only combine fractions with like denominators. Therefore, before we add or subtract fractions with unlike denominators, we must change the fractions to equivalent fractions with common denominators.

To Combine Rational Expressions with Unlike Denominators:

1. Change all of the fractions to equivalent fractions with common denominators.
2. Add or subtract the numerators, and keep the common denominator.
3. Reduce when possible.

➡ Example 1

If $x \neq 0$, then $\dfrac{3}{x} + \dfrac{4}{5x} - \dfrac{1}{x} = ?$

(A) $\dfrac{6}{x}$

(B) $\dfrac{6}{5x}$

(C) $\dfrac{8}{5x}$

(D) $\dfrac{14}{5x}$

Solution

First, consider what $x \neq 0$ means. Put simply, x is not equal to 0 because a variable, in this case x, cannot be 0 if it results in a denominator of 0. Fractions are another way of writing division, and dividing by 0 is not allowed in mathematics as it is considered undefined. Do not do a problem such as this any differently whether or not the question states $x \neq 0$. Basically, you can ignore that statement.

Now, let's start the problem by changing each fraction to an equivalent fraction with a common denominator. Since we have only two different denominators to work with, x and $5x$, we can use $5x$ as a common denominator.

$$\frac{3 \bullet 5}{x \bullet 5} + \frac{4}{5x} - \frac{1 \bullet 5}{x \bullet 5} = \frac{15}{5x} + \frac{4}{5x} - \frac{5}{5x}$$

Finally, combine the numerators, and keep the common denominator.

$$\frac{15}{5x} + \frac{4}{5x} - \frac{5}{5x} = \frac{19}{5x} - \frac{5}{5x} = \frac{14}{5x}$$

Answer: Therefore, the correct answer is choice D.

➡ Example 2

Simplify $\dfrac{2}{3b} + \dfrac{5}{3a}$.

Solution

Before we can add these rational expressions together, we need to change the fractions to equivalent fractions with common denominators.

$$\frac{2 \bullet a}{3b \bullet a} + \frac{5 \bullet b}{3a \bullet b} = \frac{2a}{3ab} + \frac{5b}{3ab}$$

Next, we add the numerators, and keep the common denominator.

$$\frac{2a}{3ab} + \frac{5b}{3ab} = \frac{2a + 5b}{3ab}$$

In the numerator, $2a$ cannot be added to $5b$, so the final simplified rational expression is $\dfrac{2a + 5b}{3ab}$.

Answer: $\dfrac{2a + 5b}{3ab}$

LESSON 14.2B—SKILLS CHECK

Directions: Simplify each rational expression.

1. $\dfrac{2}{x} + \dfrac{1}{3x}$

2. $\dfrac{5}{2y} - \dfrac{3}{4y}$

3. $\dfrac{7}{n} + \dfrac{2}{3n} - \dfrac{2}{15n}$

4. $\dfrac{a}{b} + \dfrac{b}{a}$

5. $\dfrac{3d}{c} - \dfrac{d}{3c} + \dfrac{5d}{c}$

(Answers are on page 331.)

LESSON 14.2—ACCUPLACER CHALLENGE

Directions: For each of the questions below, choose the best answer from the four choices given.

1. If $p \neq 0$, then $\dfrac{8}{3p} + \dfrac{4}{3p} =$

   (A) $\dfrac{12}{9p}$

   (B) $\dfrac{4}{p}$

   (C) $\dfrac{12}{6p}$

   (D) $\dfrac{2}{p}$

4. If $m \neq 0$, then $\dfrac{4}{m} + \dfrac{2}{m^2} =$

   (A) $\dfrac{6}{m^2}$

   (B) $\dfrac{6}{m^3}$

   (C) $\dfrac{2m+4}{m^2}$

   (D) $\dfrac{4m+2}{m^2}$

2. If $x \neq -4$, then $\dfrac{4x-1}{x+4} - \dfrac{2x-5}{x+4} =$

   (A) $\dfrac{2x+4}{x+4}$

   (B) $\dfrac{2x+1}{x+1}$

   (C) $\dfrac{2x-6}{x+4}$

   (D) $\dfrac{2x-4}{x+4}$

5. If $a > -1$, then $\dfrac{5}{a+1} + \dfrac{3}{a+2} =$

   (A) $\dfrac{8}{2a+3}$

   (B) $\dfrac{9}{a+2}$

   (C) $\dfrac{8a+13}{a^2+3a+2}$

   (D) $\dfrac{8a+24}{a^2+3a+2}$

3. If $x \neq 0$, then $\dfrac{y}{x} + \dfrac{4y}{x} - \dfrac{2y}{4x} =$

   (A) $\dfrac{11y}{2x}$

   (B) $\dfrac{3y}{2x}$

   (C) $\dfrac{9y}{2x}$

   (D) $-\dfrac{3y}{2x}$

(Answers are on page 332.)

> To complete a set of practice questions that reviews all of Chapter 14,
> and to complete a set of practice questions that covers all of Unit 2, go to:
> *barronsbooks.com/TP/accuplacer/math29qw/*

Lesson 14.1—Skills Check (page 322)

1. **3** Divide the numerator and the denominator by $(x + 3)$, their common factor. Then simplify the expression.

$$\frac{3(x+3)}{x+3} = \frac{3\overset{1}{\cancel{(x+3)}}}{\underset{1}{\cancel{x+3}}} = 3$$

2. $\dfrac{3x-2}{x+2}$ Factor the numerator and the denominator. Then divide the numerator and the denominator by 4, their common factor. Finally, simplify the expression.

$$\frac{12x-8}{4x+8} = \frac{4(3x-2)}{4(x+2)} = \frac{\overset{1}{\cancel{4}}(3x-2)}{\underset{1}{\cancel{4}}(x+2)} = \frac{3x-2}{x+2}$$

3. **4** The numerator here contains like terms, so we can first combine the like terms. Then divide $20p^2$ by $5p^2$. Finally, simplify the expression.

$$\frac{25p^2-5p^2}{5p^2} = \frac{20p^2}{5p^2} = \frac{\overset{4}{\overset{\cancel{}}{\cancel{20p^2}}}}{\underset{1}{\underset{\cancel{}}{\cancel{5p^2}}}} \overset{1}{\underset{1}{}} = \frac{4}{1} = 4$$

4. $5m^3 - 4m^2$ The denominator here is $6m$, and the largest factor of any quantity is itself, in this case $6m$. This also happens to be a factor common to all the terms in the numerator. Therefore, we can start by factoring $6m$ from the numerator. Then we can divide both the numerator and the denominator by $6m$, their common factor.

$$\frac{30m^4-24m^3}{6m} = \frac{6m\left(5m^3-4m^2\right)}{6m} = \frac{\overset{1}{\cancel{6m}}\left(5m^3-4m^2\right)}{\underset{1}{\cancel{6m}}} = 5m^3 - 4m^2$$

5. $\dfrac{k^4}{4k-3}$ Factor the denominator. Then divide the numerator and the denominator by $4k$, their common factor.

$$\frac{4k^5}{16k^2-12k} = \frac{4k^5}{4k(4k-3)} = \frac{\overset{1k^4}{\cancel{4k^5}}}{\underset{1}{\cancel{4k}}(4k-3)} = \frac{k^4}{4k-3}$$

6. $4p^5 - 16p^3$ Factor the numerator. Then divide the numerator and the denominator by $2p^3$, their common factor. Finally, simplify the expression.

$$\frac{8p^8 - 32p^6}{2p^3} = \frac{2p^3\left(4p^5 - 16p^3\right)}{2p^3} = \frac{\overset{1}{2p^3}\left(4p^5 - 16p^3\right)}{\underset{1}{2p^3}} = 4p^5 - 16p^3$$

7. $\dfrac{1}{h-3}$ Factor the denominator. Then divide the numerator and the denominator by $h+3$, their common factor. Finally, simplify the expression.

$$\frac{h+3}{h^2 - 9} = \frac{h+3}{(h+3)(h-3)} = \frac{\overset{1}{\cancel{h+3}}}{\cancel{(h+3)}(h-3)} = \frac{1}{h-3}$$

8. $x+3$ Factor the numerator. Then divide the numerator and the denominator by $x-1$, their common factor. Finally, simplify the expression.

$$\frac{x^2 + 2x - 3}{x - 1} = \frac{(x-1)(x+3)}{x-1} = \frac{\overset{1}{\cancel{(x-1)}}(x+3)}{\underset{1}{\cancel{x-1}}} = x+3$$

9. $2x+3$ Factor the numerator. Then divide the numerator and the denominator by $x+4$, their common factor. Finally, simplify the expression.

$$\frac{2x^2 + 11x + 12}{x+4} = \frac{(2x+3)(x+4)}{(x+4)} = \frac{\overset{1}{(2x+3)\cancel{(x+4)}}}{\underset{1}{\cancel{(x+4)}}} = 2x+3$$

10. $\dfrac{2x+1}{2x-3}$ Factor the numerator and the denominator. Then divide the numerator and the denominator by $x+6$, their common factor. Finally, simplify the expression.

$$\frac{2x^2 + 13x + 6}{2x^2 + 9x - 18} = \frac{(2x+1)(x+6)}{(2x-3)(x+6)} = \frac{(2x+1)\overset{1}{\cancel{(x+6)}}}{(2x-3)\underset{1}{\cancel{(x+6)}}} = \frac{2x+1}{2x-3}$$

Lesson 14.1—ACCUPLACER Challenge (page 323)

1. **(C)** Factor the numerator. Then divide the numerator and the denominator by $4p^4$, the common factor.

$$\frac{12p^8 + 8p^4}{4p^4} = \frac{4p^4\left(3p^4 + 2\right)}{4p^4} = \frac{\overset{1}{\cancel{4p^4}}\left(3p^4 + 2\right)}{\underset{1}{\cancel{4p^4}}} = 3p^4 + 2$$

We're left with $3p^4 + 2$, so the correct answer is choice C.

2. **(B)** To divide the numerator by the denominator, $2m^3$, we have to divide both terms by $2m^3$. Start by factoring $2m^3$ out of the numerator. Then divide both the numerator and the denominator by $2m^3$, the common factor.

$$\frac{12m^5 - 4m^4}{2m^3} = \frac{2m^3\left(6m^2 - 2m\right)}{2m^3} = \frac{\overset{1}{\cancel{2m^3}}\left(6m^2 - 2m\right)}{\underset{1}{\cancel{2m^3}}} = 6m^2 - 2m$$

We're left with $6m^2 - 2m$, so the correct answer is choice B.

3. **(C)** Factor both the numerator and the denominator. Then divide both the numerator and the denominator by $x + 5$, the common factor.

$$\frac{x^2 + 3x - 10}{x^2 + 2x - 15} = \frac{(x+5)(x-2)}{(x+5)(x-3)} = \frac{\overset{1}{\cancel{(x+5)}}(x-2)}{\underset{1}{\cancel{(x+5)}}(x-3)} = \frac{x-2}{x-3}$$

We're left with $\dfrac{x-2}{x-3}$, so the correct answer is choice C.

4. **(B)** Factor the numerator and the denominator. Then divide the numerator and the denominator by $p + 4$, the common factor.

$$\frac{p^2 + 2p - 8}{p^2 - 16} = \frac{(p+4)(p-2)}{(p+4)(p-4)} = \frac{\overset{1}{\cancel{(p+4)}}(p-2)}{\underset{1}{\cancel{(p+4)}}(p-4)} = \frac{p-2}{p-4}$$

We're left with $\dfrac{p-2}{p-4}$, so the correct answer is choice B.

5. **(D)** We could start by factoring the numerator and the denominator. Another option is to combine the two binomials in the numerator first. The parentheses here do not change the problem so we can start by removing them and then combining like terms.

$$\frac{\left(x^2 - 9\right) + (x+3)}{x^2 + 6x + 9} = \frac{x^2 - 9 + x + 3}{x^2 + 6x + 9} = \frac{x^2 + x - 6}{x^2 + 6x + 9}$$

Next, factor the numerator and the denominator. Then divide both the numerator and the denominator by $x + 3$, the common factor.

$$\frac{x^2 + x - 6}{x^2 + 6x + 9} = \frac{(x+3)(x-2)}{(x+3)(x+3)} = \frac{\overset{1}{\cancel{(x+3)}}(x-2)}{\underset{1}{\cancel{(x+3)}}(x+3)} = \frac{x-2}{x+3}$$

We're left with $\dfrac{x-2}{x+3}$, so the correct answer is choice D.

Lesson 14.2A—Skills Check (page 324)

1. **b** Add the numerators, and keep the common denominator. Then simplify the expression.

$$\frac{b}{2} + \frac{b}{2} = \frac{2b}{2} = \frac{\overset{1}{\cancel{2}b}}{\cancel{2}} = \frac{b}{1} = b$$

2. $\dfrac{4}{m}$ Add the numerators, and keep the common denominator.

$$\frac{2}{m} + \frac{2}{m} = \frac{4}{m}$$

3. $\dfrac{3y}{x}$ Subtract the numerators, and keep the common denominator. Then simplify the expression.

$$\frac{11y}{3x} - \frac{2y}{3x} = \frac{9y}{3x} = \frac{\overset{3}{\cancel{9}y}}{\cancel{3}x} = \frac{3y}{x}$$

4. $\dfrac{-7p+8}{2p}$ Subtract the numerators, and keep the common denominator. Then simplify the expression.

$$\frac{3p+1}{2p} - \frac{10p-7}{2p} = \frac{3p - 10p + 1 - (-7)}{2p} = \frac{-7p+8}{2p}$$

5. **5** Combine the numerators, and keep the common denominator.

$$\frac{5x+9}{x+2} + \frac{x+4}{x+2} - \frac{x+3}{x+2} = \frac{5x + x - x + 9 + 4 - 3}{x+2} = \frac{5x+10}{x+2}$$

Then simplify the expression by first factoring the numerator and then dividing both the numerator and the denominator by $x + 2$, their common factor.

$$\frac{5x+10}{x+2} = \frac{5(x+2)}{x+2} = \frac{5\overset{1}{\cancel{(x+2)}}}{\underset{1}{\cancel{x+2}}} = 5$$

Lesson 14.2B—Skills Check (page 326)

1. $\dfrac{7}{3x}$ Change all the fractions to equivalent fractions with common denominators.

$$\frac{2}{x} + \frac{1}{3x} = \frac{2 \bullet 3}{x \bullet 3} + \frac{1}{3x} = \frac{6}{3x} + \frac{1}{3x}$$

Then, add the numerators, and keep the common denominator.

$$\frac{6}{3x} + \frac{1}{3x} = \frac{7}{3x}$$

2. $\dfrac{7}{4y}$ Change all the fractions to equivalent fractions with common denominators.

$$\frac{5}{2y} - \frac{3}{4y} = \frac{5 \cdot 2}{2y \cdot 2} - \frac{3}{4y} = \frac{10}{4y} - \frac{3}{4y}$$

Then, subtract the numerators, and keep the common denominator.

$$\frac{10}{4y} - \frac{3}{4y} = \frac{7}{4y}$$

3. $\dfrac{113}{15n}$ Change all the fractions to equivalent fractions with common denominators.

$$\frac{7}{n} + \frac{2}{3n} - \frac{2}{15n} = \frac{7 \cdot 15}{n \cdot 15} + \frac{2 \cdot 5}{3n \cdot 5} - \frac{2}{15n} = \frac{105}{15n} + \frac{10}{15n} - \frac{2}{15n}$$

Then, combine the numerators, and keep the common denominator.

$$\frac{105}{15n} + \frac{10}{15n} - \frac{2}{15n} = \frac{113}{15n}$$

4. $\dfrac{a^2 + b^2}{ab}$ Change all the fractions to equivalent fractions with common denominators.

$$\frac{a}{b} + \frac{b}{a} = \frac{a \cdot a}{b \cdot a} + \frac{b \cdot b}{a \cdot b} = \frac{a^2}{ab} + \frac{b^2}{ab}$$

Then, add the numerators, and keep the common denominator.

$$\frac{a^2}{ab} + \frac{b^2}{ab} = \frac{a^2 + b^2}{ab}$$

5. $\dfrac{23d}{3c}$ Change all the fractions to equivalent fractions with common denominators.

$$\frac{3d}{c} - \frac{d}{3c} + \frac{5d}{c} = \frac{3d \cdot 3}{c \cdot 3} - \frac{d}{3c} + \frac{5d \cdot 3}{c \cdot 3} = \frac{9d}{3c} - \frac{d}{3c} + \frac{15d}{3c}$$

Then, combine the numerators, and keep the common denominator.

$$\frac{9d}{3c} - \frac{d}{3c} + \frac{15d}{3c} = \frac{23d}{3c}$$

Lesson 14.2—ACCUPLACER Challenge (page 327)

1. **(B)** Add the numerators, and keep the common denominator. Then reduce.

$$\frac{8}{3p} + \frac{4}{3p} = \frac{12}{3p} = \frac{\overset{4}{\cancel{12}}}{\underset{1}{\cancel{3}}p} = \frac{4}{p}$$

2. **(A)** Subtract the numerators, and keep the common denominator.

$$\frac{4x-1}{x+4} - \frac{2x-5}{x+4} = \frac{4x-2x-1-(-5)}{x+4} = \frac{2x+4}{x+4}$$

3. **(C)** Change all the fractions to equivalent fractions with common denominators.

$$\frac{y}{x} + \frac{4y}{x} - \frac{2y}{4x} = \frac{y \cdot 4}{x \cdot 4} + \frac{4y \cdot 4}{x \cdot 4} - \frac{2y}{4x} = \frac{4y}{4x} + \frac{16y}{4x} - \frac{2y}{4x}$$

Combine the numerators, and keep the common denominator. Then reduce.

$$\frac{4y}{4x} + \frac{16y}{4x} - \frac{2y}{4x} = \frac{18y}{4x} = \frac{\overset{9}{\cancel{18}}y}{\underset{2}{\cancel{4}}x} = \frac{9y}{2x}$$

4. **(D)** Change all the fractions to equivalent fractions with common denominators.

$$\frac{4}{m} + \frac{2}{m^2} = \frac{4 \cdot m}{m \cdot m} + \frac{2}{m^2} = \frac{4m}{m^2} + \frac{2}{m^2}$$

Then, add the numerators, and keep the common denominator.

$$\frac{4m}{m^2} + \frac{2}{m^2} = \frac{4m+2}{m^2}$$

5. **(C)** Change all the fractions to equivalent fractions with common denominators.

$$\frac{5}{a+1} + \frac{3}{a+2} = \frac{5(a+2)}{(a+1)(a+2)} + \frac{3(a+1)}{(a+1)(a+2)} = \frac{5a+10}{a^2+3a+2} + \frac{3a+3}{a^2+3a+2}$$

Then, add the numerators, and keep the common denominator.

$$\frac{5a+10}{a^2+3a+2} + \frac{3a+3}{a^2+3a+2} = \frac{5a+3a+10+3}{a^2+3a+2} = \frac{8a+13}{a^2+3a+2}$$

UNIT THREE
College-Level Mathematics

Graphs in the Coordinate Plane

15

On the ACCUPLACER College-Level Mathematics test, you should expect to see problems that ask you to interpret graphs and understand the key features of linear graphs in particular. To make sure that you're successful when solving these problems, we will review the coordinate plane and examine the graphs of several different lines and curves.

> **In this chapter, we will work to master**
>
> - plotting points on the coordinate plane,
> - graphing lines using a table of values,
> - defining and understanding slope,
> - graphing lines using slope-intercept form,
> - comparing the equations of parallel and perpendicular lines, and
> - graphing horizontal and vertical lines.

LESSON 15.1—THE COORDINATE PLANE

The **coordinate plane** is what we use to plot coordinates and graph equations. It is formed by the intersection of a horizontal number line, called the x-axis, and a vertical number line, called the y-axis. The point at which the x-axis and the y-axis intersect is called the **origin**. The coordinate plane is usually represented as a grid, like the one that follows. Notice that the origin is in the center of the plane.

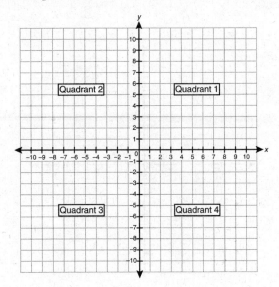

Along the x-axis, the values of x increase as we move to the right. On the y-axis, the values of y increase as we move up. The positive values of x lie to the right of the y-axis, whereas the negative values of x lie to the left of the y-axis. The positive values of y lie above the x-axis, whereas the negative values of y lie below the x-axis.

The coordinate plane consists of four **quadrants**. Notice that the first quadrant consists of places on the plane where both x and y are positive numbers. The second quadrant consists of negative x-values and positive y-values. The third quadrant is made up of points where both the x- and y-values are negative. Finally, the fourth quadrant consists of positive x-values and negative y-values. To summarize:

First Quadrant	Second Quadrant	Third Quadrant	Fourth Quadrant
Positive x	Negative x	Negative x	Positive x
Positive y	Positive y	Negative y	Negative y

We represent points in the coordinate plane using **ordered pairs**. Ordered pairs are two numbers indicating where the point is located on the coordinate plane. They are written as (x, y), where the first number indicates the x-value of the point, and the second number indicates its y-value. The x- and y-values given in an ordered pair tell us how far, and in which direction, to move *from the origin* along the vertical and horizontal axes in order to locate a point. If you feel uncomfortable plotting points, it's always a good idea to begin at the origin, which is the ordered pair $(0, 0)$.

➡ Example 1

Locate the point $(3, 2)$ on the coordinate plane.

Solution

Ordered pairs are of the format (x, y), so the x-value for this point is 3, and the y-value is 2. Both of these values are positive, so the point will be located in the first quadrant.

To locate a point, start at the origin, or $(0, 0)$. The x-value, 3, tells us to move three units to the right. From there, the y-value tells us to move 2 units up.

Answer:

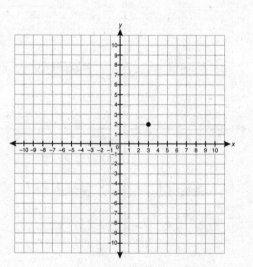

➡ Example 2

Mark the point (–5, –2) on the coordinate plane.

Solution

The *x*-value in this ordered pair is –5, which tells us that we need to move 5 units to the left, starting at the origin. Now, because the *y*-value is –2, we need to move 2 units down. Mark the point as shown on the graph below, and notice that it lies in the third quadrant.

Answer:

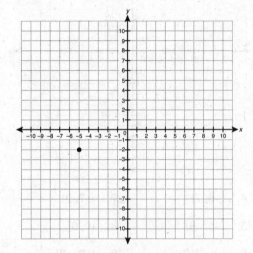

Often, an ordered pair will contain a zero, such as (0, 5) or (–1, 0). An *x*-value of zero tells us that we do not move to the left or right along the *x*-axis, and a *y*-value of zero tells us that we do not move up or down along the *y*-axis.

➡ Example 3

Locate the point (0, 5) on the coordinate plane.

Solution

The *x*-value is 0, so we do not move to the left or right. The *y*-value tells us to move 5 units up, and mark the point as shown below.

Answer:

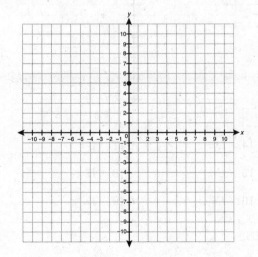

Directions: Mark the following points on the coordinate plane below.

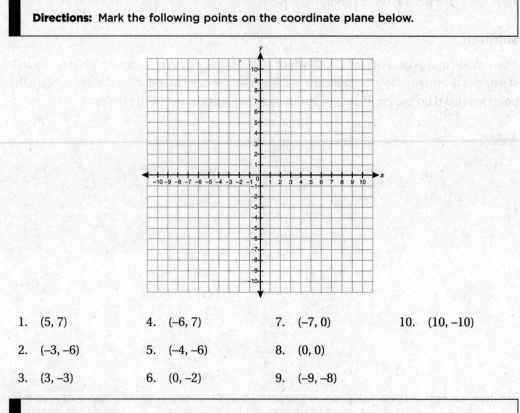

1. (5, 7)
2. (–3, –6)
3. (3, –3)

4. (–6, 7)
5. (–4, –6)
6. (0, –2)

7. (–7, 0)
8. (0, 0)
9. (–9, –8)

10. (10, –10)

Directions: Give the ordered pair for each point on the coordinate plane below.

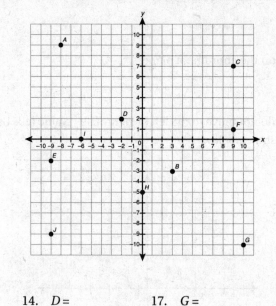

11. *A* =
12. *B* =
13. *C* =

14. *D* =
15. *E* =
16. *F* =

17. *G* =
18. *H* =
19. *I* =

20. *J* =

(Answers are on page 358.)

LESSON 15.2—GRAPHING A LINE USING A TABLE OF VALUES

The following are examples of equations whose graphs are straight, diagonal lines.

$$y = x \qquad y = 3x + 1 \qquad 6x + 2y = y + 3x + 10$$

For each of these, there are an infinite number of possible values for x and y that will make them true. Even though these equations are very different, all of their graphs will be diagonal lines because they all contain a separate x, a separate y, and no visible exponents. If the variables x or y in an equation have an exponent other than 1, the resulting graph will be a curve.

One easy way to graph a line is to use a table of values. When using a table of values, you can begin by choosing any x-value, substituting it into the equation that you want to graph, and then determining the corresponding y-value.

➡ **Example 1**

Graph the line $y = 2x - 4$ using a table of values.

Solution

On the graph of a line, each x-value will correspond to exactly one y-value. Therefore, we make a table that looks like this:

x	$y = 2x - 4$

It's always a good idea to start with $x = 0$ because the math is easier. After that, you can choose any x-values you wish, and then solve for y. Let's fill in the chart for $x = 2$ and $x = 4$.

x	$y = 2x - 4$
0	$y = 2(0) - 4 = -4$
2	$y = 2(2) - 4 = 0$
4	$y = 2(4) - 4 = 4$

This generates the coordinates (0, –4), (2, 0), and (4, 4).

Now that we know a few points on the graph of the line, we can plot them on the coordinate grid and connect them. Any point on the graph of the line is a possible solution for $y = 2x - 4$. For example, the ordered pairs (1, –2) and (3, 2) are located on this line, which means that they are possible solutions to the equation. Since this line continues on toward positive and negative infinity, there are an infinite number of possible values for (x, y) that would make this equation true. For this reason, the graph of the line has arrows at both ends to indicate that the line continues in both directions.

Answer:

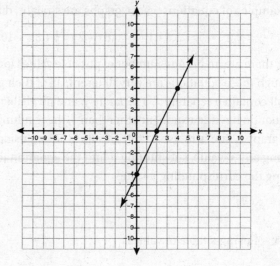

Let's look at some of the characteristics of the graph of $y = 2x - 4$. Moving from left to right along the x-axis, the line increases toward infinity. This means that the graph has a positive slope. The **slope** of a line describes its steepness and direction. You'll also notice that the graph crosses both the x-axis and the y-axis. It crosses the x-axis at the point (2, 0); this is called the ***x*-intercept**. The point where it crosses the y-axis, (0, –4), is called the **y-intercept**. All diagonal lines will have an x-intercept and a y-intercept.

➡ Example 2

Using a table of values, graph the line $y = -\dfrac{1}{2}x + 3$.

Solution

As in the previous example, choose values for x, and then evaluate to find y. Notice that if we choose multiples of 2 for x, we won't have any fractional answers for y. Let's use 0, 2, and 4 again. It's worth noting that you don't *have* to use these values for x; you could use any values. However, it's always good to keep the numbers small and simple so that you don't make mistakes.

x	$y = -\dfrac{1}{2}x + 3$
0	$y = -\dfrac{1}{2}(0) + 3 = 3$
2	$y = -\dfrac{1}{2}(2) + 3 = 2$
4	$y = -\dfrac{1}{2}(4) + 3 = 1$

After plugging in 0, 2, and 4 for x, we find the points (0, 3), (2, 2), and (4, 1). Plot these points on the graph and connect them. As we move from left to right along the graph, we travel downhill so to speak. That is, the slope is negative. Notice that it crosses the y-axis at (0, 3) and the x-axis at (6, 0).

Answer:

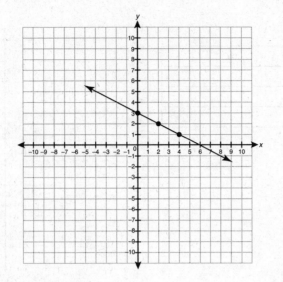

LESSON 15.2—SKILLS CHECK

Directions: Complete the table of values for each linear equation. Then graph each line on the coordinate plane provided.

1. $y = x + 5$

x	y
0	
1	
2	

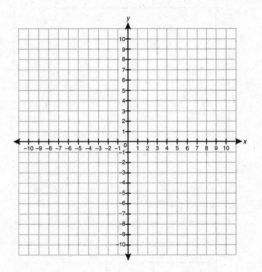

2. $y = -2x$

x	y
−2	
0	
2	

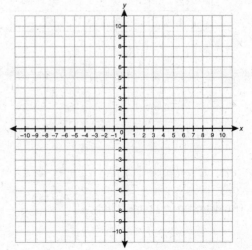

3. $y = \dfrac{1}{2}x - 8$

x	y
4	
6	
8	

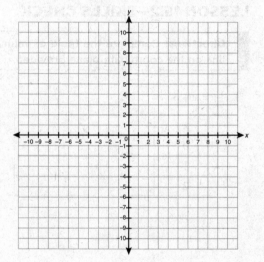

4. $y = 4x - 6$

x	y
0	
1	
2	

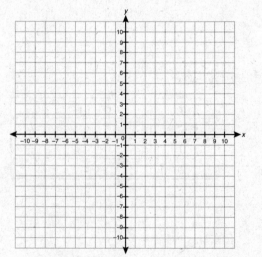

5. $y = -3x + 8$

x	y
0	
2	
4	

(Answers are on page 358.)

LESSON 15.3—UNDERSTANDING AND CALCULATING SLOPE

All the lines we have graphed so far are diagonal, and they either increased or decreased as we move from left to right. Some were steep, and some were relatively flat. This measure of a line's steepness and direction is called its **slope**. The slope of a line is the ratio of its rise, or vertical change, to its run, or horizontal change. A line's slope can be positive, negative, zero, or undefined. Knowing the slope of a line is highly valuable when visualizing the shape of a graph. If we know the slope and *any point on the graph*, we will be able to accurately draw or sketch the entire graph.

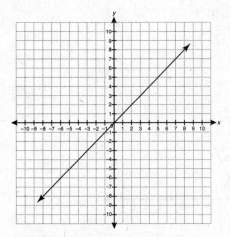

Positive Slope: The line increases, or rises, as we move along the graph from left to right.

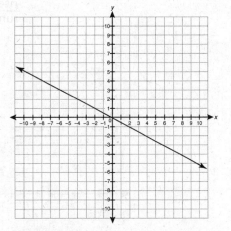

Negative Slope: The line decreases, or drops, as we move along the graph from left to right.

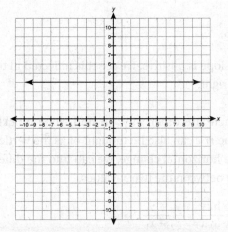

Zero Slope: Lines with zero slope are horizontal. They have no vertical change.

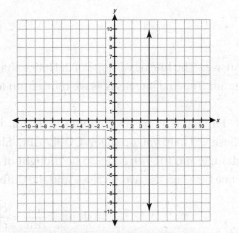

Undefined Slope: Lines with an undefined slope are vertical. They have no horizontal change.

➡ Example 1

Find the slope of a line that passes through the points (–1, –3) and (2, 3).

Solution

The slope of this line can be found by plotting the two points, connecting them with a line, and then finding the ratio of the line's **rise** to its **run**. In this case, we have a rise of 6 units and a run of 3 units. Set up the ratio as:

$$\frac{\text{rise}}{\text{run}} = \frac{6}{3} = 2$$

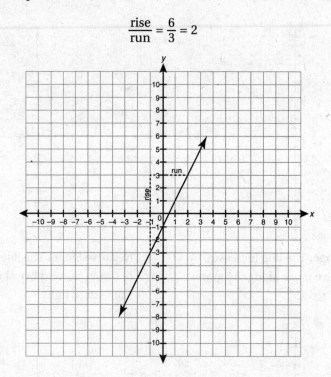

Answer: The line has a slope of 2. Notice that the slope is positive. We see this reflected in the graph. The line increases as we move from left to right.

Finding a slope by graphing is useful, but there are other ways to find slope quickly; these will help you on the ACCUPLACER. Since the *rise* can also be thought of as the vertical change, and the *run* can also be thought of as the horizontal change, we can find the difference between the two *y*-values and the difference between the two *x*-values.

$$\text{slope} = \frac{\text{rise}}{\text{run}} = \frac{\text{change in } y}{\text{change in } x} = \frac{y_2 - y_1}{x_2 - x_1}$$

When you sit down to take the ACCUPLACER, this formula won't be given to you; you'll need to have it memorized. The ordered pair (x_2, y_2) refers to the coordinates of the second point, and (x_1, y_1) refers to the coordinates of the first point. Try using this formula in Example 1. You should get the same slope that we got the first time. Let's look at another example.

➡ Example 2

Find the slope of a line that passes through the points (–2, –3) and (4, 5).

Solution

In terms of the slope formula, remember that the first point is (x_1, y_1) and the second point is (x_2, y_2).

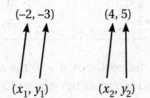

Now plug these values into the formula, and evaluate.

$$\text{slope} = \frac{y_2 - y_1}{x_2 - x_1} = \frac{5 - (-3)}{4 - (-2)} = \frac{8}{6} = \frac{4}{3}$$

We have an improper fraction for the slope. Slope is almost never written as a mixed number, so leave the fraction improper.

Answer: The slope is $\frac{4}{3}$.

➡ Example 3

Find the slope of the line that passes through the points $(2a, 9b)$ and $(-a, 6b)$.

Solution

Even though this slope has variables in it, the same formula applies.

$$\text{slope} = \frac{y_2 - y_1}{x_2 - x_1} = \frac{6b - 9b}{-a - 2a} = \frac{-3b}{-3a} = \frac{b}{a}$$

Answer: The slope is $\frac{b}{a}$.

You can also find the slope if you are given a table of values that form the graph of a line. Let's look at an example.

➡ Example 4

Find the slope of the line represented by the table below.

x	y
0	12
3	3
6	–6
9	–15

Solution

To find the slope, we have to find the ratio of the line's vertical change to its horizontal change. If given a table, we just need to look at how much the y-values change and how much the x-values change, and then set up a ratio.

x	y
0	12
3	3
6	−6
9	−15

$+3$ between x-values; $−9$ between y-values

The y-values are consistently decreasing by 9, and the x-values are consistently increasing by 3. We know that this table represents the graph of a straight line because the x- and y-values go up or down by the same amount every time.

To calculate the slope, we just need to set up the ratio $\dfrac{\text{change in } y}{\text{change in } x}$.

$$\frac{\text{change in } y}{\text{change in } x} = \frac{-9}{3} = -3$$

Answer: The slope of the line represented by this table is −3. This means that this line will be decreasing sharply as we move along the graph from left to right.

LESSON 15.3—SKILLS CHECK

Directions: Find the slope of the line that passes through each pair of coordinates.

1. (2, 4) and (3, 8)   3. (−9, −12) and (−7, −2)   5. (−3, 9) and (3, 12)

2. (0, 0) and (−3, 5)   4. (3, 8) and (−2, −7)

Directions: Find the slope of the line that is represented by each table of values.

6.

x	y
−2	−4
−1	0
0	4
1	8

7.

x	y
0	15
3	0
6	−15
9	−30

8.

x	y
−5	−11
0	−9
5	−7
10	−5

(Answers are on page 360.)

LESSON 15.4—SLOPE-INTERCEPT FORM OF A LINE

An equation whose graph is a straight line can take many forms, as long as it contains a separate x-variable term, a separate y-variable term, and no visible exponents, but the standard form for writing the equation of a line is called **slope-intercept form**. You will need to have a clear understanding of slope-intercept form to successfully answer some of the questions on the College-Level Mathematics test.

When an equation is written in slope-intercept form, it becomes very easy to extract information about the slope and the y-intercept of the line, which in turn makes it easier to graph the line if need be. Slope-intercept form is generalized as

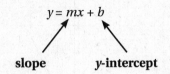

$$y = mx + b$$

slope y-intercept

For the purpose of graphing a line using this form, we only need to know the values of m and b. The line's slope is represented by m, and b represents its y-intercept (where the line crosses, or *intercepts*, the y-axis). By knowing only the slope and the y-intercept, we can accurately graph the line.

➡ Example 1

What are the slope and y-intercept of the line $y = \frac{3}{4}x - 10$?

Solution

This equation is already in the form of $y = mx + b$. Remember that m represents the slope, and b represents the y-intercept.

$$y = \frac{3}{4}x - 10$$

$$y = mx + b$$

We can see that $m = \frac{3}{4}$, and $b = -10$.

<u>Answer</u>: The slope is $\frac{3}{4}$, and the line crosses the y-axis at the point $(0, -10)$.

Now let's look at an example to see how slope-intercept form can be used to graph a line.

➡ Example 2

Graph the line $y = 2x - 4$.

Solution

This equation is already in the form of $y = mx + b$, so we can use the m and b values to graph it. Here, $m = 2$ and $b = -4$. Since the slope, m, is positive, we know that the graph will be increasing as we move from left to right. The *b* value, –4, tells us that the graph will cross the *y*-axis at the point (0, –4).

Begin by placing a point at the *y*-intercept (0, –4). Remember that the slope is the ratio of the line's *rise* to its *run*: $\frac{2}{1}$. From the *y*-intercept, we go up two units and then one unit to the right before marking our next point at (1, –2). We can repeat this process as many times as necessary before connecting the dots. The graph of the line shows us all the possible solutions for *x* and *y*.

Answer:

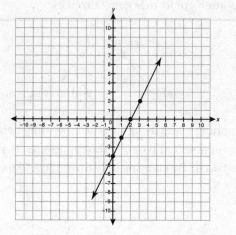

➡ Example 3

Write the equation of a line that passes through the origin and the point (2, –6).

Solution

To write the equation of this line in slope-intercept form, we need to know the slope *m* and the *y*-intercept *b*. The prompt already gives us the *y*-intercept. If the line passes through the origin, then it crosses the *y*-axis at the point (0, 0). Now we can find the slope using the formula $m = \frac{y_2 - y_1}{x_2 - x_1}$.

$$m = \frac{y_2 - y_1}{x_2 - x_1} = \frac{-6 - 0}{2 - 0} = \frac{-6}{2} = -3$$

Finally, we plug in –3 in place of *m* and 0 in place of *b*.

$$y = mx + b$$
$$y = -3x + 0$$
$$y = -3x$$

Answer: The equation of the line that passes through the origin and the point (2, –6) is $y = -3x$.

Sometimes, on the ACCUPLACER, you will be asked about the characteristics of a line, but the equation you are given will not be in slope-intercept form. In this case, you will need to use the rules of algebra to put the equation into slope-intercept form. Being able to reorganize an equation into slope-intercept form is valuable in that you can then easily answer questions about the line's characteristics.

➥ Example 4

What are the slope and y-intercept of the line formed by the equation $-5x + 4y = 3y - 10$?

Solution

Our goal is to reorganize the equation so that it takes the form $y = mx + b$, so we want to gather all of the ys on the left side of the equation, and move everything else to the right side.

$$
\begin{array}{r}
-5x + 4y = 3y - 10 \\
\underline{-3y \quad -3y} \\
-5x + y = -10 \\
\underline{+5x \qquad +5x} \\
y = 5x - 10
\end{array}
$$

In slope-intercept form, the equation is $y = 5x - 10$.

Answer: This line has a slope of 5, and it crosses the y-axis at $(0, -10)$.

➥ Example 5

Find the slope and y-intercept of the line formed by the equation $4(-2x + y) = -24$.

Solution

To put this equation into slope-intercept form, we need to isolate y on the left side of the equal sign. Begin by using the distributive property. Notice that, when we divide by 4, we have to divide every term in the equation by 4.

$$
\begin{array}{r}
4(-2x + y) = -24 \\
-8x + 4y = -24 \\
\underline{+8x \qquad +8x} \\
4y = 8x - 24 \\
\dfrac{4y}{4} = \dfrac{8x}{4} - \dfrac{24}{4} \\
y = 2x - 6
\end{array}
$$

The equation of this line is $y = 2x - 6$.

Answer: The slope of this line is 2, and it crosses the y-axis at the point $(0, -6)$.

LESSON 15.4—SKILLS CHECK

> **Directions:** Put the following equations in slope-intercept form, and then identify m and b.

1. $-3x + y = 5$

2. $8x + y + 7 = 14$

3. $5y - 2x = 8x + 25$

4. $y - 15 = 3(-2x - 1)$

5. $2x + 2y = 8$

6. $-3(x + y) = 2(x + 9)$

(Answers are on page 361.)

LESSON 15.4—ACCUPLACER CHALLENGE

> **Directions:** For each of the questions below, choose the best answer from the five choices given.

1. Consider a circle with its center at the origin and a radius of 5. How many points of intersection will the circle have with the line $y = x + 8$?

   (A) 0
   (B) 1
   (C) 2
   (D) 3
   (E) More than 3

2. Which of the following lines passes through the points $(-3, 4)$ and $(1, 6)$?

   (A) $y = 2x + 12$
   (B) $y = x + 2$
   (C) $y = \frac{1}{2}x + 5\frac{1}{2}$
   (D) $y = -\frac{1}{2}x + 2$
   (E) $y = -2x - 4$

3. At which point will the line $y = 4x - 6$ intersect the line $y = -2x$?

   (A) $(1, 2)$
   (B) $(-1, 2)$
   (C) $(1, -2)$
   (D) $(-1, -2)$
   (E) The lines will not intersect.

4. Which of the following lines has the steepest slope?

   (A) $y = \frac{5}{2}x - 5$
   (B) $y = \frac{8}{3}x$
   (C) $y = \frac{11}{4}x + 6$
   (D) $y = \frac{4}{3}x + 1$
   (E) $y = \frac{10}{3}x - 5$

5. Consider the line below that passes through the points (a, b) and (c, d). Which of the following choices represents the slope of this line?

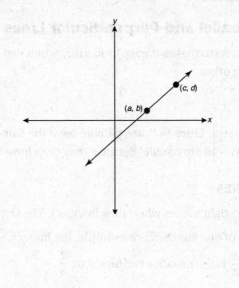

(A) $\dfrac{d - c}{b - a}$

(B) $\dfrac{d - b}{c - a}$

(C) $\dfrac{c - a}{d - b}$

(D) $\dfrac{d - a}{c - b}$

(E) $\dfrac{c - b}{d - a}$

6. What is the equation of the line shown below?

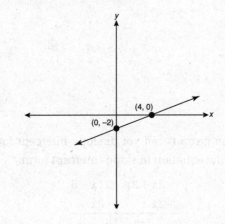

(A) $y = \dfrac{1}{2}x + 4$

(B) $y = 2x + 4$

(C) $y = \dfrac{1}{2}x + 2$

(D) $y = \dfrac{1}{2}x - 2$

(E) $y = 2x - 2$

(Answers are on page 362.)

LESSON 15.5—PARALLEL, PERPENDICULAR, HORIZONTAL, AND VERTICAL LINES

Lesson 15.5A—Parallel and Perpendicular Lines

Knowing slope-intercept form makes it easy to identify when two or more lines are parallel or perpendicular to each other.

PARALLEL LINES

Parallel lines never intersect. Lines that are parallel have the same slope. For example, the lines $y = 3x + 4$ and $y = 3x - 10$ are parallel because they both have the same slope ($m = 3$).

PERPENDICULAR LINES

Perpendicular lines form right angles when they intersect. The slopes of perpendicular lines are negative reciprocals of one another. For example, the line $y = \frac{2}{3}x$ is perpendicular to the line $y = -\frac{3}{2}x$ because $-\frac{3}{2}$ is the negative reciprocal of $\frac{2}{3}$.

On the ACCUPLACER, you may be asked to identify the equation of a line that is parallel or perpendicular to a given equation. Let's look at a couple of examples.

➥ Example 1

Which of the following lines is parallel to the line $2x + 3y = 11x - 6$?

(A) $y = -3x + 6$

(B) $y = 3x - 12$

(C) $y = \frac{1}{3}x + 9$

(D) $y = -\frac{1}{3}x + 1$

(E) None of these

Solution

Unfortunately, the equation given is not yet in slope-intercept form, so it's difficult to tell what its slope is. First, put the equation in slope-intercept form.

$$2x + 3y = 11x - 6$$
$$\underline{-2x \qquad\quad -2x}$$
$$\frac{3y}{3} = \frac{9x - 6}{3}$$
$$y = 3x - 2$$

In slope-intercept form, this line is written as $y = 3x - 2$. It has a slope of $m = 3$. The only line in the answer choices that also has a slope of 3 is choice B, $y = 3x - 12$.

Answer: Choice B is correct.

Notice that these lines have different y-intercepts but the same slope. This is how we know that they are parallel. Now let's look at a sample question that would ask you to identify perpendicular lines.

➡ Example 2

Which of the following choices has two lines that are perpendicular to one another?

(A) $y = 4x + 3$ and $y = -4x - 2$

(B) $y = 2x + 7$ and $y = \frac{1}{2}x + 4$

(C) $y = \frac{1}{5}x - 11$ and $y = -5x + 9$

(D) $y = -7x + 6$ and $y = -\frac{1}{7}x + 3$

(E) $y = \frac{2}{3}x + 1$ and $y = \frac{3}{2}x + 3$

Solution

If two lines are perpendicular, their slopes are negative reciprocals of one another. Let's break down the answer choices one by one.

The first line in choice A has a slope of 4. The negative reciprocal of $\frac{4}{1}$ is $-\frac{1}{4}$. The second line in choice A does not have the correct slope. The first line in choice B has a slope of 2. The negative reciprocal of $\frac{2}{1}$ is $-\frac{1}{2}$, which means that the second line does not have the correct slope. In choice C, the slope of the first line is $\frac{1}{5}$, which has a negative reciprocal of $-\frac{5}{1}$, or –5. Since the slopes of the two lines in choice C are negative reciprocals of one another, choice C is correct. Notice that in choices D and E, the slopes are *reciprocals* of one another, but the sign does not change.

<u>Answer</u>: Choice C is correct.

Lesson 15.5B—Horizontal and Vertical Lines

The equations of horizontal and vertical lines do not follow slope-intercept form as diagonal lines do. Their equations are much simpler.

HORIZONTAL LINES

The graph of a horizontal line is $y = c$, where c is a constant or number. This means that, for any point on the graph, y will be equal to some number. If we mark several points on the graph where $y = 3$ and connect them, we will have a horizontal line. The slope of all horizontal lines is zero.

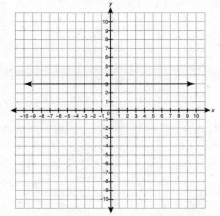

Above is the horizontal line $y = 3$.

VERTICAL LINES

The graph of a vertical line is $x = c$, where c is a constant. This means that, at all points on the graph, x will be equal to some number. If we mark several points on the graph where $x = 4$ and connect them, we will have a vertical line. The slope of a vertical line is said to be undefined. Remember that slope equals $\dfrac{\text{vertical change}}{\text{horizontal change}}$. Vertical lines have infinite vertical change and zero horizontal change. Therefore, the slope of a vertical line equals $\dfrac{\text{some constant}}{0}$. We cannot divide by 0, so we therefore say that the slope of a vertical line is undefined.

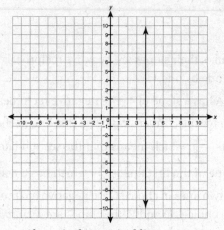

Above is the vertical line $x = 4$.

➡ Example 1

Graph the line $x = -5$ and the line $y = 2$.

Solution

The first line is in the format $x = c$, so it will be a vertical line. From the origin, move left to $x = -5$, and then sketch a vertical line.

The second line will be a horizontal line because it has the format $y = c$. Starting at the origin, move up to $y = 2$, and sketch a horizontal line.

Notice that the lines meet at the point $(-5, 2)$, the same x- and y-values as those in the equations we started with.

Answer:

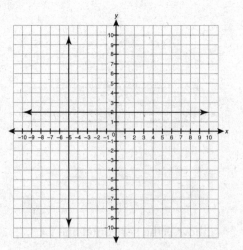

LESSON 15.5—ACCUPLACER CHALLENGE

Directions: For each of the questions below, choose the best answer from the five choices given.

1. At which point will the line $x = 6$ intersect the line $y = -1$?

   (A) $(-1, 6)$
   (B) $(6, -1)$
   (C) $(1, 6)$
   (D) $(6, 1)$
   (E) The lines do not intersect.

2. Which of the following lines is parallel to the line $y = \frac{1}{2} x + 7$?

   (A) $2x - 2y = 0$
   (B) $x + 2y = 0$
   (C) $x - 2y = 0$
   (D) $2x - y = 0$
   (E) $x - y = 0$

3. Consider a line, $y = mx + b$, where $m = \frac{4}{3}$. Which of the following lines is perpendicular to this line?

   (A) $y = \frac{3}{4} x + 4$
   (B) $y = -\frac{4}{3} x + 5$
   (C) $y = -3x - 4$
   (D) $y = -4x + 3$
   (E) $y = -\frac{3}{4} x + 6$

4. At which coordinate will the line $y = 8$ intersect the line $y = \frac{1}{4} x + 6$?

   (A) $(12, 8)$
   (B) $(8, 12)$
   (C) $(8, 8)$
   (D) $(-12, 8)$
   (E) $(-8, 8)$

5. Where will the line $4x - y = 6$ intersect the line $3y = 12x + 5$?

   (A) $(4, 6)$
   (B) $\left(4, \frac{5}{3}\right)$
   (C) $(6, 4)$
   (D) $(3, 12)$
   (E) They will not intersect.

6. Consider a circle with its center at the point $(2, 5)$ and a radius of 4. How many points of intersection will it have with the line $x = -2$?

   (A) 0
   (B) 1
   (C) 2
   (D) 3
   (E) More than 3

(Answers are on page 363.)

To complete a set of practice questions that reviews all of Chapter 15, go to:
barronsbooks.com/TP/accuplacer/math29qw/

Lesson 15.1—Skills Check (page 340)

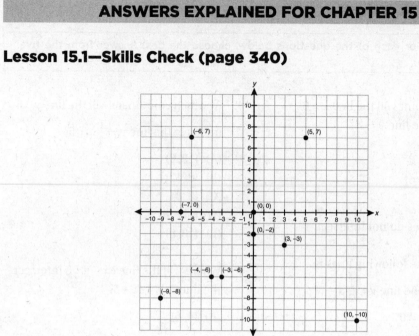

1.–10. The first value in each ordered pair denotes the *x*-coordinate, and the second value denotes the *y*-coordinate. Each ordered pair represents only one point on the coordinate plane.

11. **(–8, 9)**	14. **(–2, 2)**	17. **(10, –10)**	20. **(–9, –9)**
12. **(3, –3)**	15. **(–9, –2)**	18. **(0, –5)**	
13. **(9, 7)**	16. **(9, 1)**	19. **(–6, 0)**	

Lesson 15.2—Skills Check (pages 343–344)

1.

x	y
0	**5**
1	**6**
2	**7**

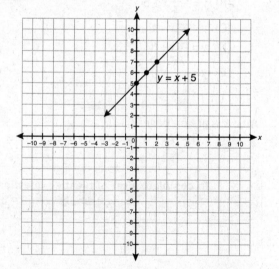

x	y
0	$y = 0 + 5 = 5$
1	$y = 1 + 5 = 6$
2	$y = 2 + 5 = 7$

2.

x	y
−2	**4**
0	**0**
2	**−4**

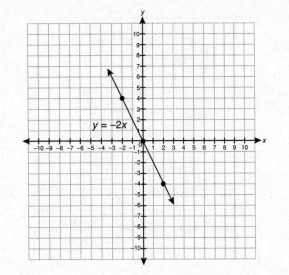

x	y
−2	$y = -2(-2) = 4$
0	$y = -2(0) = 0$
2	$y = -2(2) = -4$

3.

x	y
4	**−6**
6	**−5**
8	**−4**

x	y
4	$y = \frac{1}{2}(4) - 8 = -6$
6	$y = \frac{1}{2}(6) - 8 = -5$
8	$y = \frac{1}{2}(8) - 8 = -4$

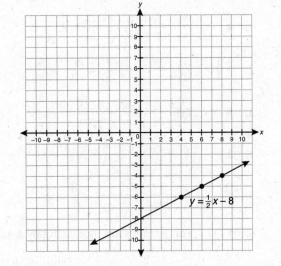

4.

x	y
0	**−6**
1	**−2**
2	**2**

x	y
0	$y = 4(0) - 6 = -6$
1	$y = 4(1) - 6 = -2$
2	$y = 4(2) - 6 = 2$

5.

x	y
0	**8**
2	**2**
4	**−4**

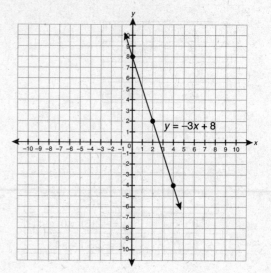

$y = -3x + 8$

x	y
0	$y = -3(0) + 8 = 8$
2	$y = -3(2) + 8 = 2$
4	$y = -3(4) + 8 = -4$

Lesson 15.3—Skills Check (page 348)

1. **4** $\dfrac{y_2 - y_1}{x_2 - x_1} = \dfrac{8 - 4}{3 - 2} = \dfrac{4}{1} = 4$

2. $-\dfrac{5}{3}$ $\dfrac{y_2 - y_1}{x_2 - x_1} = \dfrac{5 - 0}{-3 - 0} = \dfrac{5}{-3} = -\dfrac{5}{3}$

3. **5** $\dfrac{y_2 - y_1}{x_2 - x_1} = \dfrac{-2 - (-12)}{-7 - (-9)} = \dfrac{10}{2} = 5$

4. **3** $\dfrac{y_2 - y_1}{x_2 - x_1} = \dfrac{-7 - 8}{-2 - 3} = \dfrac{-15}{-5} = 3$

5. $\dfrac{1}{2}$ $\dfrac{y_2 - y_1}{x_2 - x_1} = \dfrac{12 - 9}{3 - (-3)} = \dfrac{3}{6} = \dfrac{1}{2}$

6. **4**   The y-values are increasing by 4, while the x-values are increasing by 1.

$$\frac{\text{change in } y}{\text{change in } x} = \frac{4}{1} = 4$$

7. **−5**   The y-values are decreasing by 15, while the x-values are increasing by 3.

$$\frac{\text{change in } y}{\text{change in } x} = \frac{-15}{3} = -5$$

8. $\dfrac{2}{5}$ The y-values are increasing by 2, while the x values are increasing by 5.

$$\frac{\text{change in } y}{\text{change in } x} = \frac{2}{5}$$

Lesson 15.4—Skills Check (page 351)

1. $y = 3x + 5$
   $m = 3$
   $b = 5$

   $-3x + y = 5$
   $\underline{+3x \qquad +3x}$
   $y = 3x + 5$

   $m = 3,\ b = 5$

2. $y = -8x + 7$
   $m = -8$
   $b = 7$

   $8x + y + 7 = 14$
   $\underline{\qquad -7 \quad -7}$
   $8x + y = 7$
   $\underline{-8x \qquad -8x}$
   $y = -8x + 7$

   $m = -8,\ b = 7$

3. $y = 2x + 5$
   $m = 2$
   $b = 5$

   $5y - 2x = 8x + 25$
   $\underline{+2x \quad +2x}$
   $\dfrac{5y}{5} = \dfrac{10x + 25}{5}$
   $y = 2x + 5$

   $m = 2,\ b = 5$

4. $y = -6x + 12$
   $m = -6$
   $b = 12$

   $y - 15 = 3(-2x - 1)$
   $y - 15 = -6x - 3$
   $\underline{+15 \qquad +15}$
   $y = -6x + 12$

   $m = -6,\ b = 12$

5. $y = -x + 4$
   $m = -1$
   $b = 4$

   $2x + 2y = 8$
   $\underline{-2x \qquad -2x}$
   $\dfrac{2y}{2} = \dfrac{-2x + 8}{2}$
   $y = -x + 4$

   $m = -1,\ b = 4$

6. $y = -\dfrac{5}{3}x - 6$

   $m = -\dfrac{5}{3}$

   $b = -6$

   $-3(x + y) = 2(x + 9)$
   $-3x - 3y = 2x + 18$
   $\underline{+3x \qquad +3x}$
   $\dfrac{-3y}{-3} = \dfrac{5x + 18}{-3}$
   $y = -\dfrac{5}{3}x - 6$

   $m = -\dfrac{5}{3},\ b = -6$

Lesson 15.4—ACCUPLACER Challenge (pages 352–353)

1. **(A)** The line $y = x + 8$ will cross the y-axis at the point $(0, 8)$, which lies above the circle. It will also cross the x-axis at the point $(-8, 0)$, which lies to the left of the circle. A quick sketch, like the one below, will show that the line and the circle will never intersect.

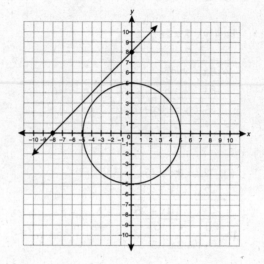

2. **(C)** All five of these lines have different slopes, so we only need to confirm the slope of the line that passes through these points in order to choose an answer.

$$\frac{y_2 - y_1}{x_2 - x_1} = \frac{6 - 4}{1 - (-3)} = \frac{2}{4} = \frac{1}{2}$$

Only choice C has a slope of $\frac{1}{2}$, so this is the correct choice.

3. **(C)** Treat the lines like a system of equations. Since both are in the form "$y =$," you can set $4x - 6$ equal to $-2x$, and then solve.

$$4x - 6 = -2x$$
$$\underline{-4x \qquad\qquad -4x}$$
$$\frac{-6}{-6} = \frac{-6x}{-6}$$
$$1 = x$$

Now plug $x = 1$ into either of the two original equations.

$$y = -2x$$
$$y = -2(1)$$
$$y = -2$$

The lines intersect at the point $(1, -2)$.

4. **(E)** All the slopes are positive, and the one with the greatest value will be the steepest. To compare, raise each fraction so that they all have a common denominator of 12.

$$\frac{5}{2} \times \frac{6}{6} = \frac{30}{12} \qquad \frac{8}{3} \times \frac{4}{4} = \frac{32}{12} \qquad \frac{11}{4} \times \frac{3}{3} = \frac{33}{12} \qquad \frac{4}{3} \times \frac{4}{4} = \frac{16}{12} \qquad \frac{10}{3} \times \frac{4}{4} = \frac{40}{12}$$

$\frac{10}{3}$ has the greatest value, so the line $y = \frac{10}{3}x - 5$ has the steepest slope.

5. **(B)** To find the slope of this line, use the formula $m = \dfrac{y_2 - y_1}{x_2 - x_1}$. Let $(a, b) = (x_1, y_1)$ and $(c, d) = (x_2, y_2)$.

$$\frac{y_2 - y_1}{x_2 - x_1} = \frac{d - b}{c - a}$$

6. **(D)** This line crosses the y-axis at -2. This means that $b = -2$. Next, find the slope.

$$\frac{y_2 - y_1}{x_2 - x_1} = \frac{0 - (-2)}{4 - 0} = \frac{2}{4} = \frac{1}{2}$$

The slope is $\dfrac{1}{2}$, and the y-intercept is -2. This means that the equation of the line is $y = \dfrac{1}{2}x - 2$.

Lesson 15.5—ACCUPLACER Challenge (page 357)

1. **(B)** The line $x = 6$ is vertical, so its x-value is 6 everywhere. The line $y = -1$ is horizontal, so its y-value is -1 everywhere. Therefore, the only point they have in common would occur when $x = 6$ and $y = -1$, or $(6, -1)$.

2. **(C)** The line in the question has a slope of $m = \dfrac{1}{2}$. Only choice C also has a slope of $m = \dfrac{1}{2}$.

$$\begin{aligned} x - 2y &= 0 \\ +2y \quad &+2y \\ \hline \frac{x}{2} &= \frac{2y}{2} \\ \frac{1}{2}x &= y \end{aligned}$$

3. **(E)** Perpendicular lines have slopes that are negative reciprocals. If a line has a slope of $m = \dfrac{4}{3}$, a perpendicular line would have a slope of $m = -\dfrac{3}{4}$.

4. **(C)** Since we know that $y = 8$, plug in 8 in place of y in the equation $y = \dfrac{1}{4}x + 6$. This will give the x-coordinate of the point where the two lines intersect.

$$\begin{aligned} y &= \frac{1}{4}x + 6 \\ 8 &= \frac{1}{4}x + 6 \\ -6 \quad &\quad -6 \\ \hline 2 &= \frac{1}{4}x \\ 4(2) &= \left(\frac{1}{4}x\right)4 \\ 8 &= x \end{aligned}$$

The lines will intersect at the point $(8, 8)$.

5. **(E)** Neither line is in slope-intercept form. Use algebra to rearrange the terms in the equations.

$$4x - y = 6$$
$$\underline{+y \quad +y}$$
$$4x = y + 6$$
$$\underline{-6 \quad\quad -6}$$
$$4x - 6 = y$$

$$\frac{3y}{3} = \frac{12x + 5}{3}$$
$$y = 4x + \frac{5}{3}$$

The two lines have the same slope and different y-intercepts. This means they are parallel and will never intersect.

6. **(B)** The line $x = -2$ is vertical. If the circle's center is located at $(2, 5)$ and it has a radius of 4, then the circle will also pass through the point $(-2, 5)$, which is also on the line $x = -2$. This is the only point they have in common.

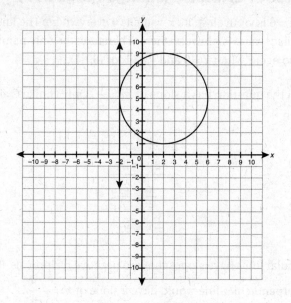

Solving Quadratic Equations

16

Until now, we have only worked on solving equations with a single solution. For example, if $x + 5 = 8$, then x must be 3. This is the one and only quantity that makes this equation true. However, **quadratic equations**, equations that contain a variable to the second power as the highest power in the equation, usually have two solutions. For example, $p^2 = 25$ has two solutions, 5 and –5. This is because both of the following are true:

$$5^2 = 25 \text{ because } 5 \times 5 = 25 \quad and \quad (-5)^2 = 25 \text{ because } -5 \times -5 = 25$$

The solutions to the equation $p^2 = 25$ could be written as $p = \{-5, 5\}$ or $p = \pm 5$. The symbol $\pm$ is read as "plus or minus."

Quadratic equations are a fundamental part of many college algebra courses, not to mention science and engineering, so the ACCUPLACER will surely test your ability to solve questions involving quadratics.

In this chapter, we will work to master

- solving quadratic equations by factoring, and
- solving quadratic equations using the quadratic formula.

LESSON 16.1—SOLVING QUADRATIC EQUATIONS BY FACTORING

We generally solve quadratic equations of the standard form $ax^2 + bx + c = 0$. In standard form, 0 is on one side of the equation, and all the other terms are on the other side of the equation. The letter a is the coefficient of the x^2 term, the letter b is the coefficient of the x term, and c is the constant, which is just a regular number. For example, the equation $3x^2 + 5x - 2 = 0$ is a quadratic equation in standard form, where $a = 3$, $b = 5$, and $c = -2$.

The **solutions** to a quadratic equation are the values, or value, that make the equation equal 0. These values are also referred to as **roots**, and they are sometimes referred to as **zeros**.

For example, the solutions to the quadratic equation $x^2 - 3x + 2 = 0$ are $x = 2$ or $x = 1$. This is because both values make the equation equal 0 when *separately* substituted into the equation, as shown below.

Substituting 2 for x in $x^2 - 3x + 2 = 0$	Substituting 1 for x in $x^2 - 3x + 2 = 0$
$2^2 - 3(2) + 2 = 0$	$1^2 - 3(1) + 2 = 0$
$4 - 6 + 2 = 0$	$1 - 3 + 2 = 0$
$-2 + 2 = 0$	$-2 + 2 = 0$
$0 = 0$	$0 = 0$

Both numbers make the equation true, so $x = 2$ or $x = 1$ are solutions.

Note that x cannot be two values at the same time in the same equation. This is why we made a point of substituting 2 for x and then separately substituting 1 for x. The following false statements would have resulted if we had substituted both 2 for x and 1 for x in the same equation at the same time.

Incorrectly substituting 2 and 1 for x at the same time into $x^2 - 3x + 2 = 0$

$2^2 - 3(1) + 2 = 0$	$1^2 - 3(2) + 2 = 0$
$4 - 3 + 2 = 0$	$1 - 6 + 2 = 0$
$1 + 2 = 0$	$-5 + 2 = 0$
$3 = 0$	$-3 = 0$
✗ False	✗ False

Lesson 16.1A—The Zero Product Property

The **zero product property** is as follows:

> If $ab = 0$, then either $a = 0$ or $b = 0$.

Before we solve quadratic equations by factoring, we will first use the zero product property to solve equations that are already factored.

To Solve Equations Using the Zero Product Property:

1. Separately set each factor equal to 0.
2. Solve both equations.

➡ Example 1

Solve $(x + 6)(x + 5) = 0$.

Solution

Write each binomial as a separate equation that is equal to 0.

$$(x + 6)(x + 5) = 0$$

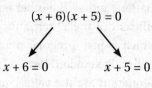

$$x + 6 = 0 \qquad x + 5 = 0$$

Then solve both equations.

$x + 6 = 0$	$x + 5 = 0$
$x = -6$	$x = -5$

To check the solutions, substitute -6 for x and then separately substitute -5 for x.

Substitute -6 for x in $(x + 6)(x + 5) = 0$	Substitute -5 for x in $(x + 6)(x + 5) = 0$
$(-6 + 6)(-6 + 5) = 0$	$(-5 + 6)(-5 + 5) = 0$
$(0)(-1) = 0$	$(1)(0) = 0$
$0 = 0$ ✔	$0 = 0$ ✔

Answer: The solution to $(x + 6)(x + 5) = 0$ is $x = -6$ *or* $x = -5$, which could also be written as $x = \{-6, -5\}$. When a solution is written like this, it is called a **solution set**.

➡ Example 2

Solve $(p - 2)(p - 3) = 0$.

Solution

Write each binomial as separate equations, and then solve.

$$p - 2 = 0 \qquad\qquad p - 3 = 0$$
$$p = 2 \qquad\qquad\qquad p = 3$$

$$p = \{2, 3\}$$

Check by separately substituting 2 for p and then 3 for p.

Substitute 2 for p in $(p - 2)(p - 3) = 0$	Substitute 3 for p in $(p - 2)(p - 3) = 0$
$(2 - 2)(2 - 3) = 0$	$(3 - 2)(3 - 3) = 0$
$(0)(-1) = 0$	$(1)(0) = 0$
$0 = 0$ ✔	$0 = 0$ ✔

Answer: The solution is $p = 2$ *or* $p = 3$. The solution set may also be written as $p = \{2, 3\}$.

ACCUPLACER TIPS

As shown in the previous examples, the solutions to the factored equation $(x + 6)(x + 5) = 0$ were $x = \{-6, -5\}$, and the solutions to the factored equation $(p - 2)(p - 3) = 0$ were $p = \{2, 3\}$. You may notice that the solutions are the opposites of the numbers found in the original equations. A common mistake is to choose the numbers that are the same as these original numbers as solutions. Here's what we mean.

Which of the following are solutions to $(m - 3)(m + 4) = 0$?

(A) $m = \{-3, 4\}$
(B) $m = \{-3, -4\}$
(C) $m = \{3, -4\}$
(D) $m = \{3, 4\}$
(E) $m = \{0, 0\}$

The common error would be to choose choice A, $m = \{-3, 4\}$, because those are the same numbers that we see in the original equation $(m - 3)(m + 4) = 0$. However, the correct answer is choice C, $m = \{3, -4\}$. These numbers will make the equation true because a number combined with its opposite equals 0. That is, $(3 - 3) = 0$ and $(-4 + 4) = 0$.

LESSON 16.1A—SKILLS CHECK

> **Directions**: Solve the following equations.

1. $(x + 2)(x + 3) = 0$

2. $(n - 2)(n - 3) = 0$

3. $(y + 5)(y - 5) = 0$

4. $(x - 8)(x + 10) = 0$

5. $(p + 12)(p + 12) = 0$

6. $(2x + 3)(-3x - 2) = 0$

7. $(4x - 1)(5x + 2) = 0$

8. $(9 - 2n)(4 + 4n) = 0$

(Answers are on page 383.)

Lesson 16.1B—Solving Quadratic Equations by Factoring

Now that we know about the zero product property and what is meant by a solution to a quadratic equation, let's dive into solving quadratic equations by factoring. To do so, you will already have to have mastered factoring quadratic expressions.

To Solve a Quadratic Equation by Factoring:

1. If it is not already done, set the equation equal to 0 to fit the form $ax^2 + bx + c = 0$.
2. Factor the quadratic expression into two binomials.
3. Use the zero product property to set each binomial equal to 0.
4. Separately solve each binomial.
5. Check by substituting your solutions back into the original equation to see if 0 is obtained.

➡ Example 1

Solve $x^2 + 6x + 8 = 0$.

Solution

First, factor the quadratic expression.

$$(x + 4)(x + 2) = 0$$

Separately set each binomial equal to 0.

$$x + 4 = 0 \qquad x + 2 = 0$$

Solve for x.

$$x = -4 \qquad x = -2$$

$$x = \{-4, -2\}$$

Check by separately substituting –4 for x and –2 for x into $x^2 + 6x + 8 = 0$.

Substitute –4 for x in $x^2 + 6x + 8 = 0$	Substitute –2 for x in $x^2 + 6x + 8 = 0$
$(-4)^2 + 6(-4) + 8 = 0$	$(-2)^2 + 6(-2) + 8 = 0$
$16 - 24 + 8 = 0$	$4 - 12 + 8 = 0$
$-8 + 8 = 0$	$-8 + 8 = 0$
$0 = 0$ ✔	$0 = 0$ ✔

Answer: $x = \{-4, -2\}$

➡ Example 2

Find the solutions to $x^2 = 7x - 12$.

Solution

Here the quadratic equation is not in the form $ax^2 + bx + c = 0$, so we need to first subtract $(7x - 12)$ from both sides.

$$x^2 - (7x - 12) = 7x - 12 - (7x - 12)$$
$$x^2 - 7x + 12 = 0$$

Next, we factor the quadratic expression.

$$(x - 4)(x - 3) = 0$$

Then, separately set each binomial equal to 0, and solve for x.

$$x - 4 = 0 \qquad x - 3 = 0$$
$$x = 4 \qquad x = 3$$

$$x = \{3, 4\}$$

Check by separately substituting 3 for x and 4 for x into $x^2 = 7x - 12$.

Substitute 3 for x in $x^2 = 7x - 12$	Substitute 4 for x in $x^2 = 7x - 12$
$(3)^2 = 7(3) - 12$	$(4)^2 = 7(4) - 12$
$9 = 21 - 12$	$16 = 28 - 12$
$9 = 9$ ✔	$16 = 16$ ✔

Answer: $x = \{3, 4\}$

LESSON 16.1B—SKILLS CHECK

Directions: Solve the following equations by factoring.

1. $x^2 + 9x + 8 = 0$

2. $x^2 + 2x = 8$

3. $x^2 - 9x + 18 = 0$

4. $2x^2 - 5x - 3 = 0$

5. $5x^2 = 13x + 6$

(Answers are on page 383.)

ACCUPLACER TIPS

Using the answer choices, problems like these can be solved by working backwards. Try the following example.

For which of the following equations are $x = 6$ and $x = -3$ both solutions?

(A) $x^2 - 36 = 0$
(B) $x^2 - 9 = 0$
(C) $x^2 - 3x - 18 = 0$
(D) $x^2 + 3x - 18 = 0$
(E) $x^2 - 6 = 0$

An alternative way to solve this problem is to plug the solutions $x = 6$ and $x = -3$ into each equation given as an answer choice. The equation that is made true by both solutions will be the correct answer.

Choice	Substituting 6 for x	Substituting –3 for x
A. $x^2 - 36 = 0$	$6^2 - 36 = 0$ $36 - 36 = 0$ ✔	$(-3)^2 - 36 = 0$ $9 - 36 = 0$ $-27 = 0$ ✗
B. $x^2 - 9 = 0$	$6^2 - 9 = 0$ $36 - 9 = 0$ $27 = 0$ ✗	$(-3)^2 - 9 = 0$ $9 - 9 = 0$ $0 = 0$ ✔
C. $x^2 - 3x - 18 = 0$	$6^2 - 3(6) - 18 = 0$ $36 - 18 - 18 = 0$ $18 - 18 = 0$ $0 = 0$ ✔	$(-3)^2 - 3(-3) - 18 = 0$ $9 + 9 - 18 = 0$ $18 - 18 = 0$ $0 = 0$ ✔
D. $x^2 + 3x - 18 = 0$	$6^2 + 3(6) - 18 = 0$ $36 + 18 - 18 = 0$ $54 - 18 = 0$ $36 = 0$ ✗	$(-3)^2 + 3(-3) - 18 = 0$ $9 - 9 - 18 = 0$ $0 - 18 = 0$ $-18 = 0$ ✗
E. $x^2 - 6 = 0$	$6^2 - 6 = 0$ $36 - 6 = 0$ $30 = 0$ ✗	$(-3)^2 - 6 = 0$ $9 - 6 = 0$ $3 = 0$ ✗

Only choice C was made true by $x = 6$ and $x = -3$, so this is the correct answer.

LESSON 16.1—ACCUPLACER CHALLENGE

> **Directions**: For each of the questions below, choose the best answer from the five choices given.

1. What is the solution set of $g^2 - 3g - 10 = 0$?

   (A) $g = \{5, -2\}$
   (B) $g = \{2, -5\}$
   (C) $g = \{3, -10\}$
   (D) $g = \{3, 10\}$
   (E) $g = \{2, -10\}$

2. The length of Tamara's rectangular garden is four feet more than the width. The area of her garden is 480 square feet. Which of the following equations could be used to find the width of Tamara's garden in feet?

   (A) $4w + 8 = 480$
   (B) $w^2 - 4w + 480 = 0$
   (C) $w^2 + 4w - 480 = 0$
   (D) $w^2 - 4w = 480$
   (E) $w^2 + 4w + 480 = 0$

3. For which of the following equations are $x = 7$ and $x = -7$ both solutions?

   (A) $x^2 + 14x - 49 = 0$
   (B) $x^2 - 14x - 49 = 0$
   (C) $x^2 - 14x + 49 = 0$
   (D) $x^2 - 49 = 0$
   (E) $x^2 + 49 = 0$

4. What is the product of the roots of the equation $x^2 + 5x - 36 = 0$?

   (A) -36
   (B) -5
   (C) 13
   (D) 36
   (E) 180

5. If $2x^2 + 5x = 12$, then one of the possible values of $5x^2 + 2x$ is

   (A) 12
   (B) 24
   (C) 72
   (D) 88
   (E) 125

(Answers are on page 384.)

LESSON 16.2—SOLVING QUADRATIC EQUATIONS USING THE QUADRATIC FORMULA

Not all quadratic expressions can be factored easily. For example, to factor and solve $x^2 + 5x + 8 = 0$, we would need to find two numbers that multiply to 8 and add to 5. There are no whole numbers that will do this. Factoring just isn't the way to tackle this problem. Luckily, there is another way. It's called the quadratic formula, and it looks like this:

The Quadratic Formula

If $ax^2 + bx + c = 0$, then: $x = \dfrac{-b \pm \sqrt{b^2 - 4ac}}{2a}$

Use the quadratic formula to solve quadratic equations when factoring is too complicated or difficult. Actually, if you prefer, you can use the quadratic formula to solve quadratic equation problems that can be solved by factoring too. In other words, you can always use the quadratic formula to solve any quadratic equation.

To Solve a Quadratic Equation Using the Quadratic Formula:

1. If it is not already done, set the equation equal to 0 to fit the form $ax^2 + bx + c = 0$.
2. Identify a, b, and c, and substitute the values into the quadratic formula.
3. Following the order of operations (PEMDAS), first evaluate $2a$ in the denominator and $\sqrt{b^2 - 4ac}$ in the numerator.
4. Split the equation into separate equations, one for $+\sqrt{b^2 - 4ac}$ and one for $-\sqrt{b^2 - 4ac}$.
5. Simplify the expression.

Before we start, remember that $\pm$ is read as "plus or minus" and is used in situations where we have two possible solutions. For example, the equation $x^2 = 25$ has two possible solutions, 5 or –5. We know this because $5^2 = 25$ and $(-5)^2 = 25$ as well. To solve the equation $x^2 = 25$, we can say that $x = \sqrt{25}$. Normally, when we think of a square root, we are looking for one solution, the positive solution 5, which is called the **principal root** of 25. The **negative square root** of 25 could be written as $-\sqrt{25} = -5$. To account for both possibilities, we write $\pm\sqrt{25} = \pm 5$. This says that the principal and negative roots of 25 are 5 and –5.

Now, let's start with a problem that we could also solve by factoring.

➡ Example 1

Solve $x^2 + 5x + 6 = 0$.

Solution

First, identify a, b, and c, and substitute the values into the quadratic formula.

$$a = 1, \ b = 5, \text{ and } c = 6$$

$$x = \frac{-5 \pm \sqrt{5^2 - 4(1)(6)}}{2(1)}$$

Evaluate $2a$ in the denominator and $\sqrt{b^2 - 4ac}$ in the numerator.

$$x = \frac{-5 \pm \sqrt{25 - 24}}{2}$$

$$x = \frac{-5 \pm \sqrt{1}}{2}$$

$$x = \frac{-5 \pm 1}{2}$$

Split the equation into separate equations.

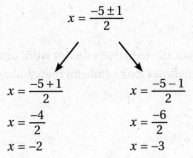

$$x = \frac{-5 \pm 1}{2}$$

$$x = \frac{-5 + 1}{2} \qquad x = \frac{-5 - 1}{2}$$

$$x = \frac{-4}{2} \qquad x = \frac{-6}{2}$$

$$x = -2 \qquad x = -3$$

<u>Answer</u>: $x = \{-2, -3\}$

➡ Example 2

Solve $2x^2 - x = 3$.

Solution

Start out by setting the equation equal to 0 by subtracting 3 from both sides.

$$2x^2 - x - 3 = 3 - 3$$

$$2x^2 - x - 3 = 0$$

Identify a, b, and c, and substitute the values into the quadratic formula.

$$a = 2,\ b = -1,\text{ and } c = -3$$

$$x = \frac{-(-1) \pm \sqrt{(-1)^2 - 4(2)(-3)}}{2(2)}$$

Evaluate $2a$ in the denominator and $\sqrt{b^2 - 4ac}$ in the numerator.

$$x = \frac{1 \pm \sqrt{1 + 24}}{4}$$

$$x = \frac{1 \pm \sqrt{25}}{4}$$

$$x = \frac{1 \pm 5}{4}$$

Split the equation into separate equations.

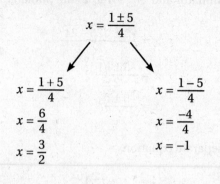

$$x = \frac{1 \pm 5}{4}$$

$$x = \frac{1+5}{4} \qquad x = \frac{1-5}{4}$$

$$x = \frac{6}{4} \qquad x = \frac{-4}{4}$$

$$x = \frac{3}{2} \qquad x = -1$$

<u>Answer:</u> $x = \left\{ \frac{3}{2}, -1 \right\}$

Often, the solutions to quadratic equations do not work out to be nice, neat fractions or whole numbers. They can sometimes look a little bit messy, like the ones in the next example.

➥ Example 3

Find the roots of $2x^2 + 3x - 8 = 0$.

Solution

The **roots** are the values that make the equation equal to 0; they are the solutions, in other words.

First, identify a, b, and c, and substitute the values into the quadratic formula.

$$a = 2, \; b = 3, \text{ and } c = -8$$

$$x = \frac{-3 \pm \sqrt{3^2 - 4(2)(-8)}}{2(2)}$$

Evaluate $2a$ in the denominator and $\sqrt{b^2 - 4ac}$ in the numerator.

$$x = \frac{-3 \pm \sqrt{9 + 64}}{4}$$

$$x = \frac{-3 \pm \sqrt{73}}{4}$$

Split the equation into separate equations.

$$x = \frac{-3 \pm \sqrt{73}}{4}$$

$$x = \frac{-3 + \sqrt{73}}{4} \qquad x = \frac{-3 - \sqrt{73}}{4}$$

<u>Answer:</u> The roots are $\left\{ \frac{-3 + \sqrt{73}}{4}, \frac{-3 - \sqrt{73}}{4} \right\}$. Notice how similar these two solutions are.

The only difference is the operation in front of the square root sign in the numerator.

LESSON 16.2—SKILLS CHECK

Directions: Find the roots for each equation using the quadratic formula.

1. $2x^2 - 8x - 10 = 0$

2. $3x^2 + 3x = 6$

3. $3x^2 + 11x - 4 = 0$

4. $x^2 + 4x + 2 = 0$

5. $3n^2 = 3 + 2n$

(Answers are on page 385.)

LESSON 16.2—ACCUPLACER CHALLENGE

Directions: For each of the questions below, choose the best answer from the five choices given.

1. What is one root of $2x^2 - 4x - 5 = 0$?

   (A) $1 - \sqrt{14}$

   (B) $\dfrac{2 + \sqrt{14}}{2}$

   (C) $1 + \sqrt{7}$

   (D) $\dfrac{-2 + \sqrt{14}}{2}$

   (E) $1 - \sqrt{6}$

2. For which of the following equations are $x = 4$ and $x = -3$ both solutions?

   (A) $x^2 + x - 12 = 0$
   (B) $x^2 - 7x + 12 = 0$
   (C) $x^2 - x - 12 = 0$
   (D) $x^2 + 2x - 24 = 0$
   (E) $x^2 + 2x + 3 = 0$

3. For which of the following equations is $\dfrac{-3 + \sqrt{13}}{2}$ a solution?

   (A) $x^2 + 3x - 1 = 0$
   (B) $x^2 + 3x + 1 = 0$
   (C) $x^2 - 3x - 1 = 0$
   (D) $x^2 - 3x + 1 = 0$
   (E) $2x^2 + 3x + 1 = 0$

4. If $3x^2 + 2x - 4 = 0$, then one of the possible values of $\left(x + \dfrac{1}{9}\right)$ is

(A) $\dfrac{\sqrt{13}}{9}$

(B) $\dfrac{-2 + 3\sqrt{13}}{9}$

(C) $\dfrac{-1 + \sqrt{13}}{3}$

(D) $\dfrac{-4 + 3\sqrt{13}}{9}$

(E) $\dfrac{-4 - 3\sqrt{13}}{9}$

5. A rectangular room has an area of 100 square feet. If the width of the room is 10 feet more than the length, what is the length of the room in feet?

(A) $-10\sqrt{5}$

(B) $-5 - 5\sqrt{5}$

(C) $-5 + 5\sqrt{5}$

(D) $-10 + 10\sqrt{5}$

(E) $-10 - 10\sqrt{5}$

(Answers are on page 388.)

LESSON 16.3—THE GRAPHS OF QUADRATIC FUNCTIONS

The graph of a quadratic function forms a shape called a parabola. A **parabola** is a U-shaped curve. On the ACCUPLACER, you will need to know how to use quadratic equations to make inferences about the shapes of graphs. For example, we have spent time in this chapter looking at how to find the zeros of quadratic functions. On the graph of a parabola, these zeros are where the graph crosses the x-axis—in other words, where $y = 0$.

One benchmark parabola is the graph of $y = x^2$. We can see from the figure below that the graph of $y = x^2$ only touches the x-axis at one place, the origin. Therefore, we could say that it has one real zero.

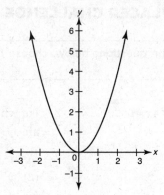

Let's look at another example of how we can use zeros to understand the shape of a parabola.

➡ Example 1

Where will the graph of $y = x^2 - 9$ cross the x-axis?

Solution

To find out where this graph crosses the x-axis, we need to find the zeros of the equation. This means that we should set y equal to 0 and then solve the quadratic equation by factoring or by using the quadratic formula.

$$y = x^2 - 9$$
$$0 = x^2 - 9$$
$$0 = (x + 3)(x - 3)$$

<u>Answer</u>: Solving each binomial for x, we find that $x = -3$ and $x = 3$. This tells us that the parabola crosses that x-axis at the points $(-3, 0)$ and $(3, 0)$.

Quadratic equations can give us more key information than just the location of the zeros. One other feature we can quickly identify is whether or not the parabola opens upward or downward.

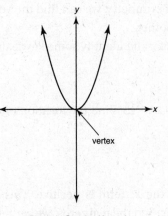

Parabola Opening Upward

A parabola that opens upward will have a positive coefficient of the x^2 term. For example, all of the following graphs would open upward:

$$y = x^2 \qquad y = 5x^2 - 36 \qquad y = x^2 + 29x + 100$$

Looking at the image of the parabola opening upward, we can see that there is a lowest point that the parabola reaches called its **minimum**. This point is also referred to as the parabola's **vertex**.

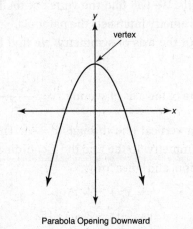

Parabola Opening Downward

A parabola that opens downward will have a negative coefficient of the x^2 term. These parabolas would open downward:

$$y = -x^2 \qquad y = 9 - x^2 \qquad y = -3x^2 + 7x - 10$$

Looking at the image of the parabola opening downward, we can see that there is a highest point that the parabola reaches called its **maximum**. This point is also referred to as the parabola's **vertex**.

All upward or downward facing parabolas are perfectly symmetrical across a vertical line called an **axis of symmetry**. The axis of symmetry runs perfectly down the middle of the parabola through the vertex.

On the ACCUPLACER, you may be asked to find the vertex of a quadratic equation. For quadratic equations with only one root, the vertex and the one real zero will be the same point. This is the one place where the parabola touches the x-axis. For quadratic equations with two roots, we must first find the axis of symmetry exactly halfway between the two real zeros. Once we find the axis of symmetry, we can find the vertex where the axis of symmetry and the parabola cross one another.

Let's look at another parabola, and identify some key features of its graph.

➡ Example 2

Consider the graph of $y = x^2 + 2x - 15$. Find the location of the zeros of the graph as well as its vertex.

Solution

Notice that the coefficient of the x^2 term is positive. This means that the graph will be a parabola that opens upward. To find the real zeros, we set y equal to 0, and solve by factoring or by using the quadratic formula. This quadratic equation happens to be factorable (as many on the ACCUPLACER will be).

$$y = x^2 + 2x - 15$$
$$0 = x^2 + 2x - 15$$
$$0 = (x + 5)(x - 3)$$

Setting each binomial equal to zero and solving, we find that $x = -5$ and $x = 3$. This means that the graph crosses the x-axis at $(-5, 0)$ and $(3, 0)$. The axis of symmetry will pass through the middle of these two points. We will find the vertex at the bottom, or minimum, of the parabola where the axis of symmetry intersects the parabola.

To find the x-coordinate of the axis of symmetry, we find the middle of the two zeros, $x = -5$ and $x = 3$.

$$x\text{-coordinate of the axis of symmetry} = \frac{-5 + 3}{2} = \frac{-2}{2} = -1$$

The axis of symmetry is a vertical line through $x = -1$. The vertex will have the same x-coordinate as the axis of symmetry. To the find the y-coordinate of the vertex, we can plug $x = -1$ into the original equation, and solve for y.

$$y = x^2 + 2x - 15$$
$$y = (-1)^2 + 2(-1) - 15$$
$$y = 1 + (-2) - 15$$
$$y = -1 - 15$$
$$y = -16$$

The vertex is located at the point $(-1, -16)$.

Answer: As shown below, the graph opens upward and has a vertex at the point (–1, –16). It has real zeros at the points (–5, 0) and (3, 0).

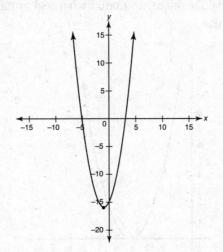

Example 3

For what real numbers x is $x^2 - 8x + 16$ negative?

(A) $-4 < x < 4$

(B) $x < -4 \ or \ x > 4$

(C) $x = -4 \ or \ x = 4$

(D) $0 < x < 8$

(E) For no real numbers x

Solution

The wording of this problem is tricky. We're looking for values of x that will make the expression $x^2 - 8x + 16$ result in a negative number.

It's difficult to determine this without testing out a few things. First, let's start by solving the equation $x^2 - 8x + 16 = 0$ to determine the real numbers x that make this expression equal 0. This time, let's use the quadratic formula.

Identify a, b, and c, and substitute the values into the quadratic formula.

$$a = 1, \ b = -8, \text{ and } c = 16$$

$$x = \frac{-(-8) \pm \sqrt{(-8)^2 - 4(1)(16)}}{2(1)}$$

$$x = \frac{8 \pm \sqrt{64 - 64}}{2}$$

$$x = \frac{8 \pm 0}{2} = \frac{8}{2} = 4$$

This quadratic equation has only one real zero, or solution, at (4, 0). Remember that if a quadratic equation has only one real zero, then its graph only intersects the x-axis at that one single point. Therefore, the one solution, or 0, is also the location of the vertex. We also know that, because this equation has a positive coefficient of x^2, its graph is a parabola that opens upward. This means that the lowest point of the parabola touches the x-axis at (4, 0), but it never goes below the x-axis! Therefore, no matter what real number value we enter for x, it will always produce a positive y.

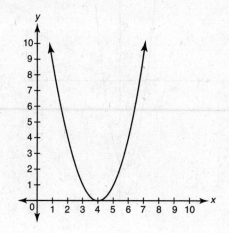

Answer: As a result, we must conclude that no real numbers x will result in a negative value for $y = x^2 - 8x + 16$, so the answer is choice E.

➥ Example 4

For what real numbers x will $25 - x^2$ be positive?

(A) $-5 < x < 5$

(B) $-5 > x \ or \ 5 < x$

(C) $x = -5 \ and \ x = 5$

(D) $0 < x < 25$

(E) For no real numbers x

Solution

Notice that the coefficient of the x^2 term is negative. This means that the graph will be a parabola that opens downward. Let's find its zeros by factoring.

$$y = 25 - x^2$$
$$0 = 25 - x^2$$
$$0 = (5 - x)(5 + x)$$

Solving $5 - x = 0$ and $5 + x = 0$, we find that the quadratic equation has real zeros at $x = -5$ and $x = 5$. We can find the x-coordinate of the vertex by first finding the middle of the two zeros.

$$x\text{-coordinate of the vertex} = \frac{-5 + 5}{2} = \frac{0}{2} = 0$$

Now find the y-coordinate of the vertex.

$$y = 25 - x^2$$
$$y = 25 - (0)^2$$
$$y = 25$$

The vertex is located at the point (0, 25). This means that this equation reaches its maximum when $x = 0$ and then decreases as x increases or decreases. We also know that it crosses the x-axis when $x = -5$ or $x = 5$. This means that every x-value in between—but not including—those two points will produce a positive value for y. The graph of this equation is shown below.

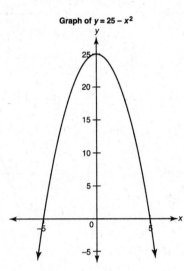

Graph of $y = 25 - x^2$

Answer: Since this equation will only produce positive values when x is greater than -5 but less than 5, choice A is correct.

LESSON 16.3—SKILLS CHECK

Directions: For each quadratic equation, identify the real zeros and the x-value of the vertex.

1. $y = 1 - x^2$

3. $y = -x^2 + 2x + 35$

5. $y = (x - 8)^2$

2. $y = x^2 + 11x - 26$

4. $y = -x^2 + 81$

(Answers are on page 390.)

LESSON 16.3—ACCUPLACER CHALLENGE

Directions: For each of the questions below, choose the best answer from the five choices given.

1. For which real numbers x will $x^2 + 6x + 9$ be greater than 0?

(A) $x > -3$
(B) $x < -3$
(C) $x = -3$
(D) $x \neq -3$
(E) For no real numbers x

2. Which of the following choices is the graph of $y = 4 - x^2$?

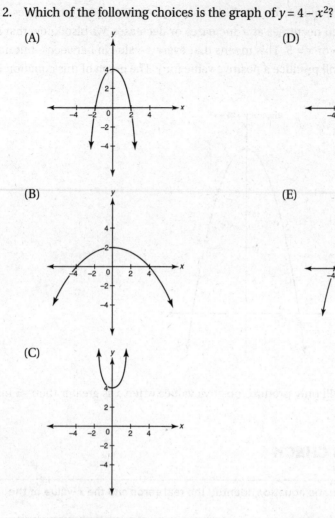

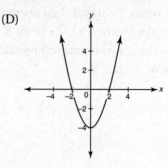

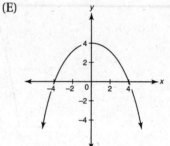

3. On which interval is $x^2 - x - 20$ negative?

(A) $-20 < x < 20$

(B) $-5 < x < 4$

(C) $-4 < x < 5$

(D) $-5 < x < -4$

(E) There is no interval for which $x^2 - x - 20$ is negative.

4. What is the location of the vertex of the parabola $y = x^2 - 8x + 12$?

(A) $(-4, 4)$

(B) $(0, 4)$

(C) $(0, -4)$

(D) $(4, 4)$

(E) $(4, -4)$

5. Which equation has real zeros at the points $(-7, 0)$ and $(4, 0)$?

(A) $y = x^2 - 28$

(B) $y = x^2 - 11x + 28$

(C) $y = x^2 + 3x - 28$

(D) $y = x^2 - 3x - 28$

(E) $y = x^2 + 3x + 28$

(Answers are on page 391.)

To complete a set of practice questions that reviews all of Chapter 16, go to:
barronsbooks.com/TP/accuplacer/math29qw/

Lesson 16.1A—Skills Check (page 368)

1. **$x = -2$ or $x = -3$**   Write each binomial as a separate equation equal to 0.

   $(x + 2) = 0$ $(x + 3) = 0$

   Separately solve each equation.

   $x = -2$ $x = -3$

2. **$n = 2$ or $n = 3$**   Write each binomial as a separate equation equal to 0.

   $(n - 2) = 0$ $(n - 3) = 0$

   Separately solve each equation.

   $n = 2$ $n = 3$

3. **$y = -5$ or $y = 5$**   Write each binomial as a separate equation equal to 0.

   $(y + 5) = 0$ $(y - 5) = 0$

   Separately solve each equation.

   $y = -5$ $y = 5$

4. **$x = 8$ or $x = -10$**   Write each binomial as a separate equation equal to 0.

   $(x - 8) = 0$ $(x + 10) = 0$

   Separately solve each equation.

   $x = 8$ $x = -10$

5. **$p = -12$**   Both binomials are the same, so there is only one solution to find. Setting both binomials equal to 0 here gives us the same result.

   $(p + 12) = 0$ $(p + 12) = 0$

There is only one solution to obtain. Solve for p.

$$p = -12$$

6. **$x = -\dfrac{3}{2}$ or $x = -\dfrac{2}{3}$**   Write each binomial as a separate equation equal to 0.

   $(2x + 3) = 0$ $(-3x - 2) = 0$

   Solving each equation separately, we find that:

   $x = -\dfrac{3}{2}$ $x = -\dfrac{2}{3}$

7. **$x = \dfrac{1}{4}$ or $x = -\dfrac{2}{5}$**   Write each binomial as a separate equation equal to 0.

   $(4x - 1) = 0$ $(5x + 2) = 0$

   Solving each equation separately, we find that:

   $x = \dfrac{1}{4}$ $x = -\dfrac{2}{5}$

8. **$n = \dfrac{9}{2}$ or $n = -1$**   Write each binomial as a separate equation equal to 0.

   $(9 - 2n) = 0$ $(4 + 4n) = 0$

   Solving each equation separately, we find that:

   $n = \dfrac{9}{2}$ $n = -1$

Lesson 16.1B—Skills Check (page 369)

1. **$x = -8$ or $x = -1$**

   $$x^2 + 9x + 8 = 0$$

   Factor the quadratic expression into two binomials.

   $$(x + 8)(x + 1) = 0$$

Write each binomial as a separate equation equal to 0.

$x + 8 = 0$ $x + 1 = 0$

Separately solve each equation.

$x = -8$ $x = -1$

2. $x = -4$ *or* $x = 2$

$$x^2 + 2x = 8$$

Write the equation in $ax^2 + bx + c = 0$ form.

$$x^2 + 2x - 8 = 0$$

Factor the quadratic expression into two binomials.

$$(x + 4)(x - 2) = 0$$

Write each binomial as a separate equation equal to 0.

$$x + 4 = 0 \qquad x - 2 = 0$$

Separately solve each equation.

$$x = -4 \qquad x = 2$$

3. $x = 6$ *or* $x = 3$

$$x^2 - 9x + 18 = 0$$

Factor the quadratic expression into two binomials.

$$(x - 6)(x - 3) = 0$$

Write each binomial as a separate equation equal to 0.

$$x - 6 = 0 \qquad x - 3 = 0$$

Separately solve each equation.

$$x = 6 \qquad x = 3$$

4. $x = -\dfrac{1}{2}$ *or* $x = 3$

$$2x^2 - 5x - 3 = 0$$

Factor the quadratic expression into two binomials.

$$(2x + 1)(x - 3) = 0$$

Write each binomial as a separate equation equal to 0.

$$2x + 1 = 0 \qquad x - 3 = 0$$

Separately solve each equation.

$$x = -\frac{1}{2} \qquad x = 3$$

5. $x = -\dfrac{2}{5}$ *or* $x = 3$

$$5x^2 = 13x + 6$$

Write the equation in $ax^2 + bx + c = 0$ form.

$$5x^2 - 13x - 6 = 0$$

Factor the quadratic expression into two binomials.

$$(5x + 2)(x - 3) = 0$$

Write each binomial as a separate equation equal to 0.

$$5x + 2 = 0 \qquad x - 3 = 0$$

Separately solve each equation.

$$x = -\frac{2}{5} \qquad x = 3$$

Lesson 16.1—ACCUPLACER Challenge (page 371)

1. **(A)** To find the solution set, start by factoring the quadratic expression.

$$g^2 - 3g - 10 = 0$$
$$(g - 5)(g + 2) = 0$$

Then write each binomial as a separate equation equal to 0, and solve.

$$(g - 5) = 0 \qquad\qquad (g + 2) = 0$$
$$g = 5 \qquad\qquad g = -2$$
$$g = \{5, -2\}$$

2. **(C)** Using l to represent the length of Tamara's garden, and w to represent the width of her garden, we could write the equation $l \times w = 480$ to represent the area of her garden in square feet. We also know that the length of her garden is four feet more than the width, so we can state that $l = w + 4$. We can then rewrite the equation as $(w + 4) \times w = 480$. This then becomes $w(w + 4) = 480$, and then $w^2 + 4w = 480$. In order

to solve a quadratic equation, we normally put the equation into standard form, $ax^2 + bx + c = 0$. In standard form, $w^2 + 4w = 480$ becomes $w^2 + 4w - 480 = 0$, which is choice C.

3. **(D)** We need to find a quadratic equation that has –7 and 7 as solutions. These numbers are opposites, and opposites solve quadratic equations that are the difference of two squares equal to 0. The only answer choice that fits the form $a^2 - b^2 = 0$ is choice D. To test to see if these solutions solve $x^2 - 49 = 0$, we separately substitute 7 for x and –7 for x.

$$7^2 - 49 = 0 \qquad\qquad (-7)^2 - 49 = 0$$
$$49 - 49 = 0 \checkmark \qquad\qquad 49 - 49 = 0 \checkmark$$

4. **(A)** The roots of the equation are the solutions to the equation when the equation is equal to 0. To solve the equation, we factor, and then solve.

$$x^2 + 5x - 36 = 0$$
$$(x + 9)(x - 4) = 0$$

$$x + 9 = 0 \qquad\qquad x - 4 = 0$$
$$x = -9 \qquad\qquad x = 4$$

5. **(C)** Start by getting the equation $2x^2 + 5x = 12$ in standard form to find the value of x.

$$2x^2 + 5x - 12 = 0$$

Factor, and solve the equation.

$$(2x - 3)(x + 4) = 0$$

$$2x - 3 = 0 \qquad\qquad x + 4 = 0$$
$$x = \frac{3}{2} \qquad\qquad x = -4$$

Then, separately substitute $\frac{3}{2}$ and –4 for x into the expression $5x^2 + 2x$.

$$5\left(\frac{3}{2}\right)^2 + 2\left(\frac{3}{2}\right) = 5\left(\frac{9}{4}\right) + \frac{6}{2} = \frac{45}{4} + 3 = 11\frac{1}{4} + 3 = 14\frac{1}{4}$$

$$5(-4)^2 + 2(-4) = 5(16) - 8 = 80 - 8 = 72$$

Both $14\frac{1}{4}$ and 72 are correct answers, but only 72 is an answer choice.

Lesson 16.2—Skills Check (page 375)

1. $x = \{-1, 5\}$

$$2x^2 - 8x - 10 = 0$$

Identify a, b, and c, and substitute the values into the quadratic formula.

$$a = 2,\ b = -8,\ c = -10$$

$$x = \frac{-(-8) \pm \sqrt{(-8)^2 - 4(2)(-10)}}{2(2)}$$

$$x = \frac{8 \pm \sqrt{64 + 80}}{4}$$

$$x = \frac{8 \pm \sqrt{144}}{4}$$

Split both roots into two separate equations.

$$x = \frac{8+12}{4} \qquad x = \frac{8-12}{4}$$

$$x = \frac{20}{4} \qquad x = \frac{-4}{4}$$

$$x = 5 \qquad x = -1$$

2. $x = \{-2, 1\}$ Subtract 6 from both sides to get the equation equal to 0.

$$3x^2 + 3x - 6 = 6 - 6$$
$$3x^2 + 3x - 6 = 0$$

Identify a, b, and c, and substitute the values into the quadratic formula.

$$a = 3, \ b = 3, \ c = -6$$

$$x = \frac{-3 \pm \sqrt{3^2 - 4(3)(-6)}}{2(3)}$$

$$x = \frac{-3 \pm \sqrt{9 + 72}}{6}$$

$$x = \frac{-3 \pm \sqrt{81}}{6}$$

Split both roots into two separate equations.

$$x = \frac{-3+9}{6} \qquad x = \frac{-3-9}{6}$$

$$x = \frac{6}{6} \qquad x = \frac{-12}{6}$$

$$x = 1 \qquad x = -2$$

3. $x = \left\{-4, \dfrac{1}{3}\right\}$

$$3x^2 + 11x - 4 = 0$$

Identify a, b, and c, and substitute the values into the quadratic formula.

$$a = 3, \ b = 11, \ c = -4$$

$$x = \frac{-11 \pm \sqrt{11^2 - 4(3)(-4)}}{2(3)}$$

$$x = \frac{-11 \pm \sqrt{121 + 48}}{6}$$

$$x = \frac{-11 \pm \sqrt{169}}{6}$$

Split both roots into two separate equations.

$$x = \frac{-11+13}{6} \qquad x = \frac{-11-13}{6}$$

$$x = \frac{2}{6} \qquad x = \frac{-24}{6}$$

$$x = \frac{1}{3} \qquad x = -4$$

4. $x = \left\{ -2 - \sqrt{2}, \ -2 + \sqrt{2} \right\}$

$$x^2 + 4x + 2 = 0$$

Identify a, b, and c, and substitute the values into the quadratic formula.

$$a = 1, \ b = 4, \ c = 2$$

$$x = \frac{-4 \pm \sqrt{4^2 - 4(1)(2)}}{2(1)}$$

$$x = \frac{-4 \pm \sqrt{16 - 8}}{2}$$

$$x = \frac{-4 \pm \sqrt{8}}{2}$$

$$x = \frac{-4 \pm 2\sqrt{2}}{2}$$

Split both roots into two separate equations.

$$x = \frac{-4 + 2\sqrt{2}}{2} \qquad x = \frac{-4 - 2\sqrt{2}}{2}$$

$$x = -2 + \sqrt{2} \qquad x = -2 - \sqrt{2}$$

5. $x = \left\{ \dfrac{1 - \sqrt{10}}{3}, \ \dfrac{1 + \sqrt{10}}{3} \right\}$ Subtract $(3 + 2n)$ from both sides to get the equation equal to 0.

$$3n^2 - (3 + 2n) = 3 + 2n - (3 + 2n)$$

$$3n^2 - 2n - 3 = 0$$

Identify a, b, and c, and substitute the values into the quadratic formula.

$$a = 3, \ b = -2, \ c = -3$$

$$x = \frac{-(-2) \pm \sqrt{(-2)^2 - 4(3)(-3)}}{2(3)}$$

$$x = \frac{2 \pm \sqrt{4 + 36}}{6}$$

$$x = \frac{2 \pm \sqrt{40}}{6}$$

$$x = \frac{2 \pm 2\sqrt{10}}{6}$$

Split both roots into two separate equations.

$$x = \frac{2 - 2\sqrt{10}}{6} \qquad x = \frac{2 + 2\sqrt{10}}{6}$$

$$x = \frac{1 - \sqrt{10}}{3} \qquad x = \frac{1 + \sqrt{10}}{3}$$

Lesson 16.2—ACCUPLACER Challenge (pages 375–376)

1. **(B)** To find a root of $2x^2 - 4x - 5 = 0$, first identify a, b, and c, and then substitute the values into the quadratic formula.

$$a = 2,\ b = -4,\ \text{and}\ c = -5$$

$$x = \frac{-(-4) \pm \sqrt{(-4)^2 - 4(2)(-5)}}{2(2)}$$

$$x = \frac{4 \pm \sqrt{16 + 40}}{4}$$

$$x = \frac{4 \pm \sqrt{56}}{4}$$

Split both roots into two separate equations.

$$x = \frac{4 - \sqrt{56}}{4} \qquad x = \frac{4 + \sqrt{56}}{4}$$

$$x = \frac{4 - 2\sqrt{14}}{4} \qquad x = \frac{4 + 2\sqrt{14}}{4}$$

$$x = \frac{2 - \sqrt{14}}{2} \qquad x = \frac{2 + \sqrt{14}}{2}$$

2. **(C)** Knowing that $x = 4$ and $x = -3$ are both solutions to one of the answer choices, which are all quadratic equations equal to 0, we can work backwards to determine that these solutions must also solve $(x - 4)(x + 3) = 0$. We can then multiply the two binomials $(x - 4)(x + 3)$ to find that this expression is equivalent to $x^2 - x - 12$. Therefore, the correct answer is choice C. Using the quadratic formula to find the solution of $x^2 - x - 12 = 0$ supports this.

3. **(A)** One method to solve this problem would be to use the quadratic formula on each of the answer choices to find the one that has $\frac{-3 + \sqrt{13}}{2}$ as a solution. However, using the **discriminant**, which is the $\sqrt{b^2 - 4ac}$ part of the quadratic formula, will help us determine which of the equations will result in a solution with $\sqrt{13}$. All the choices have $a = 1$, or $a = 2$, $b = \pm3$, and $c = \pm1$. Working with those, we can determine that $a = 1$ and $c = -1$, because $\sqrt{(\pm3)^2 - 4(1)(-1)} = \sqrt{13}$. Then, looking to the -3 in the numerator of the solution $\frac{-3 + \sqrt{13}}{2}$, we know that $b = 3$, because the formula calls for $-b$ in the numerator. That leads us to choice A. To be sure, we can plug the values of a, b, and c into the quadratic formula. Doing so will support that choice A is the correct answer.

4. **(B)** We must first start out by finding the two possible values of x by using the quadratic formula to solve $3x^2 + 2x - 4 = 0$.

$$x = \frac{-2 \pm \sqrt{2^2 - 4(3)(-4)}}{2(3)}$$

$$x = \frac{-2 \pm \sqrt{4 + 48}}{6}$$

$$x = \frac{-2 \pm \sqrt{52}}{6}$$

$$x = \frac{-2 \pm 2\sqrt{13}}{6}$$

$$x = \frac{-1+\sqrt{13}}{3} \quad or \quad x = \frac{-1-\sqrt{13}}{3}$$

Then, separately substitute both solutions into $\left(x + \frac{1}{9}\right)$.

$$\left(x + \frac{1}{9}\right) = \frac{-1+\sqrt{13}}{3} + \frac{1}{9} \qquad \left(x + \frac{1}{9}\right) = \frac{-1-\sqrt{13}}{3} + \frac{1}{9}$$

$$= \frac{-3+3\sqrt{13}}{9} + \frac{1}{9} \qquad\qquad = \frac{-3-3\sqrt{13}}{9} + \frac{1}{9}$$

$$= \frac{-2+3\sqrt{13}}{9} \qquad\qquad\quad = \frac{-2-3\sqrt{13}}{9}$$

Although there are two correct outcomes, only choice B has one of them as an answer choice.

5. **(C)** The area of a rectangle is found using the formula area = length × width. Here the width is 10 feet more than the length, so we can call the length l and the width $l + 10$. It follows then that we can write the following equation to determine the area of the rectangular room in square feet.

$$l \times (l + 10) = 100$$
$$l^2 + 10l = 100$$
$$l^2 + 10l - 100 = 0$$

This quadratic equation is not easily solved by factoring, so we'll go ahead and use the quadratic formula to find l, the length of the room in feet.

$$l = \frac{-10 \pm \sqrt{10^2 - 4(1)(-100)}}{2(1)}$$

$$l = \frac{-10 \pm \sqrt{100 + 400}}{2}$$

$$l = \frac{-10 \pm \sqrt{500}}{2}$$

$$l = \frac{-10 \pm 10\sqrt{5}}{2}$$

$$l = -5 \pm 5\sqrt{5}$$

Using the quadratic formula to solve this equation results in two possible answers. However, $-5 - 5\sqrt{5}$ will result in a negative outcome, and length cannot be negative, so we reject this answer. The only valid choice is choice C.

Lesson 16.3—Skills Check (page 381)

1. **The real zeros are $x = 1$ and $x = -1$. The vertex is at $x = 0$.**

   To find the real zeros, substitute 0 in place of y, factor, and solve each binomial for x.

   $$0 = 1 - x^2$$
   $$0 = (1 - x)(1 + x)$$

   $$1 - x = 0 \qquad\qquad 1 + x = 0$$
   $$x = 1 \qquad\qquad\qquad x = -1$$

   The real zeros are $x = -1$ and $x = 1$. The x-value of the vertex is the midpoint of the real zeros.

   $$x\text{-value of the vertex} = \frac{-1+1}{2} = \frac{0}{2} = 0$$

2. **The real zeros are $x = -13$ and $x = 2$. The vertex is at $x = -5\frac{1}{2}$.**

   To find the real zeros, substitute 0 in place of y, factor, and solve each binomial for x.

   $$0 = x^2 + 11x - 26$$
   $$0 = (x + 13)(x - 2)$$

   $$x + 13 = 0 \qquad\qquad x - 2 = 0$$
   $$x = -13 \qquad\qquad\qquad x = 2$$

   The real zeros are $x = -13$ and $x = 2$. The x-value of the vertex is the midpoint of the real zeros.

   $$x\text{-value of the vertex} = \frac{-13+2}{2} = \frac{-11}{2} = -5\frac{1}{2}$$

3. **The real zeros are $x = 7$ and $x = -5$. The vertex is at $x = 1$.**

   To find the real zeros, substitute 0 in place of y, factor, and solve each binomial for x.

   $$0 = -x^2 + 2x + 35$$
   $$-1(0) = -1(-x^2 + 2x + 35)$$
   $$0 = x^2 - 2x - 35$$
   $$0 = (x - 7)(x + 5)$$

   $$x - 7 = 0 \qquad\qquad x + 5 = 0$$
   $$x = 7 \qquad\qquad\qquad x = -5$$

   The real zeros are $x = 7$ and $x = -5$. The x-value of the vertex is the midpoint of the real zeros.

   $$x\text{-value of the vertex} = \frac{7+(-5)}{2} = \frac{2}{2} = 1$$

4. **The real zeros are $x = 9$ and $x = -9$. The vertex is at $x = 0$.**

   To find the real zeros, substitute 0 in place of y, factor, and solve each binomial for x.

   $$0 = -x^2 + 81$$
   $$0 = 81 - x^2$$
   $$0 = (9 - x)(9 + x)$$

$9 - x = 0$	$9 + x = 0$
$x = 9$	$x = -9$

   The real zeros are $x = 9$ and $x = -9$. The x-value of the vertex is the midpoint of the real zeros.

   $$x\text{-value of the vertex} = \frac{9 + (-9)}{2} = \frac{0}{2} = 0$$

5. **The real zero is at $x = 8$. The vertex is at $x = 8$.**

   There is only one real zero for this equation.

   $$x - 8 = 0$$
   $$x = 8$$

   This means that the vertex must also be located at $x = 8$.

Lesson 16.3—ACCUPLACER Challenge (pages 381-382)

1. **(D)** First find the real zeros by setting $x^2 + 6x + 9$ equal to 0, and then factor.

   $$x^2 + 6x + 9 = 0$$
   $$(x + 3)(x + 3) = 0$$

   Setting $x + 3 = 0$, we find that there is only one real zero at $x = -3$. Since the coefficient of the x^2 term is positive, the parabola opens upward, so the real zero ($x = -3$) is the minimum value. Every other value of x will be greater than 0.

2. **(A)** Find the real zeros and the vertex of $y = 4 - x^2$ by setting y equal to 0 and factoring.

   $$0 = 4 - x^2$$
   $$0 = (2 - x)(2 + x)$$

   Solving both binomials for x, we find that the real zeros are at $x = 2$ and $x = -2$. The coefficient of the x^2 term is negative, so the parabola will open downward. Only choice A has both of these characteristics.

3. **(C)** First find the real zeros by setting $x^2 - x - 20$ equal to 0, and then factor.

   $$x^2 - x - 20 = 0$$
   $$(x - 5)(x + 4) = 0$$

   Solving both binomials for x, we find that the real zeros are at $x = 5$ and $x = -4$. The coefficient of the x^2 term is positive, so the parabola will open upward. It will be negative, or below the x-axis, at all x-values in between -4 and 5.

4. **(E)** First find the real zeros by setting $x^2 - 8x + 12$ equal to 0, and then factor.

$$x^2 - 8x + 12 = 0$$
$$(x-6)(x-2) = 0$$

Solving each binomial for x, we find that the real zeros are $x = 6$ and $x = 2$. The x-value of the vertex will be in the middle of these.

$$x\text{-value of the vertex} = \frac{6+2}{2} = \frac{8}{2} = 4$$

To find the y-coordinate of the vertex, plug $x = 4$ into the original equation.

$$y = x^2 - 8x + 12$$
$$y = (4)^2 - 8(4) + 12$$
$$y = 16 - 32 + 12$$
$$y = -4$$

The vertex is at the point (4, –4).

5. **(C)** The only equation that will produce the real zeros $x = -7$ and $x = 4$ is $y = x^2 + 3x - 28$. Set y equal to 0, factor, and solve each binomial for x.

$$y = x^2 + 3x - 28$$
$$y = (x+7)(x-4)$$

Solving each binomial for x, we get the zeros $x = -7$ and $x = 4$.

Solving Absolute Value Equations, Radical Equations, and Other Complex Equations

17

Let's review the kinds of equations that we've solved so far. We have looked at how to find solutions to one-variable equations and inequalities. We have examined systems of equations that involve solving for two variables. We have examined two key ways of solving quadratic equations. In this chapter, we are going to look at three more kinds of equations that you can expect to see on the College-Level Mathematics test of the ACCUPLACER.

In this chapter, we will work to master

- solving absolute value equations,
- solving equations with radicals/roots, and
- solving systems of equations that involve lines and curves.

LESSON 17.1—SOLVING ABSOLUTE VALUE EQUATIONS

You'll remember from Chapter 8 that the term **absolute value** refers to the distance between a number and zero on the number line. For example, the absolute value of –4 is 4. Absolute value is signified by two tall bars around a number or expression, like this:

$$|{-4}| = 4$$

It's also important to remember that absolute value can never be negative. The College-Level Mathematics test takes the concept of absolute value a step further and asks you to solve equations that involve absolute value. Consider the example below.

➡ Example 1

Solve the equation $|x - 4| = 10$.

Solution

We know that the absolute value of the expression on the left side of the equation must be equal to 10. This means that $x - 4$ can be equal to 10 or –10. Both "cases" will have an absolute value of 10.

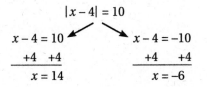

Answer: The solutions are $x = 14$ and $x = -6$. We know this is true because if we plug either 14 or –6 into the original expression on the left side of the equal sign, we get an absolute value of 10.

Absolute value equations, like polynomial equations, can have multiple solutions. On the ACCUPLACER, though, you will really only need to solve simple absolute value equations that result in two solutions.

To Solve an Absolute Value Equation:

1. Isolate the absolute value part of the equation.
2. Create two equations, and remove the absolute value bars.
3. Change all the signs of the terms on the other side of the equation in the second case.
4. Solve both equations.

Let's look at another example.

➥ Example 2

Find the solutions to the equation $|a - 7| - 12 = -4$.

Solution

First, we need to isolate the absolute value part of the equation by adding 12 to both sides.

$$|a - 7| - 12 = -4$$
$$\underline{ +12 +12}$$
$$|a - 7| = 8$$

Next, create two equations. In the first equation, leave the sign of +8 the same. In the second equation, change it to –8. Solve each equation.

$$|a - 7| = 8$$

$$a - 7 = 8 \qquad\qquad a - 7 = -8$$
$$\underline{+7 \ \ +7} \qquad\qquad \underline{+7 \ \ +7}$$
$$a = 15 \qquad\qquad\quad a = -1$$

Answer: The solution is the set $a = \{-1, 15\}$. Like last time, we need to plug both of these solutions back into the original equation to make sure that they are actually correct. Both $|-1 - 7|$ and $|15 - 7|$ are equal to 8, so both solutions are correct.

⇒ Example 3

What is the solution set to the equation $|2x + 3| = x + 6$?

Solution

First, create two equations. Notice that on the right side of the equation, we have to change the signs of both terms in the second case.

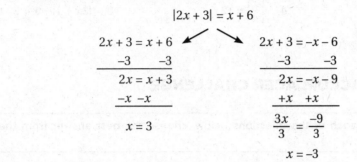

We have a solution set of $x = 3$ and $x = -3$. Plug both into the original equation to make sure that they satisfy it.

$$|2x + 3| = x + 6 \qquad\qquad |2x + 3| = x + 6$$
$$|2(3) + 3| = 3 + 6 \qquad\qquad |2(-3) + 3| = -3 + 6$$
$$|6 + 3| = 9 \qquad\qquad |-6 + 3| = 3$$
$$|9| = 9 \qquad\qquad |-3| = 3$$
$$9 = 9 \qquad\qquad 3 = 3$$

Both of these check out, so our solutions are correct. This may not always be the case, though, so be sure to always double-check your solutions!

Answer: The solution set is $x = \{-3, 3\}$.

Let's try one last example before we move on.

⇒ Example 4

Find the solutions to $|5n + 12| = -32$.

Solution

Look carefully at what this equation says. It says that if we find 5 times n, add 12, and then find the absolute value, the answer will be a *negative* 32. This can't be possible because we know that absolute value can never be negative. In this case, there are no real solutions that will satisfy this equation.

Answer: The set of solutions is the empty set, or $n = \varnothing$, or { }.

The big takeaway from this example is that, if you end up with an absolute value expression on one side of the equal sign and a negative number on the other side of the equal sign, then there are no real solutions to that equation.

LESSON 17.1—SKILLS CHECK

Directions: Find the solution set for each absolute value equation.

1. $|x| = 49$

2. $|x - 11| = 11$

3. $|5x - 5| = 35$

4. $\left|\dfrac{x}{4}\right| = 12$

5. $3|2x + 7| = 45$

6. $|2x + 1| = x + 2$

(Answers are on page 406.)

LESSON 17.1—ACCUPLACER CHALLENGE

Directions: For each of the questions below, choose the best answer from the five choices given.

1. How many real solutions exist for the equation $-3|x + 5| = 24$?

   (A) 0
   (B) 1
   (C) 2
   (D) 3
   (E) More than 3

2. What is the solution set for the absolute value equation $|2x - 3| = 7$?

   (A) $x = -2$ only
   (B) $x = 5$ only
   (C) $x = -2$ and $x = 5$
   (D) $x = 2$ and $x = -5$
   (E) $x = -5$ only

3. Which of the following choices is a solution to the equation $|4x + 1| = 2x + 3$?

   (A) 0
   (B) -1
   (C) $\dfrac{2}{3}$
   (D) $-\dfrac{2}{3}$
   (E) No real solutions exist.

4. Find the solution set for the equation $6 - 5|x + 7| = -19$.

   (A) $x = \{2, 12\}$
   (B) $x = \{-2, 12\}$
   (C) $x = \{2, -12\}$
   (D) $x = \{-2, -12\}$
   (E) None of these

5. Which of the following choices is the solution set to $\left|\dfrac{1}{2}x + 4\right| = x + 2$?

   (A) $x = -4$
   (B) $x = 4$
   (C) $x = \{-4, 4\}$
   (D) $x = 2$
   (E) $x = \{\ \}$

(Answers are on page 408.)

LESSON 17.2—SOLVING RADICAL EQUATIONS

Simply put, **radical equations** are equations that involve square root signs over one or more expressions contained in the equation. As we've seen before when solving equations, we need to keep the equation balanced by performing the same operation to the expressions on both sides of the equal sign. Radical equations involve squaring both sides of the equation in order to eliminate the radical sign. As we'll see, we need to be careful and always check our solutions.

To Solve a Radical Equation:

1. Isolate the radical on one side of the equation.
2. Square both sides of the equation to eliminate the radical.
3. Solve the equation by elimination or by factoring.
4. Check your solution(s) by plugging the solution(s) into the original equation.

This last step isn't just a recommendation when solving these kinds of equations; it's a requirement. As you'll see in some of the examples and practice problems in this chapter, sometimes you get what is called an **extraneous solution**. This happens when you perform all the correct operations, but your answer doesn't check out. Let's look at some examples.

➡ Example 1

Solve $\sqrt{x-8} = 2$.

Solution

Here, the radical is already by itself on one side of the equal sign. To eliminate the radical sign, we need to square both sides of the equation.

$$\left(\sqrt{x-8}\right)^2 = (2)^2$$
$$x - 8 = 4$$
$$\underline{+8 \quad +8}$$
$$x = 12$$

Now that we have a solution, we need to plug it back into the original equation to make sure that it works.

$$\sqrt{x-8} = 2$$
$$\sqrt{12-8} = 2$$
$$\sqrt{4} = 2$$
$$2 = 2$$

Answer: The solution checks out, so the solution is $x = 12$.

➥ Example 2

Solve the equation $3\sqrt{2x-6} - 24 = 0$.

Solution

First, we need to isolate the part of the equation underneath the radical.

$$3\sqrt{2x-6} - 24 = 0$$
$$\underline{\phantom{3\sqrt{2x-6}} + 24 + 24}$$
$$\frac{3\sqrt{2x-6}}{3} = \frac{24}{3}$$
$$\sqrt{2x-6} = 8$$

Next, square both sides of the equation and solve for x.

$$\left(\sqrt{2x-6}\right)^2 = (8)^2$$
$$2x - 6 = 64$$
$$\underline{ + 6 \quad + 6}$$
$$\frac{2x}{2} = \frac{70}{2}$$
$$x = 35$$

Finally, check the solution in the original equation to make sure that it works.

$$3\sqrt{2x-6} - 24 = 0$$
$$3\sqrt{2(35)-6} - 24 = 0$$
$$3\sqrt{70-6} - 24 = 0$$
$$3\sqrt{64} - 24 = 0$$
$$3(8) - 24 = 0$$
$$24 - 24 = 0$$
$$0 = 0$$

Answer: The solution checks out, so $x = 35$.

Let's look at an example of a radical equation where the solution *doesn't* work.

➥ Example 3

Solve $\sqrt{4x+9} + 12 = 7$.

Solution

First, isolate the radical part of the equation.

$$\sqrt{4x+9} + 12 = 7$$
$$\underline{\phantom{\sqrt{4x+9}} - 12 \quad -12}$$
$$\sqrt{4x+9} = -5$$

Square both sides of the equation, and solve for x.

$$\left(\sqrt{4x+9}\right)^2 = (-5)^2$$
$$4x + 9 = 25$$
$$\underline{-9 \quad -9}$$
$$\frac{4x}{4} = \frac{16}{4}$$
$$x = 4$$

We get a solution of $x = 4$. However, look what happens when we check it.

$$\sqrt{4x+9} + 12 = 7$$
$$\sqrt{4(4)+9} + 12 = 7$$
$$\sqrt{16+9} + 12 = 7$$
$$\sqrt{25} + 12 = 7$$
$$5 + 12 = 7$$
$$17 \neq 7$$

We get an untrue statement! This means that our solution doesn't actually satisfy the equation and that there are, in fact, no real solutions to this radical equation. This example demonstrates why it's so important to take the time to check your solution!

Answer: There are no real solutions.

➡ Example 4

Solve the radical equation $\sqrt{2x+8} = x + 4$.

Solution

Since the radical is already isolated on one side of the equal sign, square both sides.

$$\left(\sqrt{2x+8}\right)^2 = (x+4)^2$$
$$2x + 8 = x^2 + 8x + 16$$

Notice that, this time, we have a quadratic equation. We should combine all the like terms, and set the equation equal to 0. We can then factor to find possible solutions.

$$2x + 8 = x^2 + 8x + 16$$
$$\underline{-8 \qquad\qquad -8}$$
$$2x = x^2 + 8x + 8$$
$$\underline{-2x \qquad -2x}$$
$$0 = x^2 + 6x + 8$$

Upon factoring, we find that $x^2 + 6x + 8 = (x + 4)(x + 2) = 0$. This gives us the possible solutions $x = -4$ and $x = -2$. Plug them back into the original equation to check them.

$$\sqrt{2x + 8} = x + 4 \qquad\qquad \sqrt{2x + 8} = x + 4$$
$$\sqrt{2(-4) + 8} = -4 + 4 \qquad \sqrt{2(-2) + 8} = -2 + 4$$
$$\sqrt{-8 + 8} = 0 \qquad\qquad \sqrt{-4 + 8} = 2$$
$$\sqrt{0} = 0 \qquad\qquad\qquad \sqrt{4} = 2$$
$$0 = 0 \qquad\qquad\qquad\quad 2 = 2$$

Both solutions check out! Be careful though—this won't always be the case.

Answer: $x = \{-4, -2\}$

LESSON 17.2—SKILLS CHECK

Directions: Find the possible solutions for each of the equations below. Be sure to check your solutions!

1. $\sqrt{x} = 25$

2. $\sqrt{x + 5} = 7$

3. $\sqrt{x - 12} = 3$

4. $2\sqrt{3x - 4} = 8$

5. $-4 + \sqrt{x + 10} = -2$

6. $\sqrt{6x - 3} + 3 = 0$

7. $\sqrt{5x - 6} = x$

8. $\sqrt{3x + 1} = x - 3$

(Answers are on page 410.)

LESSON 17.2—ACCUPLACER CHALLENGE

Directions: For each of the questions below, choose the best answer from the five choices given.

1. Which of the following choices represents the solution set for

   $\sqrt{|x - 3|} = 1$?

   (A) $x = 4$
   (B) $x = \{2, 4\}$
   (C) $x = \{4, -4\}$
   (D) $x = 2$
   (E) $x = \{\ \}$

2. Which of the following choices is a possible solution to $15 + \sqrt{2x + 5} = 10$?

   (A) -10
   (B) 10
   (C) -5
   (D) 5
   (E) No real solutions exist.

3. Solve the equation $\sqrt{2x + 1} - x = 1$.

   (A) $x = 1$
   (B) $x = \{-1, 1\}$
   (C) $x = 0$
   (D) $x = \{0, 1\}$
   (E) No real solutions exist.

4. Which of the following choices represents the solution set for the equation $x - 5 = \sqrt{x + 1}$?

(A) 3 only

(B) 8 only

(C) 3 and 8

(D) −3 and 8

(E) No real solutions exist.

5. Find the solution set for

$$\sqrt{x^2 - 14x + 49} = 9.$$

(A) $x = \{-2, 16\}$

(B) $x = -2$

(C) $x = 16$

(D) $x = \{2, -16\}$

(E) No real solutions exist.

(Answers are on page 413.)

LESSON 17.3—SOLVING SYSTEMS OF EQUATIONS INVOLVING LINES AND CURVES

The ACCUPLACER College-Level Mathematics test requires you to have background knowledge of the shapes of different curves and how they behave in the coordinate plane. We already looked at the graphs of curves and lines previously, but now we need to see how we can find points of intersection between two or more graphs using algebra. Let's look at some examples.

➡ Example 1

At which points will the line $y = -2x + 8$ intersect the parabola $y = x^2 - 7x + 12$?

Solution

The line could intersect the parabola in 0, 1, or 2 places. To understand why, it helps to visualize a line and a parabola on a graph and how they could intersect.

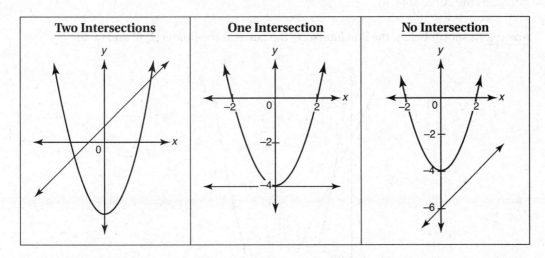

To find the possible points where the graphs met, we can set them equal to each other since they are both already solved for y.

$$-2x + 8 = x^2 - 7x + 12$$

Notice that when we combine all the like terms, this will be a quadratic equation.

$$\begin{array}{r} -2x + 8 = x^2 - 7x + 12 \\ \underline{-8 \qquad\qquad\quad -8} \\ -2x = x^2 - 7x + 4 \\ \underline{+2x \qquad\quad +2x} \\ 0 = x^2 - 5x + 4 \end{array}$$

Now we have a factorable trinomial, so we can find the values of x.

$$x^2 - 5x + 4$$
$$(x - 1)(x - 4)$$

$$x - 1 = 0 \qquad x - 4 = 0$$
$$x = 1 \qquad\quad x = 4$$

Now we know that the line intersects the curve where $x = 1$ and where $x = 4$. To find the actual coordinates, though, we have to separately plug each of these values into either of the original equations in order to find the y-coordinate. Since it requires fewer calculations, let's use the linear equation $y = -2x + 8$.

When $x = 1$

$$y = -2x + 8$$
$$y = -2(1) + 8$$
$$y = -2 + 8$$
$$y = 6$$

When $x = 4$

$$y = -2x + 8$$
$$y = -2(4) + 8$$
$$y = -8 + 8$$
$$y = 0$$

When $x = 1$, $y = 6$, so the line intersects the curve at $(1, 6)$. When $x = 4$, $y = 0$, so the line also intersects the curve at $(4, 0)$.

Answer: As shown below, the line intersects the curve at the points $(1, 6)$ and $(4, 0)$.

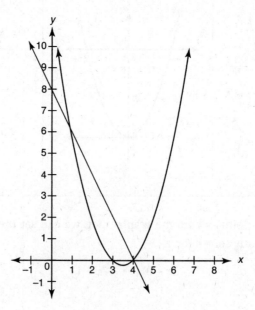

➡ Example 2

Find the points where the parabola $y = x^2$ intersects the circle $x^2 + y^2 = 20$.

Solution

In the case of circles, the graphs will usually be provided for you. This particular circle is centered at the origin and has a radius equal to $\sqrt{20} = 2\sqrt{5}$. We know that the graph of $y = x^2$ is a parabola that has one real zero at the origin. This means that the graphs will intersect in two places.

We can solve this equation by substitution. The equation of the parabola tells us that $y = x^2$. We can substitute y in place of x^2 in the equation of the circle.

$$x^2 + y^2 = 20$$
$$(y) + y^2 = 20$$

Now we can set this equation equal to zero, and factor or use the quadratic equation to find possible values of y.

$$y^2 + y = 20$$
$$y^2 + y - 20 = 0$$
$$(y + 5)(y - 4) = 0$$

This means that the possible values for y are –5 and 4. However, we can rule out $y = -5$ as a possibility. Here's why: If we plug –5 into either of the original equations, we get a statement that doesn't have a real solution for x.

$$y = x^2$$
$$-5 = x^2$$

No real number raised to the second power will produce –5. This would give us a complex number, which will be covered in Chapter 21. Also, if we plug $y = -5$ into the equation of the circle:

$$x^2 + y^2 = 20$$
$$x^2 + (-5)^2 = 20$$
$$x^2 + 25 = 20$$
$$x^2 = -5$$

We are back at the same untrue statement. When we plug $y = 4$ into either equation, though, we find possible values of x.

$$y = x^2$$
$$4 = x^2$$
$$\sqrt{4} = \sqrt{x^2}$$
$$\pm 2 = x$$

Now we know that our points of intersection occur where $x = -2$ and $x = 2$, and we only have one possible y-value: 4.

Answer: As shown below, the parabola intersects the circle at the points (–2, 4) and (2, 4).

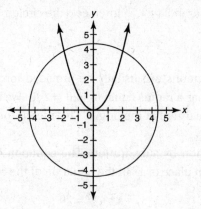

➡ **Example 3**

At what points will the curve $y = x(x^2 + 2)$ intersect the curve $y = x^3 + 6$?

Solution

Even though we haven't spent time looking at the curves of cubic functions, we can still find the points of intersection by setting the two curves equal to each other and solving for x.

$$x(x^2 + 2) = x^3 + 6$$

First, use the distributive property to eliminate the parentheses.

$$x(x^2 + 2) = x^3 + 6$$
$$x^3 + 2x = x^3 + 6$$

Notice how the x^3 term on each side of the equation will cancel out, leaving us with something much simpler.

$$
\begin{array}{r}
x^3 + 2x = x^3 + 6 \\
\underline{-x^3 \qquad\quad -x^3} \\
\dfrac{2x}{2} = \dfrac{6}{2} \\
x = 3
\end{array}
$$

We know that the graphs will only intersect once, when $x = 3$. To find the point where they actually meet, plug 3 in place of x in either equation.

$$y = x^3 + 6$$
$$y = (3)^3 + 6$$
$$y = 27 + 6$$
$$y = 33$$

Answer: As shown below, the graphs intersect at the point (3, 33).

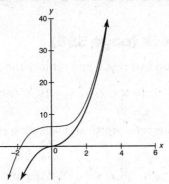

LESSON 17.3—ACCUPLACER CHALLENGE

Directions: For each of the questions below, choose the best answer from the five choices given.

1. At which value of x will the parabola $y = 2x^2 + 6x - 7$ intersect the parabola $y = 2x^2 + 4x + 5$?

   (A) 3
   (B) 4
   (C) 5
   (D) 6
   (E) They will not intersect.

2. At how many points will the line $y = 12$ intersect the parabola $y = 3x^2$?

   (A) 0
   (B) 1
   (C) 2
   (D) 3
   (E) More than 3

3. At which points will the line $y = x + 7$ intersect the curve $y = (x - 2)(x - 1)$?

   (A) (6, -1) and (12, 5)
   (B) (-1, 6) and (5, 12)
   (C) (-1, 8) and (5, 9)
   (D) (1, 12) and (5, 6)
   (E) They will not intersect.

4. At which value of x will the graph of $y = x^3 - 2x + c$ intersect the graph of $y = x^3$?

   (A) c
   (B) $-c$
   (C) $2c$
   (D) $-2c$
   (E) $\dfrac{c}{2}$

5. At how many points will the parabola $y = x^2 - 3$ intersect the circle $x^2 + y^2 = 9$?

   (A) 0
   (B) 1
   (C) 2
   (D) 3
   (E) More than 3

(Answers are on page 416.)

To complete a set of practice questions that reviews all of Chapter 17, go to:
barronsbooks.com/TP/accuplacer/math29qw/

Lesson 17.1—Skills Check (page 396)

1. **$x = 49$ and $x = -49$** Set up two cases, and remove the absolute value bars.

$$|x| = 49 \qquad\qquad |x| = -49$$
$$x = 49 \qquad\qquad\quad x = -49$$

The two possible solutions for x are 49 and –49. Plug each into the original equation.

$$|x| = 49 \qquad\qquad |x| = 49$$
$$|49| = 49 \qquad\qquad |-49| = 49$$
$$49 = 49 \qquad\qquad\quad 49 = 49$$

Both 49 and –49 are solutions.

2. **$x = 22$ and $x = 0$** Set up two cases, and remove the absolute value bars.

$$|x - 11| = 11 \qquad\qquad |x - 11| = -11$$
$$x - 11 = 11 \qquad\qquad\quad x - 11 = -11$$
$$\underline{+11 \ +11} \qquad\qquad\quad \underline{+11 \ \ +11}$$
$$x = 22 \qquad\qquad\qquad\quad x = 0$$

Check both solutions by plugging them into the original equation.

$$|x - 11| = 11 \qquad\qquad |x - 11| = 11$$
$$|22 - 11| = 11 \qquad\qquad |0 - 11| = 11$$
$$|11| = 11 \qquad\qquad\quad |-11| = 11$$
$$11 = 11 \qquad\qquad\qquad 11 = 11$$

Both solutions check out.

3. **$x = 8$ and $x = -6$** Set up two cases, and remove the absolute value bars.

$$|5x - 5| = 35 \qquad\qquad |5x - 5| = -35$$
$$5x - 5 = 35 \qquad\qquad\quad 5x - 5 = -35$$
$$\underline{+5 \ \ +5} \qquad\qquad\qquad \underline{+5 \quad +5}$$
$$\frac{5x}{5} = \frac{40}{5} \qquad\qquad\qquad \frac{5x}{5} = \frac{-30}{5}$$
$$x = 8 \qquad\qquad\qquad\quad x = -6$$

Check both solutions by plugging them into the original equation.

$$|5x - 5| = 35 \qquad\qquad |5x - 5| = 35$$
$$|5(8) - 5| = 35 \qquad\qquad |5(-6) - 5| = 35$$
$$|40 - 5| = 35 \qquad\qquad |-30 - 5| = 35$$
$$|35| = 35 \qquad\qquad\qquad |-35| = 35$$
$$35 = 35 \qquad\qquad\qquad\quad 35 = 35$$

Both solutions check out.

4. **$x = 48$ and $x = -48$** Set up two cases, and remove the absolute value bars.

$$\left|\frac{x}{4}\right| = 12 \qquad\qquad \left|\frac{x}{4}\right| = -12$$

$$\frac{x}{4} = 12 \qquad\qquad\quad \frac{x}{4} = -12$$

$$\frac{4}{1}\left(\frac{x}{4}\right) = 4(12) \qquad \frac{4}{1}\left(\frac{x}{4}\right) = 4(-12)$$

$$x = 48 \qquad\qquad\quad x = -48$$

Check both solutions by plugging them into the original equation.

$$\left|\frac{x}{4}\right| = 12 \qquad\qquad \left|\frac{x}{4}\right| = 12$$

$$\left|\frac{48}{4}\right| = 12 \qquad\qquad \left|\frac{-48}{4}\right| = 12$$

$$|12| = 12 \qquad\qquad |-12| = 12$$

$$12 = 12 \qquad\qquad\quad 12 = 12$$

Both solutions check out.

5. **$x = 4$ and $x = -11$** First, isolate the absolute value part of the equation.

$$\frac{3|2x + 7|}{3} = \frac{45}{3}$$

$$|2x + 7| = 15$$

Next, set up two cases, and remove the absolute value bars.

$$|2x + 7| = 15 \qquad\qquad |2x + 7| = -15$$

$$2x + 7 = 15 \qquad\qquad 2x + 7 = -15$$

$$\underline{-7 \quad -7} \qquad\qquad \underline{-7 \quad -7}$$

$$\frac{2x}{2} = \frac{8}{2} \qquad\qquad\quad \frac{2x}{2} = \frac{-22}{2}$$

$$x = 4 \qquad\qquad\qquad x = -11$$

The possible solutions are $x = 4$ and $x = -11$. Check them by plugging them into the original equation.

$$3|2x + 7| = 45 \qquad\qquad 3|2x + 7| = 45$$

$$3|2(4) + 7| = 45 \qquad\qquad 3|2(-11) + 7| = 45$$

$$3|8 + 7| = 45 \qquad\qquad 3|-22 + 7| = 45$$

$$3|15| = 45 \qquad\qquad\quad 3|-15| = 45$$

$$45 = 45 \qquad\qquad\qquad 3(15) = 45$$

$$45 = 45$$

Both solutions check out.

6. **$x = 1$ and $x = -1$** Set up two cases, and solve both equations. In the second case, make sure that you change the sign of each term on the right side of the equation.

$$|2x + 1| = x + 2 \qquad\qquad |2x + 1| = -x - 2$$

$$
\begin{array}{c}
2x + 1 = x + 2 \\
\underline{-1 \qquad -1} \\
2x = x + 1 \\
\underline{-x - x} \\
x = 1
\end{array}
\qquad\qquad
\begin{array}{c}
2x + 1 = -x - 2 \\
\underline{+x \qquad +x} \\
3x + 1 = -2 \\
\underline{-1 \quad -1} \\
\dfrac{3x}{3} = \dfrac{-3}{3} \\
x = -1
\end{array}
$$

The possible solutions are $x = 1$ and $x = -1$. Check them by plugging them into the original equation.

$$
\begin{array}{c}
|2x + 1| = x + 2 \\
|2(1) + 1| = 1 + 2 \\
|2 + 1| = 1 + 2 \\
|3| = 3 \\
3 = 3
\end{array}
\qquad\qquad
\begin{array}{c}
|2x + 1| = x + 2 \\
|2(-1) + 1| = -1 + 2 \\
|-2 + 1| = -1 + 2 \\
|-1| = 1 \\
1 = 1
\end{array}
$$

Both solutions check out.

Lesson 17.1—ACCUPLACER Challenge (page 396)

1. **(A)** First, isolate the absolute value part of the equation.

$$\frac{-3|x + 5|}{-3} = \frac{24}{-3}$$
$$|x + 5| = -8$$

We can stop working here. The absolute value of a quantity cannot be equal to a negative number, so there are no real solutions.

2. **(C)** Set up two cases, and solve both equations.

$$
\begin{array}{c}
|2x - 3| = 7 \\
2x - 3 = 7 \\
\underline{+3 \ +3} \\
\dfrac{2x}{2} = \dfrac{10}{2} \\
x = 5
\end{array}
\qquad\qquad
\begin{array}{c}
|2x - 3| = -7 \\
2x - 3 = -7 \\
\underline{+3 \ +3} \\
\dfrac{2x}{2} = \dfrac{-4}{2} \\
x = -2
\end{array}
$$

The possible solutions are $x = 5$ and $x = -2$. Check both solutions by plugging them into the original equation.

$$
\begin{array}{c}
|2x - 3| = 7 \\
|2(5) - 3| = 7 \\
|10 - 3| = 7 \\
|7| = 7 \\
7 = 7
\end{array}
\qquad\qquad
\begin{array}{c}
|2x - 3| = 7 \\
|2(-2) - 3| = 7 \\
|-4 - 3| = 7 \\
|-7| = 7 \\
7 = 7
\end{array}
$$

Both solutions check out.

3. **(D)** Set up two cases, and solve each equation. In the second case, remember to change the signs of both terms on the right side of the equation.

$$|4x+1| = 2x+3$$
$$4x+1 = 2x+3$$
$$\underline{-1-1}$$
$$4x = 2x+2$$
$$\underline{-2x-2x}$$
$$\frac{2x}{2} = \frac{2}{2}$$
$$x = 1$$

$$|4x+1| = -2x-3$$
$$4x+1 = -2x-3$$
$$\underline{-1-1}$$
$$4x = -2x-4$$
$$\underline{+2x+2x}$$
$$\frac{6x}{6} = \frac{-4}{6}$$
$$x = -\frac{4}{6} = -\frac{2}{3}$$

$x = 1$ is not one of the answer choices, but $x = -\dfrac{2}{3}$ is. We still need to check the answer by plugging it into the original equation.

$$|4x+1| = 2x+3$$
$$\left|\frac{4}{1}\left(-\frac{2}{3}\right)+1\right| = \frac{2}{1}\left(-\frac{2}{3}\right)+3$$
$$\left|-\frac{8}{3}+\frac{3}{3}\right| = -\frac{4}{3}+\frac{9}{3}$$
$$\left|-\frac{5}{3}\right| = \frac{5}{3}$$
$$\frac{5}{3} = \frac{5}{3}$$

The solution checks out, so $x = -\dfrac{2}{3}$ is a solution.

4. **(D)** First, isolate the absolute value part of the equation.

$$6-5|x+7| = -19$$
$$\underline{-6-6}$$
$$\frac{-5|x+7|}{-5} = \frac{-25}{-5}$$
$$|x+7| = 5$$

Now set up two cases, and solve each equation.

$$|x+7| = 5 \qquad\qquad |x+7| = -5$$
$$x+7 = 5 \qquad\qquad\quad x+7 = -5$$
$$\underline{-7-7} \qquad\qquad \underline{-7-7}$$
$$x = -2 \qquad\qquad\quad x = -12$$

The possible solutions are $x = -2$ and $x = -12$. Check both by plugging them into the original equation.

$$6-5|x+7| = -19 \qquad\qquad 6-5|x+7| = -19$$
$$6-5|-2+7| = -19 \qquad\qquad 6-5|-12+7| = -19$$
$$6-5|5| = -19 \qquad\qquad\quad 6-5|-5| = -19$$
$$6-25 = -19 \qquad\qquad\quad 6-5(5) = -19$$
$$-19 = -19 \qquad\qquad\qquad 6-25 = -19$$
$$\qquad\qquad\qquad\qquad\qquad -19 = -19$$

The solutions are $x = -2$ and $x = -12$.

5. **(B)** Set up two cases, and solve each equation. Remember to change the sign of each term on the right side of the equation in the second case.

$$\left|\frac{1}{2}x+4\right| = x+2$$

$$\frac{1}{2}x+4 = x+2$$

$$\underline{-\frac{1}{2}x \qquad -\frac{1}{2}x}$$

$$4 = \frac{1}{2}x+2$$

$$\underline{-2 \qquad -2}$$

$$2(2) = \left(\frac{1}{2}x\right)\frac{2}{1}$$

$$4 = x$$

$$\left|\frac{1}{2}x+4\right| = -x-2$$

$$\frac{1}{2}x+4 = -x-2$$

$$\underline{+x \qquad +x}$$

$$\frac{3}{2}x+4 = -2$$

$$\underline{-4 \quad -4}$$

$$\frac{2}{3}\left(\frac{3}{2}x\right) = \left(-\frac{6}{1}\right)\frac{2}{3}$$

$$x = -\frac{12}{3} = -4$$

The possible solutions are $x = 4$ and $x = -4$. Check both solutions.

$$\left|\frac{1}{2}x+4\right| = x+2$$

$$\left|\frac{1}{2}\left(\frac{4}{1}\right)+4\right| = 4+2$$

$$|2+4| = 4+2$$

$$|6| = 6$$

$$6 = 6$$

$$\left|\frac{1}{2}x+4\right| = x+2$$

$$\left|\frac{1}{2}\left(-\frac{4}{1}\right)+4\right| = -4+2$$

$$|-2+4| = -4+2$$

$$|2| = -2$$

$$2 \neq -2$$

The only solution that checks out is $x = 4$.

Lesson 17.2—Skills Check (page 400)

1. **$x = 625$**   Square both sides of the equation.

$$\left(\sqrt{x}\right)^2 = 25^2$$

$$x = 625$$

Check the answer by plugging it into the original equation.

$$\sqrt{x} = 25$$

$$\sqrt{625} = 25$$

$$25 = 25$$

2. **$x = 44$**   Square both sides of the equation, and solve for x.

$$\left(\sqrt{x+5}\right)^2 = 7^2$$

$$x+5 = 49$$

$$\underline{-5 \quad -5}$$

$$x = 44$$

Check the answer by plugging it into the original equation.

$$\sqrt{x+5} = 7$$

$$\sqrt{44+5} = 7$$

$$\sqrt{49} = 7$$

$$7 = 7$$

3. **$x = 21$** Square both sides of the equation, and solve for x.

$$\left(\sqrt{x-12}\right)^2 = 3^2$$
$$x - 12 = 9$$
$$\underline{+12 \quad +12}$$
$$x = 21$$

Check the answer by plugging it into the original equation.

$$\sqrt{x-12} = 3$$
$$\sqrt{21-12} = 3$$
$$\sqrt{9} = 3$$
$$3 = 3$$

4. **$\dfrac{20}{3}$ or $6\dfrac{2}{3}$** First, divide both sides by 2.

$$\frac{2\sqrt{3x-4}}{2} = \frac{8}{2}$$
$$\sqrt{3x-4} = 4$$

Then square both sides, and solve for x.

$$\left(\sqrt{3x-4}\right)^2 = 4^2$$
$$3x - 4 = 16$$
$$\underline{+4 \quad +4}$$
$$\frac{3x}{3} = \frac{20}{3}$$
$$x = \frac{20}{3}$$

To check this solution, plug it into the original equation.

$$2\sqrt{3x-4} = 8$$
$$2\sqrt{\frac{3}{1}\left(\frac{20}{3}\right)-4} = 8$$
$$2\sqrt{20-4} = 8$$
$$2\sqrt{16} = 8$$
$$2(4) = 8$$
$$8 = 8$$

5. **$x = -6$** First, isolate the radical part of the equation.

$$-4 + \sqrt{x+10} = -2$$
$$\underline{+4 \qquad\qquad +4}$$
$$\sqrt{x+10} = 2$$

Now square both sides, and solve for x.

$$\left(\sqrt{x+10}\right)^2 = 2^2$$
$$x + 10 = 4$$
$$\underline{-10 \quad -10}$$
$$x = -6$$

Now plug $x = -6$ into the original equation to check the solution.

$$-4 + \sqrt{x + 10} = -2$$
$$-4 + \sqrt{-6 + 10} = -2$$
$$-4 + \sqrt{4} = -2$$
$$-4 + 2 = -2$$
$$-2 = -2$$

6. **No real solutions** First, isolate the radical part of the equation.

$$\sqrt{6x - 3} + 3 = 0$$
$$\underline{\quad\quad -3 \ -3}$$
$$\sqrt{6x - 3} = -3$$

Square both sides, and solve the equation for x.

$$\left(\sqrt{6x - 3}\right)^2 = (-3)^2$$
$$6x - 3 = 9$$
$$\underline{\quad +3 \ +3}$$
$$\frac{6x}{6} = \frac{12}{6}$$
$$x = 2$$

To check the solution, plug $x = 2$ into the original equation.

$$\sqrt{6x - 3} + 3 = 0$$
$$\sqrt{6(2) - 3} + 3 = 0$$
$$\sqrt{12 - 3} + 3 = 0$$
$$\sqrt{9} + 3 = 0$$
$$3 + 3 = 0$$
$$6 \neq 0$$

$x = 2$ does not check out so there are no real solutions.

7. $x = 3$ **and** $x = 2$ First, square both sides.

$$\left(\sqrt{5x - 6}\right)^2 = x^2$$
$$5x - 6 = x^2$$

Rearrange the equation so that it is set equal to zero. Then, factor to find possible solutions.

$$5x - 6 = x^2$$
$$\underline{-5x \quad\quad\quad -5x}$$
$$-6 = x^2 - 5x$$
$$\underline{+6 \quad\quad\quad +6}$$
$$0 = x^2 - 5x + 6$$
$$0 = (x - 3)(x - 2)$$

Solving each binomial for x, we get the solutions $x = 3$ and $x = 2$. We need to check both.

$$\sqrt{5x-6} = x \qquad\qquad \sqrt{5x-6} = x$$
$$\sqrt{5(3)-6} = 3 \qquad\qquad \sqrt{5(2)-6} = 2$$
$$\sqrt{15-6} = 3 \qquad\qquad \sqrt{10-6} = 2$$
$$\sqrt{9} = 3 \qquad\qquad\qquad \sqrt{4} = 2$$
$$3 = 3 \qquad\qquad\qquad\quad 2 = 2$$

Both solutions check out.

8. **$x = 8$**   Square both sides of the equation.

$$\left(\sqrt{3x+1}\right)^2 = (x-3)^2$$
$$3x+1 = x^2 - 6x + 9$$

Combine the variables, and set the equation equal to zero.

$$
\begin{array}{r}
3x+1 = x^2 - 6x + 9 \\
-3x \qquad\quad -3x \\
\hline
1 = x^2 - 9x + 9 \\
-1 \qquad\qquad -1 \\
\hline
0 = x^2 - 9x + 8
\end{array}
$$

Factor the resulting equation.

$$0 = x^2 - 9x + 8$$
$$0 = (x-8)(x-1)$$

Solving each binomial, we find that $x = 8$ and $x = 1$ are possible solutions. Plug each into the original equation to check them.

$$\sqrt{3x+1} = x - 3 \qquad\qquad \sqrt{3x+1} = x - 3$$
$$\sqrt{3(8)+1} = (8) - 3 \qquad\qquad \sqrt{3(1)+1} = 1 - 3$$
$$\sqrt{24+1} = 5 \qquad\qquad\qquad \sqrt{3+1} = -2$$
$$\sqrt{25} = 5 \qquad\qquad\qquad\quad \sqrt{4} = -2$$
$$5 = 5 \qquad\qquad\qquad\qquad 2 \neq -2$$

The solution $x = 8$ checks out, but $x = 1$ does not.

Lesson 17.2—ACCUPLACER Challenge (pages 400–401)

1. **(B)** First, square both sides to eliminate the radical.

$$\left(\sqrt{|x-3|}\right)^2 = 1^2$$
$$|x-3| = 1$$

Next, create two different equations, and solve both.

$$
\begin{array}{r}
x - 3 = 1 \\
+3 \quad +3 \\
\hline
x = 4
\end{array}
\qquad\qquad
\begin{array}{r}
x - 3 = -1 \\
+3 \quad +3 \\
\hline
x = 2
\end{array}
$$

The possible solutions are $x = 4$ and $x = 2$. Plug both into the original equation to check them.

$$\sqrt{|x-3|} = 1 \qquad\qquad \sqrt{|x-3|} = 1$$
$$\sqrt{|4-3|} = 1 \qquad\qquad \sqrt{|2-3|} = 1$$
$$\sqrt{|1|} = 1 \qquad\qquad \sqrt{|-1|} = 1$$
$$\sqrt{1} = 1 \qquad\qquad \sqrt{1} = 1$$
$$1 = 1 \qquad\qquad 1 = 1$$

Both solutions check out.

2. **(E)** First, isolate the radical part of the equation.

$$15 + \sqrt{2x+5} = 10$$
$$\underline{-15 \qquad\qquad -15}$$
$$\sqrt{2x+5} = -5$$

Square both sides, and solve for x.

$$\left(\sqrt{2x+5}\right)^2 = (-5)^2$$
$$2x + 5 = 25$$
$$\underline{-5 \quad -5}$$
$$\frac{2x}{2} = \frac{20}{2}$$
$$x = 10$$

Test the solution.

$$15 + \sqrt{2x+5} = 10$$
$$15 + \sqrt{2(10)+5} = 10$$
$$15 + \sqrt{20+5} = 10$$
$$15 + \sqrt{25} = 10$$
$$15 + 5 = 10$$
$$20 \neq 10$$

$x = 10$ does not check out, so there are no real solutions.

3. **(C)** First, add x to both sides to isolate the radical part of the equation.

$$\sqrt{2x+1} - x = 1$$
$$\underline{+x \quad +x}$$
$$\sqrt{2x+1} = x+1$$

Now square both sides, and set the resulting quadratic equation equal to zero.

$$\left(\sqrt{2x+1}\right)^2 = (x+1)^2$$
$$2x + 1 = x^2 + 2x + 1$$
$$\underline{-2x \qquad\qquad -2x}$$
$$1 = x^2 + 1$$
$$\underline{-1 \qquad -1}$$
$$0 = x^2$$

The only possible solution that would make x^2 equal to zero is when $x = 0$. Test the solution.

$$\sqrt{2x+1} - x = 1$$
$$\sqrt{2(0)+1} - 0 = 1$$
$$\sqrt{1} - 0 = 1$$
$$1 - 0 = 1$$
$$1 = 1$$

The solution checks out.

4. **(B)** Square both sides of the equation.

$$(x-5)^2 = \left(\sqrt{x+1}\right)^2$$
$$x^2 - 10x + 25 = x + 1$$

Set the equation equal to zero.

$$
\begin{array}{r}
x^2 - 10x + 25 = x + 1 \\
\underline{-x \qquad\quad -x} \\
x^2 - 11x + 25 = 1 \\
\underline{\qquad\qquad -1 \;\; -1} \\
x^2 - 11x + 24 = 0
\end{array}
$$

Factor.

$$x^2 - 11x + 24 = 0$$
$$(x-8)(x-3) = 0$$

Solving each of the two binomials, we find that the possible solutions are $x = 8$ and $x = 3$. Check both solutions.

$$
\begin{array}{ll}
x - 5 = \sqrt{x+1} \qquad & x - 5 = \sqrt{x+1} \\
8 - 5 = \sqrt{8+1} & 3 - 5 = \sqrt{3+1} \\
3 = \sqrt{9} & -2 = \sqrt{4} \\
3 = 3 & -2 \ne 2
\end{array}
$$

The solution $x = 8$ checks out, but $x = 3$ does not. Therefore, there is only one solution, $x = 8$.

5. **(A)** Square both sides of the equation.

$$\left(\sqrt{x^2 - 14x + 49}\right)^2 = 9^2$$
$$x^2 - 14x + 49 = 81$$

Set the equation equal to zero, and factor.

$$
\begin{array}{r}
x^2 - 14x + 49 = 81 \\
\underline{\qquad\qquad -81 \;\; -81} \\
x^2 - 14x - 32 = 0 \\
(x-16)(x+2) = 0
\end{array}
$$

Solving each binomial, we find that the possible solutions are $x = 16$ and $x = -2$. Check both solutions.

$$\sqrt{x^2 - 14x + 49} = 9 \qquad\qquad \sqrt{x^2 - 14x + 49} = 9$$
$$\sqrt{16^2 - 14(16) + 49} = 9 \qquad \sqrt{(-2)^2 - 14(-2) + 49} = 9$$
$$\sqrt{256 - 224 + 49} = 9 \qquad\qquad \sqrt{4 + 28 + 49} = 9$$
$$\sqrt{81} = 9 \qquad\qquad\qquad \sqrt{81} = 9$$
$$9 = 9 \qquad\qquad\qquad\qquad 9 = 9$$

Both solutions check out.

Lesson 17.3—ACCUPLACER Challenge (page 405)

1. **(D)** The point where the parabolas intersect will be at an ordered pair (x, y) that they both have in common. This problem only asks us to find the x-value that they share. Set the equations equal to one another.

$$2x^2 + 6x - 7 = 2x^2 + 4x + 5$$

Notice that the x^2 terms can be eliminated.

$$2x^2 + 6x - 7 = 2x^2 + 4x + 5$$
$$\underline{-2x^2 \qquad\qquad -2x^2}$$
$$6x - 7 = 4x + 5$$

Solve the resulting equation for x.

$$6x - 7 = 4x + 5$$
$$\underline{-4x \qquad\quad -4x}$$
$$2x - 7 = 5$$
$$\underline{+7 + 7}$$
$$\frac{2x}{2} = \frac{12}{2}$$
$$x = 6$$

The parabolas will intersect where $x = 6$.

2. **(C)** The line $y = 12$ is horizontal, and the graph of $y = 3x^2$ is a parabola that opens upward and has its vertex at the origin. This means that the horizontal line must intersect the parabola twice. We could also set $3x^2$ equal to 12, and solve for the possible values of x to learn more.

$$\frac{3x^2}{3} = \frac{12}{3}$$
$$x^2 = 4$$
$$\sqrt{x^2} = \sqrt{4}$$
$$x = \pm 2$$

The line will intersect the parabola at the points (2, 12) and (–2, 12).

3. **(B)** First, expand the equation $y = (x - 2)(x - 1)$ by multiplying the two binomials.

$$y = (x - 2)(x - 1)$$
$$y = x^2 - 3x + 2$$

If we set the equation of the line equal to the equation of the parabola and solve for x by factoring or by using the quadratic equation, then we will find the x-coordinate of the ordered pair that the two graphs have in common.

$$x + 7 = x^2 - 3x + 2$$
$$\underline{-x \qquad\qquad -x}$$
$$7 = x^2 - 4x + 2$$
$$\underline{-7 \qquad\qquad -7}$$
$$0 = x^2 - 4x - 5$$
$$0 = (x - 5)(x + 1)$$

Setting each binomial equal to zero and solving for x, we find that the line and the parabola intersect where $x = -1$ and $x = 5$. Choices B and C both have the correct x-coordinates, so we still need to find the y-coordinates where the line and the parabola intersect. We can find this by plugging the x-values into either of the equations.

$$y = x + 7 \qquad\qquad y = x + 7$$
$$y = -1 + 7 \qquad\qquad y = 5 + 7$$
$$y = 6 \qquad\qquad y = 12$$

The line and the curve intersect at the points $(-1, 6)$ and $(5, 12)$.

4. **(E)** To find where the graphs will meet, set them equal to one another, and solve for x. Notice that the x^3 terms will cancel out.

$$x^3 - 2x + c = x^3$$
$$\underline{-x^3 \qquad\qquad -x^3}$$
$$-2x + c = 0$$

Continue solving for x.

$$-2x + c = 0$$
$$\underline{-c \quad -c}$$
$$\frac{-2x}{-2} = \frac{-c}{-2}$$
$$x = \frac{c}{2}$$

5. **(D)** Notice that the equation of the parabola can be solved in terms of x^2. It can then be substituted into the equation of the circle.

$$y = x^2 - 3$$
$$\underline{+3 \qquad\quad +3}$$
$$y + 3 = x^2$$

Substituting $y + 3$ in place of x^2 in the equation $x^2 + y^2 = 9$, we have:

$$x^2 + y^2 = 9$$
$$(y + 3) + y^2 = 9$$

Combine terms, and solve for y by setting the equation equal to zero and factoring.

$$(y+3)+y^2 = 9$$
$$y^2 + y + 3 = 9$$
$$\underline{-9\ -9}$$
$$y^2 + y - 6 = 0$$
$$(y+3)(y-2) = 0$$

Solving each binomial for y, we find that the parabola and the circle will intersect where $y = -3$ and $y = 2$. It would seem like this means that the graphs intersect in two places, but we aren't quite finished. We need to plug both of these into either of the original equations in order to find the x-values that correspond with our y-values.

$$y = x^2 - 3 \qquad\qquad y = x^2 - 3$$
$$-3 = x^2 - 3 \qquad\qquad 2 = x^2 - 3$$
$$\underline{+3 \qquad +3} \qquad\qquad \underline{+3 \qquad +3}$$
$$0 = x^2 \qquad\qquad\qquad 5 = x^2$$
$$0 = x \qquad\qquad\qquad \sqrt{5} = \sqrt{x^2}$$
$$\qquad\qquad\qquad\qquad\quad \pm\sqrt{5} = x$$

Now we see that the parabola and the circle meet at three points: $(0, -3)$, $(\sqrt{5}, 2)$, and $(-\sqrt{5}, 2)$.

Functions

18

In mathematics, a **function** is a mathematical rule that assigns exactly one output to each input. If you don't remember hearing the term function, maybe you remember working with input/output tables in one of your math classes. When you were filling in these tables, you were actually working with functions.

> **In this chapter, we will work to master**
>
> - defining functions,
> - using different views of functions,
> - evaluating functions,
> - evaluating composite functions, and
> - finding and evaluating inverse functions.

LESSON 18.1—DEFINING AND EVALUATING FUNCTIONS

Lesson 18.1A—The Definition of a Function

You can think of a function like a machine. You take something and put it into the machine. Inside the machine, something happens to the item you put in, and, in the end, something comes out. In the case of mathematical functions, what you put in is called the **input**, what happens inside the machine is a mathematical operation, or operations, that produces an **output**. The set of inputs is called the **domain**, and the set of outputs is called the **range**.

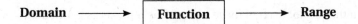

$$\text{Domain} \longrightarrow \boxed{\text{Function}} \longrightarrow \text{Range}$$

Key to the definition of a function is the requirement that each input can have only one output. If an input has multiple outputs, then the equation is not actually a function. Let's look at two different sets of values and identify whether or not they represent functions.

Function	
Input	**Output**
0	10
1	20
2	30
3	40
4	40

Not a Function	
Input	**Output**
1	10
1	20
2	30
2	40
3	50

This table represents the graph of a function because each input corresponds to *exactly one output*. The input 0 produces only the output 10, the input 1 produces only the output 20, and so on. Notice that the inputs 3 and 4 both have the same output: 40. With functions, it's okay for more than one input to have the same output.

Unlike the table on the left, in this table, two of the inputs correspond to more than one output. The definition of a function requires that each input correspond to only one output. Here, the input 1 produces two different outputs: 10 and 20. The input 2 also produces two different outputs: 30 and 40.

Functions can also be graphed using the coordinate plane. When graphing a function, we treat $f(x)$ as y. In other words, the output values correspond to the y-axis, while the input values correspond to the x-axis. Using function notation, an ordered pair on the graph of a function would be $(x, f(x))$ instead of (x, y). On the ACCUPLACER, you may need to identify the graph of a function versus the graph of a shape or a curve that is *not* a function. To do this, we use the vertical line test.

The **vertical line test** states that if a graph represents a function, we can draw a vertical line *anywhere* on the graph, and it will intersect the graph in, at most, one place. Think about what this means. A vertical line has only one x-value. If a vertical line intersects a graph in more than one place, then that graph has multiple y-values, or outputs, paired with a single input, or x-value. For example, if the vertical line $x = 3$ is drawn, and it crosses a graph at (3, 2) and then again at (3, –2), it would mean that the input, 3, has more than one output, 2 and also –2. This would violate the definition of a function. To gain more insight into this concept, let's look at an example.

➡ **Example 1**

Determine whether or not each of these two graphs represents a function.

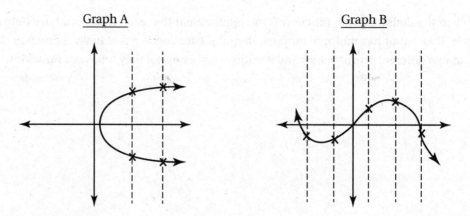

Graph A   Graph B

Solution

If the graph does represent a function, then we can drop a vertical line anywhere on the graph, and it will intersect the graph in at most one point. In the case of Graph A, if we drop a vertical line at any point to the right of the y-axis, it will intersect the graph in two places. This means that Graph A is not a function.

With Graph B, we can drop a vertical line anywhere on the graph, and it will intersect the graph only once. This means that Graph B is a function. You might notice that all diagonal and horizontal lines are functions.

Answer: Graph A is not a function. Graph B is a function.

Lesson 18.1B—Evaluating Functions

Functions are often written in a very specific form called **function notation**. If the input variable in a function is x, the output variable is written as $f(x)$. We read $f(x)$ as "f of x." In previous algebra problems, a letter preceding a quantity in parentheses indicates that we need to multiply, but this is not the case for functions. With functions, $f(x)$ simply indicates the output for a particular function of x.

On the ACCUPLACER, you might see functions that have different "names" other than $f(x)$. Most commonly, you will see $f(x)$, $g(x)$, or $h(x)$. However, the input variable might not always be x. You should expect to see n or t used as input variables as well. With this in mind, we can use the general form of a function to find the output value for any input. Let's look at an example.

➡ Example 1

Consider the function $f(x) = 2x + 1$. Find $f(5)$.

Solution

This is a function that takes an input variable x and gives an output $f(x)$. This problem gives us the input value, 5, that we should use in place of x. We read $f(5)$ as "f of 5." Now we substitute 5 into the expression $2x + 1$. Once we solve the expression, we will know the correct output value for the input 5.

Substitute 5 for x into $f(x) = 2x + 1$.

$$f(5) = 2(5) + 1$$
$$f(5) = 10 + 1$$
$$f(5) = 11$$

Answer: $f(5) = 11$

Going back to the analogy of functions as a "box," in this case, we put 5 into the box, performed some calculations, and then the output 11 came out of the box. If we kept plugging different inputs into the function, we would find that none of them have the same output.

➡ Example 2

Find $g(-3)$ for the function $g(x) = 2x^2 + 5x + 7$.

Solution

As we did in Example 1, we will find g of -3 by substituting the input -3 into the function. Begin by rewriting it as:

$$g(-3) = 2(-3)^2 + 5(-3) + 7$$

Make sure to follow the order of operations! In this case, we need to work on the exponents first, then the multiplication, and finally the addition.

$$g(-3) = 2(-3)^2 + 5(-3) + 7$$
$$g(-3) = 2(9) + 5(-3) + 7$$
$$g(-3) = 18 + (-15) + 7$$
$$g(-3) = 10$$

Answer: The input value -3 corresponds to the output value 10.

➡ Example 3

Consider the function $h(t) = \dfrac{4t^2 + 2t}{t + 5}$, where $t \neq -5$. Find $h(4)$.

Solution

Notice that this function has a restriction on the domain. If the input were -5, then the denominator of this rational expression would be equal to 0. We know that division by 0 is undefined, so this function will be undefined when $t = -5$. You should expect to see similar restrictions on the domain on the ACCUPLACER.

Plug the input value 4 into the function.

$$h(4) = \frac{4(4)^2 + 2(4)}{4 + 5}$$
$$h(4) = \frac{4(16) + 2(4)}{4 + 5}$$
$$h(4) = \frac{64 + 8}{9}$$
$$h(4) = \frac{72}{9} = 8$$

Answer: $h(4) = 8$

➡ Example 4

Which of the following functions expresses the rule for the values in the table below?

x	$f(x)$
1	2
2	6
3	22
4	56

(A) $f(x) = x^2 + 1$
(B) $f(x) = 2^x + 2$
(C) $f(x) = x^3 - 3x + 4$
(D) $f(x) = 2x - 1$
(E) $f(x) = 4x - 2$

Solution

This time, instead of being asked to find an output for a given input, we are given a set of inputs and outputs and asked to find the function rule that matches the table. To solve a problem like this on the ACCUPLACER, you should use the answer choices and work backwards.

We can eliminate choices D and E right off the bat because they are linear equations, and the outputs in the table are not increasing at a constant rate. If we try choice A, the first input and output pair match, but none of the others do.

If we try choice B, it will only work for the input 2 but none of the others. With choice C, however, all of the inputs produce the desired outputs.

$$f(x) = x^3 - 3x + 4 \qquad\qquad f(x) = x^3 - 3x + 4$$
$$f(1) = 1^3 - 3(1) + 4 \qquad\qquad f(3) = 3^3 - 3(3) + 4$$
$$f(1) = 1 - 3 + 4 \qquad\qquad f(3) = 27 - 9 + 4$$
$$f(1) = 2 \qquad\qquad f(3) = 22$$

$$f(x) = x^3 - 3x + 4 \qquad\qquad f(x) = x^3 - 3x + 4$$
$$f(2) = 2^3 - 3(2) + 4 \qquad\qquad f(4) = 4^3 - 3(4) + 4$$
$$f(2) = 8 - 6 + 4 \qquad\qquad f(4) = 64 - 12 + 4$$
$$f(2) = 6 \qquad\qquad f(4) = 56$$

Answer: Choice C is the function rule that matches this table of values.

LESSON 18.1—SKILLS CHECK

Directions: For questions 1–5, evaluate the function $f(x) = x^2 + 6x + 5$ for the given inputs.

1. $f(6)$ 3. $f(0)$ 5. $f\left(-\dfrac{3}{4}\right)$

2. $f(-7)$ 4. $f\left(\dfrac{1}{2}\right)$

Directions: For questions 6–10, evaluate the function $g(x) = \dfrac{3x+6}{x-2}$, $x \neq 2$, for the given inputs.

6. $g(3)$ 8. $g(0)$ 10. $g(8)$

7. $g(-4)$ 9. $g(-2)$

(Answers are on page 433.)

LESSON 18.1—ACCUPLACER CHALLENGE

> **Directions**: For each of the questions below, choose the best answer from the five choices given.

1. Consider the function below.
$$f(x) = \frac{3x^2 - 12}{4x - 20}$$
   For which of the following values of x will $f(x)$ be undefined?

   (A) 2
   (B) −2
   (C) 5
   (D) −5
   (E) 0

2. Which of the function rules below represents the following table?

x	$h(x)$
4	−6
8	−8
12	−10
16	−12

   (A) $h(x) = \frac{1}{2}x + 6$

   (B) $h(x) = -\frac{1}{2}x - 6$

   (C) $h(x) = 2x - 4$

   (D) $h(x) = -\frac{3}{4}x + 6$

   (E) $h(x) = -\frac{1}{2}x - 4$

(Answers are on page 434.)

3. Given the function $f(x) = x^2 - 9$, on which interval will the value of $f(x) < 0$?

   (A) $-9 < x < 9$
   (B) $-3 < x < 3$
   (C) $-3 \leq x \leq 3$
   (D) $-9 \leq x \leq 9$
   (E) $f(x)$ is always greater than or equal to 0.

4. Consider two functions $f(x) = 2^x$ and $g(x) = \frac{x^3}{4}$. Which of the following statements is true?

   (A) $f(0) < g(0)$
   (B) $f(1) = g(1)$
   (C) $f(6) < g(6)$
   (D) $f(4) = g(4)$
   (E) $f(2) < g(2)$

5. Let $h(x) = \frac{1}{x^2}$, where $x \neq 0$, and let $0 < a < b$. Which of the following is true?

   (A) $h(a) = h(b)$
   (B) $h(a) < h(b)$
   (C) $h(a) > h(b)$
   (D) $h(a) \leq h(b)$
   (E) None of these are true.

LESSON 18.2—COMPOSITE FUNCTIONS

Think of a **composite function** as a "function within a function." In the previous exercises and examples, we evaluated a single function for a given input. With composite functions, you will need to evaluate one function for an input and then use its output as the input of another function. It sounds confusing, but the process is pretty straightforward as long as you're careful.

On the ACCUPLACER, you should expect to see composite functions written in two ways:

$$f(g(x)) \qquad f \circ g(x)$$

Both of these are read as "f of g of x." To evaluate this composite function, you would first need to find $g(x)$, and then you would use that output as the input for the function f. Let's look at an example.

➡ Example 1

Let $f(x) = 6x + 8$ and $g(x) = 4x - 5$. Find $f(g(3))$.

Solution

We need to find f of g of 3. First, evaluate $g(3)$.

$$g(x) = 4x - 5$$
$$g(3) = 4(3) - 5$$
$$g(3) = 12 - 5$$
$$g(3) = 7$$

Now that we know $g(3) = 7$, we can rewrite $f(g(3))$ as $f(7)$, and evaluate the function.

$$f(x) = 6x + 8$$
$$f(7) = 6(7) + 8$$
$$f(7) = 42 + 8$$
$$f(7) = 50$$

Answer: $f(g(3)) = 50$

The order in which you evaluate composite functions matters! In other words, $f(g(x))$ is not necessarily equal to $g(f(x))$. In the previous example, $f(g(3)) = 50$. Let's check $g(f(3))$.

$$f(x) = 6x + 8$$
$$f(3) = 6(3) + 8$$
$$f(3) = 18 + 8$$
$$f(3) = 26$$

$$g(f(3)) = g(26)$$
$$g(x) = 4x - 5$$
$$g(26) = 4(26) - 5$$
$$g(26) = 104 - 5$$
$$g(26) = 99$$

We see that, in this case, $f(g(x)) \neq g(f(x))$. Make sure that you always evaluate the "inside" function first, and then work toward the outside.

➡ Example 2

Consider two functions $f(t) = \dfrac{t^2 - 1}{4}$ and $h(t) = \dfrac{3}{4}t - 2$. Evaluate $f \circ h(12)$.

Solution

First, evaluate $h(12)$ by plugging 12 in for x.

$$h(t) = \frac{3}{4}t - 2$$

$$h(12) = \frac{3}{4}\left(\frac{12}{1}\right) - 2$$

$$h(12) = \frac{36}{4} - 2$$

$$h(12) = 9 - 2$$

$$h(12) = 7$$

Since we need to find f of h of 12 and $h(12) = 7$, plug 7 into the function f.

$$f(t) = \frac{t^2 - 1}{4}$$

$$f(7) = \frac{7^2 - 1}{4}$$

$$f(7) = \frac{49 - 1}{4}$$

$$f(7) = \frac{48}{4} = 12$$

Answer: $f \circ h(12) = 12$

In the previous examples, we evaluated composite functions when given a specific input value. Some problems on the ACCUPLACER, however, will ask you to compose two functions without a particular input.

➡ **Example 3**

Let $g(x) = \frac{x+2}{3}$ and $h(x) = 3x + 1$. Find $g(h(x))$.

Solution

Notice that, this time, there isn't an input value. We can, however, still find the composition of these two functions. The problem is asking us to find $g(h(x))$, and since $h(x) = 3x + 1$, we could think of the problem as $g(3x + 1)$. This means that we plug in the function $3x + 1$ in place of the x in the function g.

$$g(h(x)) = g(3x + 1)$$

$$g(x) = \frac{x+2}{3}$$

$$g(3x + 1) = \frac{(3x + 1) + 2}{3}$$

$$g(3x + 1) = \frac{3x + 3}{3}$$

$$g(3x + 1) = \frac{3(x + 1)}{3}$$

$$g(3x + 1) = x + 1$$

Answer: $g(h(x)) = x + 1$

Example 4

$f(n) = n^2 + 3$ and $g(n) = n - 2$. Find $f(g(n))$.

Solution

Since we don't have a value for n, we need to plug in the function $g(n)$ into the function $f(n)$.

$$f(g(n)) = f(n - 2)$$
$$f(n) = n^2 + 3$$
$$f(n - 2) = (n - 2)^2 + 3$$
$$f(n - 2) = (n^2 - 4n + 4) + 3$$
$$f(n - 2) = n^2 - 4n + 7$$

Answer: $f(g(n)) = n^2 - 4n + 7$

LESSON 18.2—SKILLS CHECK

> **Directions**: For questions 1–5, use the functions *f*, *g*, and *h* below to evaluate each composite function on the given input.

$$f(x) = 4x - 12 \qquad g(x) = \frac{1}{2}x^2 \qquad h(x) = \frac{x + 10}{x + 2}$$

1. $f(g(6))$

2. $f(h(2))$

3. $h(f(0))$

4. $g(h(-4))$

5. $f(h(0))$

> **Directions**: For questions 6–10, use the functions *f*, *g*, and *h* below to find each composite function.

$$f(n) = n^2 \qquad g(n) = n - 5 \qquad h(n) = 3n + 4$$

6. $f \circ g(n)$

7. $h \circ g(n)$

8. $f \circ h(n)$

9. $h \circ f \circ g(n)$

10. $f \circ h \circ g(n)$

(Answers are on page 435.)

LESSON 18.2—ACCUPLACER CHALLENGE

> **Directions**: For each of the questions below, choose the best answer from the five choices given.

1. Let $g(x) = \dfrac{4x+1}{x+2}$ and $h(x) = x^3 - 10x + 8$. Find $g(h(3))$.

   (A) 2
   (B) 3
   (C) 4
   (D) 5
   (E) 6

2. Consider two functions $f(n) = -3n + 2$ and $g(n) = 2n + 5$. Which of the following is true?

   (A) $f \circ g(0) = g \circ f(0)$
   (B) $f \circ g(0) > g \circ f(0)$
   (C) $f \circ g(3) > g \circ f(3)$
   (D) $f \circ g(3) < g \circ f(3)$
   (E) $f \circ g(3) = g \circ f(3)$

3. Let the function $f(x) = \dfrac{x^2}{8}$ and the function $h(x) = 4x + 4$. Which of the following is equivalent to $f(h(x))$?

   (A) $\dfrac{1}{2}x^2 + 4$
   (B) $2(x + 1)^2$
   (C) $\dfrac{x+1}{2}$
   (D) $x + 1$
   (E) $2x + 12$

4. For the functions $g(t) = t^2$ and $h(t) = 6t$, which of the following is true for all t?

   (A) $g \circ h(t) > g \circ h(-t)$
   (B) $g \circ h(t) < g \circ h(-t)$
   (C) $g \circ h(t) = g \circ h(-t)$
   (D) $g \circ h(t) = h \circ g(t)$
   (E) None of these are true.

5. Consider three functions, $f(x) = x^2$, $g(x) = 3x + 1$, and $h(x) = x + 2$. Which of the following is equal to $h \circ g \circ f(x)$?

   (A) $3x^3 + x^2 + 6x + 2$
   (B) $9x^2 + 42x + 49$
   (C) $9x^2 + 12x + 9$
   (D) $3x^3 + 6x^2 + x + 2$
   (E) $3(x^2 + 1)$

(Answers are on page 438.)

LESSON 18.3—INVERSE FUNCTIONS

When we worked on solving one-variable equations, we talked a lot about "inverse" operations, and we used the word *inverse* to mean *opposite*. Inverse functions work in a similar way. Remember that a function is a process that takes an input, does some math to it, and produces an output. An **inverse function** takes the output from the original function, does some math to it, and produces the original input from the original function! Sound confusing? Think about it this way.

Consider the function $f(x) = 10x + 5$. If we evaluated the function for the input $x = 5$, we would find that $f(x) = 55$. The inverse function of $f(x)$ would need to take the input 55 and produce the output 5. Essentially, the inverse function would "undo" the work that we had done with the original function. The function we would be looking for would be called "f inverse of x," and it would look like this:

$$f^{-1}(x)$$

The –1 in the exponent position denotes that this is an inverse function. However, it is important to note that we should not confuse this notation with the typical use of a negative exponent. Here the –1 exponent indicates that $f^{-1}(x)$ is the inverse function of $f(x)$. This is *not* to say that we are raising the function to the exponent of –1.

Let's take a look at a couple of examples to see how we can find an inverse function, given an original function.

➡ Example 1

Find the inverse function of $f(x) = 10x + 5$.

Solution

Remember that $f(x)$ essentially means the same thing as y, so first, replace the $f(x)$ with y.

$$y = 10x + 5$$

Swap the positions of the variables x and y. The resulting equation should be:

$$x = 10y + 5$$

Next, solve the new equation for y.

$$x = 10y + 5$$
$$\underline{-5 \qquad\qquad -5}$$
$$\frac{x-5}{10} = \frac{10y}{10}$$
$$\frac{x-5}{10} = y$$

The last step is to replace y with $f^{-1}(x)$.

Answer: $f^{-1}(x) = \dfrac{x-5}{10}$

We initially evaluated the function $f(x) = 10x + 5$ and found that $f(5) = 55$. This means that, when we use our inverse function, we should find that $f^{-1}(55) = 5$.

$$f^{-1}(x) = \frac{x-5}{10}$$

$$f^{-1}(55) = \frac{55-5}{10}$$

$$f^{-1}(55) = \frac{50}{10}$$

$$f^{-1}(55) = 5$$

It checks out! The function $f(x) = 10x + 5$ has an inverse function of $f^{-1}(x) = \frac{x-5}{10}$.

➡ Example 2

Consider the function $f(x) = \frac{2}{3}x - 4$. Find $f^{-1}(6)$.

Solution

First, change $f(x)$ to y, and then swap the position of the variables.

$$y = \frac{2}{3}x - 4$$

$$x = \frac{2}{3}y - 4$$

Now solve for y.

$$x = \frac{2}{3}y - 4$$

$$\underline{+4 \qquad\quad +4}$$

$$x + 4 = \frac{2}{3}y$$

$$\left(\frac{3}{2}\right)(x+4) = \left(\frac{3}{2}\right)\left(\frac{2}{3}y\right)$$

$$\frac{3(x+4)}{2} = y$$

$$f^{-1}(x) = \frac{3(x+4)}{2}$$

Finally, evaluate $f^{-1}(6)$ by plugging 6 in for x in the inverse function.

$$f^{-1}(6) = \frac{3(6+4)}{2}$$

$$f^{-1}(6) = \frac{3(10)}{2}$$

$$f^{-1}(6) = \frac{30}{2}$$

$$f^{-1}(6) = 15$$

Answer: $f^{-1}(6) = 15$. This checks out because, if we plug our answer, 15, back into the original function, we get an output of 6.

$$f(x) = \frac{2}{3}x - 4$$
$$f(15) = \frac{2}{3}(15) - 4$$
$$f(15) = 10 - 4$$
$$f(15) = 6$$
$$f(15) = 6 \text{ and } f^{-1}(6) = 15$$

LESSON 18.3—SKILLS CHECK

Directions: For each of the problems below, find the inverse of the given function.

1. $f(x) = 2x + 7$

2. $g(x) = \dfrac{x + 6}{3}$

3. $h(x) = 4x$

4. $g(x) = -\dfrac{1}{2}x + 8$

5. $f(x) = \dfrac{3}{4}x - \dfrac{1}{2}$

(Answers are on page 439.)

LESSON 18.3—ACCUPLACER CHALLENGE

Directions: For each of the questions below, choose the best answer from the five choices given.

1. A function $g(x)$ has an inverse function of $g^{-1}(x)$. If $g(a) = b$, what does $g^{-1}(b)$ equal?

   (A) $g(a)$
   (B) $g^{-1}(a)$
   (C) $g^{-1}(-b)$
   (D) a
   (E) b

2. Which of the functions below has the inverse function $f^{-1}(x) = 3x + 18$?

   (A) $f(x) = \dfrac{1}{3}x - 6$

   (B) $f(x) = -3x - 18$

   (C) $f(x) = -\dfrac{1}{3}x - 6$

   (D) $f(x) = 6x - 3$

   (E) $f(x) = 18 - 3x$

3. Which of the following is the inverse of the function $f(x) = -\frac{2}{3}x + \frac{1}{4}$?

(A) $f^{-1}(x) = \dfrac{-3(4x-1)}{8}$

(B) $f^{-1}(x) = \dfrac{-3(x+1)}{8}$

(C) $f^{-1}(x) = \dfrac{-4x+1}{6}$

(D) $f^{-1}(x) = -12x + 3$

(E) $f^{-1}(x) = 12x - 3$

4. For the function $f(x) = \dfrac{4(x-1)}{5}$, find $f^{-1}(8)$.

(A) $5\dfrac{3}{5}$

(B) 44

(C) $6\dfrac{1}{5}$

(D) 9

(E) 11

5. A linear function $f(x)$ has $f(0) = -2$ and $f(1) = 5$. Which of the following is $f^{-1}(x)$?

(A) $f^{-1}(x) = \dfrac{x+2}{5}$

(B) $f^{-1}(x) = \dfrac{x+2}{7}$

(C) $f^{-1}(x) = 7x - 2$

(D) $f^{-1}(x) = 5x - 2$

(E) $f^{-1}(x) = \dfrac{x-2}{7}$

(Answers are on page 441.)

To complete a set of practice questions that reviews all of Chapter 18, go to:
barronsbooks.com/TP/accuplacer/math29qw/

Lesson 18.1—Skills Check (page 423)

1. $f(6) = 77$ Find $f(6)$ by replacing x with 6 in the function.

$$f(x) = x^2 + 6x + 5$$
$$f(6) = (6)^2 + 6(6) + 5$$
$$f(6) = 36 + 36 + 5$$
$$f(6) = 77$$

2. $f(-7) = 12$ Find $f(-7)$ by replacing x with -7 in the function.

$$f(x) = x^2 + 6x + 5$$
$$f(-7) = (-7)^2 + 6(-7) + 5$$
$$f(-7) = 49 - 42 + 5$$
$$f(-7) = 12$$

3. $f(0) = 5$ Find $f(0)$ by replacing x with 0 in the function.

$$f(x) = x^2 + 6x + 5$$
$$f(0) = (0)^2 + 6(0) + 5$$
$$f(0) = 0 + 0 + 5$$
$$f(0) = 5$$

4. $f\left(\dfrac{1}{2}\right) = 8\dfrac{1}{4}$ Find $f\left(\dfrac{1}{2}\right)$ by replacing x with $\dfrac{1}{2}$ in the function.

$$f(x) = x^2 + 6x + 5$$
$$f\left(\frac{1}{2}\right) = \left(\frac{1}{2}\right)^2 + 6\left(\frac{1}{2}\right) + 5$$
$$f\left(\frac{1}{2}\right) = \frac{1}{4} + 3 + 5$$
$$f\left(\frac{1}{2}\right) = 8\frac{1}{4}$$

5. $f\left(-\dfrac{3}{4}\right) = \dfrac{17}{16}$ Find $f\left(-\dfrac{3}{4}\right)$ by replacing x with $-\dfrac{3}{4}$ in the function.

$$f(x) = x^2 + 6x + 5$$
$$f\left(-\frac{3}{4}\right) = \left(-\frac{3}{4}\right)^2 + 6\left(-\frac{3}{4}\right) + 5$$
$$f\left(-\frac{3}{4}\right) = \frac{9}{16} - \frac{9}{2} + 5$$
$$f\left(-\frac{3}{4}\right) = \frac{9}{16} - \frac{72}{16} + \frac{80}{16}$$
$$f\left(-\frac{3}{4}\right) = \frac{17}{16}$$

6. $g(3) = 15$ Find $g(3)$ by replacing x with 3 in the function.

$$g(x) = \frac{3x + 6}{x - 2}$$
$$g(3) = \frac{3(3) + 6}{(3) - 2}$$
$$g(3) = \frac{9 + 6}{1}$$
$$g(3) = 15$$

7. $g(-4) = 1$ Find $g(-4)$ by replacing x with -4 in the function.

$$g(x) = \frac{3x + 6}{x - 2}$$
$$g(-4) = \frac{3(-4) + 6}{(-4) - 2}$$
$$g(-4) = \frac{-12 + 6}{-6}$$
$$g(-4) = \frac{-6}{-6}$$
$$g(-4) = 1$$

8. $g(0) = -3$ Find $g(0)$ by replacing x with 0 in the function.

$$g(x) = \frac{3x + 6}{x - 2}$$
$$g(0) = \frac{3(0) + 6}{(0) - 2}$$
$$g(0) = \frac{6}{-2}$$
$$g(0) = -3$$

9. **g(–2) = 0** Find $g(-2)$ by replacing x with –2 in the function.

$$g(x) = \frac{3x + 6}{x - 2}$$

$$g(-2) = \frac{3(-2) + 6}{(-2) - 2}$$

$$g(-2) = \frac{-6 + 6}{-4}$$

$$g(-2) = \frac{0}{-4}$$

$$g(-2) = 0$$

10. **g(8) = 5** Find $g(8)$ by replacing x with 8 in the function.

$$g(x) = \frac{3x + 6}{x - 2}$$

$$g(8) = \frac{3(8) + 6}{(8) - 2}$$

$$g(8) = \frac{24 + 6}{6}$$

$$g(8) = \frac{30}{6}$$

$$g(8) = 5$$

Lesson 18.1—ACCUPLACER Challenge (page 424)

1. **(C)** The function will be undefined when the denominator of the rational expression becomes zero, so we need to find the value of $4x - 20$ that is equal to zero.

$$4x - 20 = 0$$
$$\underline{+20 \quad +20}$$
$$\frac{4x}{4} = \frac{20}{4}$$
$$x = 5$$

The function will be undefined at $x = 5$ because division by zero is undefined.

2. **(E)** This is a linear function because the change in x and the change in $h(x)$ are both linear. The answer choices are also in slope-intercept form. We can find the slope of the line modeled in the function table.

$$\text{slope} = \frac{\text{change in } y}{\text{change in } x} = \frac{-2}{4} = -\frac{1}{2}$$

Two of the answer choices have this slope, so we need more information. We know that when $x = 4$, $h(x) = -6$. Test the function in choice B to see if the input 4 produces the desired output.

$$h(x) = -\frac{1}{2}x - 6$$

$$h(4) = -\frac{1}{2}(4) - 6$$

$$h(4) = -2 - 6$$

$$h(4) = -8$$

The function in choice B does not produce the desired output. Therefore, the correct choice is choice E.

$$h(x) = -\frac{1}{2}x - 4$$

$$h(4) = -\frac{1}{2}(4) - 4$$

$$h(4) = -2 - 4$$

$$h(4) = -6$$

3. **(B)** We know that this function is the graph of a parabola. We can find the real zeros by making $f(x)$ equal to 0 and then factoring to find solutions.

$$f(x) = x^2 - 9$$
$$0 = x^2 - 9$$
$$0 = (x + 3)(x - 3)$$

Setting each binomial equal to zero and solving, we find that the function has zeros at $x = 3$ and $x = -3$. We also know that it will be a parabola that opens upward and has a vertex at the point $(0, -9)$. This means that the function will be negative at all values in between, but not including, the zeros.

4. **(D)** To solve this problem, test each case.

Choice A:

$$f(x) = 2^x \qquad g(x) = \frac{x^3}{4}$$
$$f(0) = 2^0 \qquad g(0) = \frac{0^3}{4}$$
$$f(0) = 1 \qquad g(0) = 0$$

The statement $f(0) < g(0)$ is false.

Choice B:

$$f(x) = 2^x \qquad g(x) = \frac{x^3}{4}$$
$$f(1) = 2^1 \qquad g(1) = \frac{1^3}{4}$$
$$f(1) = 2 \qquad g(1) = \frac{1}{4}$$

The statement $f(1) = g(1)$ is false.

Choice C:

$$f(x) = 2^x \qquad g(x) = \frac{x^3}{4}$$
$$f(6) = 2^6 \qquad g(6) = \frac{6^3}{4}$$
$$f(6) = 64 \qquad g(6) = \frac{216}{4}$$
$$g(6) = 54$$

The statement $f(6) < g(6)$ is false.

Choice D:

$$f(x) = 2^x \qquad g(x) = \frac{x^3}{4}$$
$$f(4) = 2^4 \qquad g(4) = \frac{4^3}{4}$$
$$f(4) = 16 \qquad g(4) = \frac{64}{4}$$
$$g(4) = 16$$

This statement is true, so choice D is the correct answer.

Choice E:

$$f(x) = 2^x \qquad g(x) = \frac{x^3}{4}$$
$$f(2) = 2^2 \qquad g(2) = \frac{2^3}{4}$$
$$f(2) = 4 \qquad g(2) = \frac{8}{4}$$
$$g(2) = 2$$

The statement $f(2) < g(2)$ is false.

5. **(C)** The statement $0 < a < b$ tells us that we are only working with positive values, and that a must be less than b. We know that as the value of x increases, the value of $h(x)$ will get closer and closer to 0. Since b is the larger input value, we can conclude that $h(b)$ will always have a value closer to zero than $h(a)$ will. This means that $h(a)$ will have a greater value. Consider a test case where $a = 2$ and $b = 3$.

$$h(x) = \frac{1}{x^2} \qquad h(x) = \frac{1}{x^2}$$
$$h(a) = \frac{1}{a^2} \qquad h(b) = \frac{1}{b^2}$$
$$h(2) = \frac{1}{2^2} \qquad h(3) = \frac{1}{3^2}$$
$$h(2) = \frac{1}{4} \qquad h(3) = \frac{1}{9}$$

Here $h(a) > h(b)$. This will be true for all positive values of a and b.

Lesson 18.2—Skills Check (page 427)

1. $f(g(6)) = 60$ To evaluate $f(g(6))$, first find $g(6)$, and then use it as the input for the function f.

$$g(x) = \frac{1}{2}x^2$$
$$g(6) = \frac{1}{2}(6)^2$$
$$g(6) = \frac{1}{2}(36)$$
$$g(6) = 18$$

Now use 18 as the input value for f.

$$f(x) = 4x - 12$$
$$f(18) = 4(18) - 12$$
$$f(18) = 72 - 12$$
$$f(18) = 60$$

Therefore, $f(g(6)) = 60$.

2. $f(h(2)) = 0$ To evaluate $f(h(2))$, first find $h(2)$, and then use it as the input for the function f.

$$h(x) = \frac{x+10}{x+2}$$
$$h(2) = \frac{2+10}{2+2}$$
$$h(2) = \frac{12}{4}$$
$$h(2) = 3$$

Now use 3 as the input value for f.

$$f(x) = 4x - 12$$
$$f(3) = 4(3) - 12$$
$$f(3) = 12 - 12$$
$$f(3) = 0$$

Therefore, $f(h(2)) = 0$.

3. $h(f(0)) = \frac{1}{5}$ To evaluate $h(f(0))$, first find $f(0)$, and then use it as the input for the function h.

$$f(x) = 4x - 12$$
$$f(0) = 4(0) - 12$$
$$f(0) = 0 - 12$$
$$f(0) = -12$$

Now use −12 as the input value for h.

$$h(x) = \frac{x+10}{x+2}$$
$$h(-12) = \frac{-12+10}{-12+2}$$
$$h(-12) = \frac{-2}{-10}$$
$$h(-12) = \frac{1}{5}$$

Therefore, $h(f(0)) = \frac{1}{5}$.

4. $g(h(-4)) = \frac{9}{2}$ or $4\frac{1}{2}$ To evaluate $g(h(-4))$, first find $h(-4)$, and then use it as the input for the function g.

$$h(x) = \frac{x+10}{x+2}$$
$$h(-4) = \frac{-4+10}{-4+2}$$
$$h(-4) = \frac{6}{-2}$$
$$h(-4) = -3$$

Now use −3 as the input for the function g.

$$g(x) = \frac{1}{2}x^2$$
$$g(-3) = \frac{1}{2}(-3)^2$$
$$g(-3) = \frac{1}{2}(9)$$
$$g(-3) = \frac{9}{2} \text{ or } 4\frac{1}{2}$$

Therefore, $g(h(-4)) = \frac{9}{2}$ or $4\frac{1}{2}$.

5. $f(h(0)) = 8$ To evaluate $f(h(0))$, first find $h(0)$, and then use it as the input for the function f.

$$h(x) = \frac{x+10}{x+2}$$
$$h(0) = \frac{0+10}{0+2}$$
$$h(0) = \frac{10}{2}$$
$$h(0) = 5$$

Now use 5 as the input value for f.

$$f(x) = 4x - 12$$
$$f(5) = 4(5) - 12$$
$$f(5) = 20 - 12$$
$$f(5) = 8$$

Therefore, $f(h(0)) = 8$.

6. $f \circ g(n) = n^2 - 10n + 25$ To create the composite function $f \circ g(n)$, use $g(n)$ as the input for the function f.

$$f \circ g(n) = f(n - 5)$$
$$f(n - 5) = (n - 5)^2$$
$$f(n - 5) = n^2 - 10n + 25$$

Therefore, $f \circ g(n) = n^2 - 10n + 25$.

7. $h \circ g(n) = 3n - 11$ To create the composite function $h \circ g(n)$, use $g(n)$ as the input for the function h.

$$h \circ g(n) = h(n - 5)$$
$$h(n - 5) = 3(n - 5) + 4$$
$$h(n - 5) = 3n - 15 + 4$$
$$h(n - 5) = 3n - 11$$

Therefore, $h \circ g(n) = 3n - 11$.

8. $f \circ h(n) = 9n^2 + 24n + 16$ To create the composite function $f \circ h(n)$, use $h(n)$ as the input for the function f.

$$f \circ h(n) = f(3n + 4)$$
$$f(3n + 4) = (3n + 4)^2$$
$$f(3n + 4) = (3n + 4)(3n + 4)$$
$$f(3n + 4) = 9n^2 + 24n + 16$$

Therefore, $f \circ h(n) = 9n^2 + 24n + 16$.

9. $h \circ f \circ g(n) = 3n^2 - 30n + 79$ To create the composite function $h \circ f \circ g(n)$, first use $g(n)$ as the input for the function f.

$$f \circ g(n) = f(n - 5)$$
$$f(n - 5) = (n - 5)^2$$
$$f(n - 5) = n^2 - 10n + 25$$

$f \circ g(n) = n^2 - 10n + 25$. Now use $f \circ g(n)$ as the input for the function h.

$$h \circ f \circ g(n) = h(n^2 - 10n + 25)$$
$$h(n^2 - 10n + 25) = 3(n^2 - 10n + 25) + 4$$
$$h(n^2 - 10n + 25) = 3n^2 - 30n + 75 + 4$$
$$h(n^2 - 10n + 25) = 3n^2 - 30n + 79$$

Therefore, $h \circ f \circ g(n) = 3n^2 - 30n + 79$.

10. $f \circ h \circ g(n) = 9n^2 - 66n + 121$ To create the composite function $f \circ h \circ g(n)$, first use $g(n)$ as the input for the function h.

$$h \circ g(n) = h(n - 5)$$
$$h(n - 5) = 3(n - 5) + 4$$
$$h(n - 5) = 3n - 15 + 4$$
$$h(n - 5) = 3n - 11$$

$h \circ g(n) = 3n - 11$. Now use $h \circ g(n)$ as the input for the function f.

$$f \circ h \circ g(n) = f(3n - 11)$$
$$f(3n - 11) = (3n - 11)^2$$
$$f(3n - 11) = (3n - 11)(3n - 11)$$
$$f(3n - 11) = 9n^2 - 66n + 121$$

Therefore, $f \circ h \circ g(n) = 9n^2 - 66n + 121$.

Lesson 18.2—ACCUPLACER Challenge (page 428)

1. **(B)** First evaluate $h(3)$.

$$h(x) = x^3 - 10x + 8$$
$$h(3) = (3)^3 - 10(3) + 8$$
$$h(3) = 27 - 30 + 8$$
$$h(3) = 5$$

Now use $h(3) = 5$ as the input for the function g.

$$g(x) = \frac{4x + 1}{x + 2}$$
$$g(5) = \frac{4(5) + 1}{5 + 2}$$
$$g(5) = \frac{20 + 1}{7}$$
$$g(5) = \frac{21}{7}$$
$$g(5) = 3$$

Therefore, $g(h(3)) = 3$.

2. **(D)** Since three of the five answer choices involve evaluating and comparing the composite functions on an input value of 3, we should first evaluate $f \circ g(3)$ and $g \circ f(3)$. One of them will likely be true.

$$g(3) = 2(3) + 5 \qquad\qquad f(3) = -3(3) + 2$$
$$g(3) = 6 + 5 \qquad\qquad f(3) = -9 + 2$$
$$g(3) = 11 \qquad\qquad f(3) = -7$$

$$f \circ g(3) = f(11) \qquad\qquad g \circ f(3) = g(-7)$$
$$f(11) = -3(11) + 2 \qquad\qquad g(-7) = 2(-7) + 5$$
$$f(11) = -33 + 2 \qquad\qquad g(-7) = -14 + 5$$
$$f(11) = -31 \qquad\qquad g(-7) = -9$$

$$f \circ g(3) = -31 \qquad\qquad g \circ f(3) = -9$$

We see that $f \circ g(3) < g \circ f(3)$.

3. **(B)** To find $f(h(x))$, use $h(x)$ as the input for the function f.

$$f(h(x)) = f(4x + 4)$$
$$f(4x + 4) = \frac{(4x + 4)^2}{8}$$
$$f(4x + 4) = \frac{16x^2 + 32x + 16}{8}$$
$$f(4x + 4) = \frac{16(x^2 + 2x + 1)}{8}$$
$$f(4x + 4) = 2(x^2 + 2x + 1)$$
$$f(4x + 4) = 2(x + 1)^2$$

Therefore, $f(h(x)) = 2(x + 1)^2$.

4. **(C)** We need to evaluate $g \circ h(t)$ and $g \circ h(-t)$.

Finding $g \circ h(t)$	**Finding $g \circ h(-t)$**

$$g \circ h(t) = g(6t)$$
$$g(6t) = (6t)^2$$
$$g(6t) = 36t^2$$

$$h(t) = 6t$$
$$h(-t) = 6(-t)$$
$$h(-t) = -6t$$

$$g \circ h(-t) = g(-6t)$$
$$g(-6t) = (-6t)^2$$
$$g(-6t) = 36t^2$$

We see that $g \circ h(t)$ and $g \circ h(-t)$ are equal, so choice C is the correct answer.

5. **(E)** To find $h \circ g \circ f(x)$, first use $f(x)$ as the input for the function g.

$$g \circ f(x) = g(x^2)$$
$$g(x^2) = 3(x^2) + 1$$
$$g(x^2) = 3x^2 + 1$$

Now use $3x^2 + 1$ as the input for the function h.

$$h(3x^2 + 1) = 3x^2 + 1 + 2$$
$$h(3x^2 + 1) = 3x^2 + 3$$
$$h(3x^2 + 1) = 3(x^2 + 1)$$

Therefore, $h \circ g \circ f(x) = 3(x^2 + 1)$.

Lesson 18.3—Skills Check (page 431)

1. $f^{-1}(x) = \dfrac{x-7}{2}$ Replace $f(x)$ with y.

$$f(x) = 2x + 7$$
$$y = 2x + 7$$

Swap the position of the variables, and solve for y.

$$x = 2y + 7$$
$$\underline{ -7 \qquad -7}$$
$$\frac{x-7}{2} = \frac{2y}{2}$$
$$\frac{x-7}{2} = y$$

Replace y with $f^{-1}(x)$.

$$f^{-1}(x) = \frac{x-7}{2}$$

2. $g^{-1}(x) = 3x - 6$ Replace $g(x)$ with y.

$$g(x) = \frac{x+6}{3}$$

$$y = \frac{x+6}{3}$$

Swap the position of the variables, and solve for y.

$$x = \frac{y+6}{3}$$

$$3(x) = \left(\frac{y+6}{3}\right)3$$

$$3x = y + 6$$

$$\underline{-6 \qquad -6}$$

$$3x - 6 = y$$

Replace y with $g^{-1}(x)$.

$$y = 3x - 6$$

$$g^{-1}(x) = 3x - 6$$

3. $h^{-1}(x) = \frac{x}{4}$ Replace $h(x)$ with y.

$$h(x) = 4x$$

$$y = 4x$$

Swap the position of the variables, and solve for y.

$$\frac{x}{4} = \frac{4y}{4}$$

$$\frac{x}{4} = y$$

Replace y with $h^{-1}(x)$.

$$y = \frac{x}{4}$$

$$h^{-1}(x) = \frac{x}{4}$$

4. $g^{-1}(x) = -2x + 16$ Replace $g(x)$ with y.

$$g(x) = -\frac{1}{2}x + 8$$

$$y = -\frac{1}{2}x + 8$$

Swap the position of the variables, and solve for y.

$$x = -\frac{1}{2}y + 8$$

$$\underline{-8 \qquad -8}$$

$$-2(x - 8) = \left(-\frac{1}{2}y\right)\left(-\frac{2}{1}\right)$$

$$-2x + 16 = y$$

Replace y with $g^{-1}(x)$.

$$y = -2x + 16$$
$$g^{-1}(x) = -2x + 16$$

5. $f^{-1}(x) = \dfrac{4x+2}{3}$ Replace $f(x)$ with y.

$$f(x) = \frac{3}{4}x - \frac{1}{2}$$
$$y = \frac{3}{4}x - \frac{1}{2}$$

Swap the position of the variables, and solve for y.

$$4(x) = 4\left(\frac{3}{4}y - \frac{1}{2}\right)$$
$$4x = 3y - 2$$
$$\underline{+2 \qquad +2}$$
$$\frac{4x+2}{3} = \frac{3y}{3}$$
$$\frac{4x+2}{3} = y$$

Replace y with $f^{-1}(x)$.

$$y = \frac{4x+2}{3}$$
$$f^{-1}(x) = \frac{4x+2}{3}$$

Lesson 18.3—ACCUPLACER Challenge (pages 431–432)

1. **(D)** This is the definition of an inverse function. Remember that the inverse of a function "undoes" the work of the original function. Therefore, if the original function took an input a and produced an output b, then the inverse would take the input b and produce the output a.

2. **(A)** To find the inverse of choice A, first replace $f(x)$ with y.

$$f(x) = \frac{1}{3}x - 6$$
$$y = \frac{1}{3}x - 6$$

Swap the position of the variables, and solve for y.

$$x = \frac{1}{3}y - 6$$
$$\underline{+6 \qquad +6}$$
$$3(x+6) = \left(\frac{1}{3}y\right)3$$
$$3(x+6) = y$$
$$3x + 18 = y$$

Replace y with $f^{-1}(x)$.

$$y = 3x + 18$$
$$f^{-1}(x) = 3x + 18$$

The function $f(x) = \frac{1}{3}x - 6$ is the only one of these choices that has an inverse function of $f^{-1}(x) = 3x + 18$.

3. **(A)** To find the inverse of $f(x)$, first replace $f(x)$ with y.

$$f(x) = -\frac{2}{3}x + \frac{1}{4}$$
$$y = -\frac{2}{3}x + \frac{1}{4}$$

Swap the position of the variables, and solve for y. Notice that we multiply both sides of the equation by 12 to eliminate the fractions.

$$x = -\frac{2}{3}y + \frac{1}{4}$$
$$12(x) = \left(-\frac{2}{3}y + \frac{1}{4}\right)12$$
$$12x = -8y + 3$$
$$\underline{-3 \qquad\qquad -3}$$
$$\frac{12x - 3}{-8} = \frac{-8y}{-8}$$
$$\frac{-12x + 3}{8} = y$$
$$\frac{-3(4x - 1)}{8} = y$$

Replace y with $f^{-1}(x)$.

$$y = \frac{-3(4x - 1)}{8}$$
$$f^{-1}(x) = \frac{-3(4x - 1)}{8}$$

4. **(E)** First, find the inverse of $f(x)$. Replace $f(x)$ with y.

$$f(x) = \frac{4(x - 1)}{5}$$
$$y = \frac{4(x - 1)}{5}$$

Swap the position of the variables, and solve for y.

$$5(x) = \left(\frac{4(y - 1)}{5}\right)5$$
$$\frac{5x}{4} = \frac{4(y - 1)}{4}$$
$$\frac{5}{4}x = y - 1$$
$$\underline{+1 \qquad\qquad +1}$$
$$\frac{5}{4}x + 1 = y$$

Replace y with $f^{-1}(x)$.

$$y = \frac{5}{4}x + 1$$

$$f^{-1}(x) = \frac{5}{4}x + 1$$

Therefore:

$$f^{-1}(8) = \frac{5}{4}\left(\frac{8}{1}\right) + 1$$

$$f^{-1}(8) = \frac{40}{4} + 1$$

$$f^{-1}(8) = 10 + 1$$

$$f^{-1}(8) = 11$$

5. **(B)** For this function, we can conclude that if $f(0) = -2$, then $f^{-1}(-2) = 0$. Also, if $f(1) = 5$, then $f^{-1}(5) = 1$. This follows the definition of inverse functions. If we use the input -2 in the correct function, it should produce an output of 0. Likewise, if we use the input 5 in the correct function, it should produce an output of 1. This will only work with the function in choice B.

$$f^{-1}(x) = \frac{x+2}{7} \qquad f^{-1}(x) = \frac{x+2}{7}$$

$$f^{-1}(-2) = \frac{-2+2}{7} \qquad f^{-1}(5) = \frac{5+2}{7}$$

$$f^{-1}(-2) = \frac{0}{7} \qquad f^{-1}(5) = \frac{7}{7}$$

$$f^{-1}(-2) = 0 \qquad f^{-1}(5) = 1$$

Logarithms and Rational Exponents

19

Throughout this book, you have practiced working with exponents and solving algebraic equations. The two types of mathematical problems involving exponents that we haven't yet addressed are

When the Exponent is the Unknown

$$2^x = 8$$

When the Exponent is a Fraction

$$8^{\frac{1}{3}} = b$$

> As a means to solving problems like these and more, in this chapter, we will work to master
>
> - understanding the rules governing logarithms, and
> - working with rational exponents.

LESSON 19.1—LOGARITHMS

Lesson 19.1A—Logarithmic Form

Let's consider the following:

$$2^x = 8$$

Now, this is a fairly easy problem if you are familiar with your powers of 2. The variable is $x = 3$ because $2^3 = 2 \times 2 \times 2 = 8$. How would we solve $2^x = 20$? You will not need to accurately solve problems like these for the ACCUPLACER, but you do need to be familiar with the common mathematical notation used in such problems.

Typically, when we solve for a variable, say x, we work to isolate the variable or get the variable by itself. Even though solving $2^x = 8$ may be straightforward, it isn't written such that x is isolated. One way to isolate the variable x in an exponential equation such as $2^x = 8$ is to use logarithms. Put simply, a **logarithm** is the inverse of exponentiation (exponents). Remember, subtraction is the inverse of addition, division is the inverse of multiplication, square root is the inverse of squaring, and logarithm, log for short, is the inverse of exponentiation.

For example,

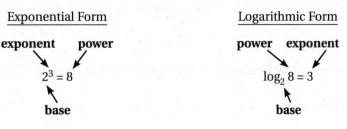

In this way, we can write $2^x = 8$ as $\log_2 8 = x$, and the exponent, the variable, has been isolated. The logarithmic equation would be read as *log base 2 of 8 equals x*.

To Write an Exponential Equation in Logarithmic Form:

1. Identify the base, exponent, and power.
2. Substitute the base, power, and exponent into the form $\log_{\text{base}}(\text{power}) = \text{exponent}$.

➡ Example 1

Write $5^x = 125$ in logarithmic form, and solve for x.

Solution

Let's first identify the base, exponent, and power.

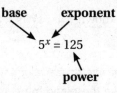

In order to isolate the variable, we use log notation.

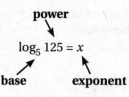

To solve the problem, we must ask ourselves, "which power of 5 is 125?" Through multiplication knowledge, or trial and error, we can determine that $5^3 = 125$, so $\log_5 125 = 3$. In other words, the third power of 5 is 125.

Answer: $\log_5 125 = x$ and $x = 3$

➡ Example 2

Write $\log 10,000 = 4$ in exponential form.

Solution

If you see a logarithm written with no visible base, the base is 10. A logarithm of base 10 is called the **common logarithm**.

Let's identify the base, exponent, and power.

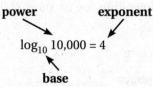

Write the base, power, and exponent in the exponential form $\text{base}^{\text{exponent}} = \text{power}$.

Answer: $10^4 = 10,000$

➡ Example 3

Solve $\log_3 81 = x$.

Solution

We need to determine which power of 3 is 81. Without a calculator, we could list the powers of 3 until we find the one that produces 81.

$$3^1 = 3$$
$$3^2 = 3 \times 3 = 9$$
$$3^3 = 3 \times 3 \times 3 = 27$$
$$3^4 = 3 \times 3 \times 3 \times 3 = 81 \checkmark$$

As we can see, the fourth power of 3 is 81. Therefore, $\log_3 81 = 4$.

Answer: $x = 4$

ACCUPLACER TIPS

Tip #1

It's easy to get confused when setting up logarithms, especially during a test. Memorize one example of a logarithm, and the power it corresponds with, and use this one example to remember how to set up others. One example that we like, and one that we recommend burning into your brain, is

$$\log_2 8 = 3 \qquad \text{goes with} \qquad 2^3 = 8$$

Tip #2

Logarithms start out pretty straightforward, and then they start to get less straightforward as we start using more than one logarithm in the same equation, as you will see later in the chapter. It's helpful to be familiar with the powers of lesser numbers in order to make solving logarithmic and exponential equations easier. On the ACCUPLACER, you will not be expected to know powers of greater numbers, but you may have to solve logarithm problems that involve powers of 2 through powers of 10. It is a good idea to memorize at least the following powers.

$2^0 = 1$	$3^0 = 1$	$4^0 = 1$	$5^0 = 1$	$6^0 = 1$	$7^0 = 1$	$8^0 = 1$	$9^0 = 1$	$10^0 = 1$
$2^1 = 2$	$3^1 = 3$	$4^1 = 4$	$5^1 = 5$	$6^1 = 6$	$7^1 = 7$	$8^1 = 8$	$9^1 = 9$	$10^1 = 10$
$2^2 = 4$	$3^2 = 9$	$4^2 = 16$	$5^2 = 25$	$6^2 = 36$	$7^2 = 49$	$8^2 = 64$	$9^2 = 81$	$10^2 = 100$
$2^3 = 8$	$3^3 = 27$	$4^3 = 64$	$5^3 = 125$	$6^3 = 216$		$8^3 = 512$	$9^3 = 729$	$10^3 = 1,000$
$2^4 = 16$	$3^4 = 81$	$4^4 = 256$	$5^4 = 625$					$10^4 = 10,000$
$2^5 = 32$		$4^5 = 1,024$						$10^5 = 100,000$
$2^6 = 64$								$10^6 = 1,000,000$
$2^7 = 128$								
$2^8 = 256$								
$2^9 = 512$								
$2^{10} = 1,024$								

LESSON 19.1A—SKILLS CHECK

> **Directions**: For questions 1–5, write each equation in logarithmic form ($\log_b a = x$).

1. $2^5 = 32$

2. $3^4 = 81$

3. $10^6 = 1{,}000{,}000$

4. $3^x = 45$

5. $7^a = b$

> **Directions**: For questions 6–10, write each equation in exponential form ($b^x = a$).

6. $\log_2 32 = 5$

7. $\log 100 = 2$

8. $\log_3 1 = 0$

9. $\log_b 1{,}000 = 3$

10. $\log_b a = 4$

> **Directions**: For questions 11–15, solve each equation.

11. $\log_2 16 = x$

12. $\log_4 64 = x$

13. $\log_{10} 100{,}000 = x$

14. $\log_b 36 = 2$

15. $\log_3 a = 5$

(Answers are on page 461.)

Lesson 19.1B—The Product Rule of Logarithms

Experience with logarithms has lead to the development of some **logarithm identities** that you will need to know for the ACCUPLACER. The first logarithm identity that you should be familiar with is

Product Rule of Logarithms

$$\log_b (a \bullet c) = \log_b a + \log_b c$$

➥ **Example 1**

Check that $\log_2 (8)(4) = \log_2 8 + \log_2 4$.

Solution

$$\log_2 (8)(4) = \log_2 8 + \log_2 4$$
$$\log_2 32 = 3 + 2$$
$$5 = 5$$

<u>Answer</u>: This checks out.

➡ Example 2

Check that $\log_{10} 10{,}000 + \log_{10} 1{,}000 = \log_{10}(10{,}000)(1{,}000)$.

Solution

$$\log_{10} 10{,}000 + \log_{10} 1{,}000 = \log_{10}(10{,}000)(1{,}000)$$
$$4 + 3 = \log_{10}(10{,}000{,}000)$$
$$7 = 7$$

Answer: This checks out.

➡ Example 3

Find the value of x that makes the equation $\log_2 8 + \log_2 x = \log_2 256$ true.

Solution

To solve this problem, we need to remember the fundamentals of the product rule.

$$\log_b (a \cdot c) = \log_b a + \log_b c$$

Looking at the equation $\log_2 8 + \log_2 x = \log_2 256$, we can determine the following about a, b, and c.

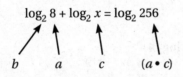

Therefore, we can determine that $b = 2$, $a = 8$, and $8 \cdot c = 256$. By substituting 8 for a and x for c, we can then write the following equation, and solve for x.

$$ac = 256$$
$$8x = 256$$
$$\frac{8x}{8} = \frac{256}{8}$$
$$x = 32$$

To make sure that this is correct, we can substitute 32 for x in the original equation, and evaluate.

$$\log_2 8 + \log_2 32 = \log_2 256$$
$$3 + 5 = \log_2 256$$
$$8 = 8$$

Answer: The solution checks out, so $x = 32$.

LESSON 19.1B—SKILLS CHECK

> **Directions**: Solve for the value of x.

1. $\log_{10}(100)(10) = \log_{10}100 + \log_{10}x$

2. $\log_2 64 = \log_2 4 + \log_2 x$

3. $\log_2 5 + \log_2 3 = \log_2 x$

4. $\log_a 2 + \log_a 4 = \log_a x$

5. $\log_b 3x = \log_b 3 + \log_b 6$

(Answers are on page 462.)

Lesson 19.1C—The Quotient Rule of Logarithms

The second logarithm identity that you should be familiar with is

Quotient Rule of Logarithms

$$\log_b\left(\frac{a}{c}\right) = \log_b a - \log_b c$$

➥ Example 1

Check that $\log_2\left(\dfrac{8}{4}\right) = \log_2 8 - \log_2 4$.

Solution

$$\log_2\left(\frac{8}{4}\right) = \log_2 8 - \log_2 4$$
$$\log_2 2 = 3 - 2$$
$$1 = 1$$

Answer: This checks out.

➥ Example 2

Check that $\log_4 64 - \log_4 16 = \log_4\left(\dfrac{64}{16}\right)$.

Solution

$$\log_4 64 - \log_4 16 = \log_4\left(\frac{64}{16}\right)$$
$$3 - 2 = \log_4 4$$
$$1 = 1$$

Answer: This checks out.

➡ Example 3

Find the value of x that makes the equation $\log_3 27 = \log_3 x - \log_3 3$ true.

Solution

To solve this problem, we need to remember the fundamentals of the quotient rule.

$$\log_b\left(\frac{a}{c}\right) = \log_b a - \log_b c$$

Looking at the equation $\log_3 27 = \log_3 x - \log_3 3$, we can determine the following about a, b, and c.

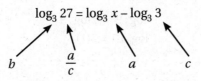

Therefore, we can determine that $b = 3$, $\frac{a}{c} = 27$, and $c = 3$. By substituting 3 for c and x for a, we can then write the following equation, and solve for x.

$$\frac{a}{c} = 27$$

$$\frac{x}{3} = 27$$

$$3 \cdot \frac{x}{3} = 27 \cdot 3$$

$$x = 81$$

To make sure that this is correct, we can substitute 81 for x in the original equation, and evaluate.

$$\log_3 27 = \log_3 81 - \log_3 3$$
$$3 = 4 - 1$$
$$3 = 3$$

Answer: The solution checks out, so $x = 81$.

LESSON 19.1C—SKILLS CHECK

> **Directions**: Solve for the value of x.

1. $\log_{10} 10,000 - \log_{10} 100 = \log_{10} x$

2. $\log_2 32 - \log_2 8 = \log_2 x$

3. $\log_y 16 - \log_y 2 = \log_y x$

4. $\log_b 8 = \log_b 24 - \log_b x$

5. $\log_a 3 = \log_a x - \log_a 3$

(Answers are on page 463.)

Lesson 19.1D—The Power Rule of Logarithms

The third logarithm identity that you should be familiar with is

Power Rule of Logarithms

$$\log_b a^c = c(\log_b a)$$

➥ Example 1

Check that $\log_9 9^2 = 2\log_9 9$.

Solution

$$\log_9 9^2 = 2\log_9 9$$
$$\log_9 81 = 2(1)$$
$$2 = 2$$

Answer: This checks out.

➥ Example 2

Check that $\log_2 8^3 = 3\log_2 8$.

Solution

$$\log_2 8^3 = 3\log_2 8$$
$$\log_2 512 = 3(3)$$
$$9 = 9$$

Answer: This checks out.

➥ Example 3

Find the value of x that makes the equation $\log_4 x = 2\log_4 16$ true.

Solution

To solve this problem, we need to remember the fundamentals of the power rule.

$$\log_b a^c = c(\log_b a)$$

Looking at the equation $\log_4 x = 2\log_4 16$, we can determine the following about a, b, and c.

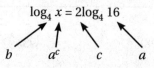

Therefore, we can determine that $b = 4$, $a^c = x$, $c = 2$, and $a = 16$. By substituting 16 for a and 2 for c, we can then write the following equation and solve for x.

$$x = a^c$$
$$x = 16^2$$
$$x = 256$$

To make sure that this is correct, we can substitute 256 for x in the original equation, and evaluate.

$$\log_4 256 = 2\log_4 16$$
$$4 = 2(2)$$
$$4 = 4$$

Answer: The solution checks out, so $x = 256$.

LESSON 19.1D—SKILLS CHECK

Directions: For questions 1–5, solve for the value of x.

1. $\log_2 5^3 = x \log_2 5$

2. $2\log_{10} 10 = \log_{10} 10^x$

3. $\log_2 16 = 2\log_2 x$

4. $\log_b x = 3\log_b 5$

5. $\log_b 64 = 3\log_b x$

(Answers are on page 464.)

LESSON 19.1—ACCUPLACER CHALLENGE

Directions: For each of the questions below, choose the best answer from the five choices given.

1. If $\log_2 a = 4$, then what does a equal?

    (A) $\dfrac{1}{2}$

    (B) $\sqrt{2}$

    (C) 2

    (D) 8

    (E) 16

2. What is the function $y = 3^x$ equivalent to?

    (A) $\log_x y = 3$
    (B) $\log_x 3 = y$
    (C) $\log_3 x = y$
    (D) $\log_3 y = x$
    (E) $\log_y 3 = x$

3. Which of the following expressions is NOT equivalent to $\log_b 64$?

    (A) $2\log_b 8$
    (B) $2\log_b 32$
    (C) $\log_b 16 + \log_b 4$
    (D) $\log_b 128 - \log_b 2$
    (E) $\log_b 8 + \log_b 8$

4. For which value of y is $\log_b y = x$ undefined?

    (A) 1.5
    (B) 1
    (C) 0
    (D) 1.7598
    (E) π

5. $3\log_4 a - \log_4 b =$

    (A) $2\log_4 (a - b)$

    (B) $\log_4\left(\dfrac{a^3}{b}\right)$

    (C) $\log_4 (3a - b)$

    (D) $\log_4 (a^3 - b)$

    (E) $3\log_4\left(\dfrac{a}{b}\right)$

(Answers are on page 465.)

LESSON 19.2—RATIONAL EXPONENTS

Lesson 19.2A—Writing and Interpreting Rational Exponents

In this section, we will explore what happens when we have an exponent that is a fraction. First, remember that when we multiply powers with the same base, we can simplify the expression by keeping the common base and adding the exponents. For example,

$$2^2 \times 2^3 = 2^{2+3} = 2^5$$

If that's true, then

$$16^{\frac{1}{2}} \times 16^{\frac{1}{2}} = 16^{\frac{1}{2}+\frac{1}{2}} = 16^1 = 16$$

We just multiplied $16^{\frac{1}{2}}$ by $16^{\frac{1}{2}}$ and got 16. The strange thing is that multiplying 4 by 4 also equals 16. In other words, $16^{\frac{1}{2}} = 4$ and $\sqrt{16} = 4$. Therefore:

$$16^{\frac{1}{2}} = \sqrt{16}$$

As a rule, we can state the following:

$$\boxed{a^{\frac{1}{2}} = \sqrt{a}}$$

➡ Example 1

$100^{\frac{1}{2}} = ?$

Solution

$$100^{\frac{1}{2}} = \sqrt{100} = 10$$

<u>Answer:</u> 10 (Remember, $10^2 = 100$.)

➡ Example 2

$45^{\frac{1}{2}} = ?$

Solution

$$45^{\frac{1}{2}} = \sqrt{45} = \sqrt{9 \bullet 5} = 3\sqrt{5}$$

<u>Answer:</u> $3\sqrt{5}$

Notice that, in Example 2, 45 is not a perfect square so we only simplified the radical. Remember, you might not have a calculator on the ACCUPLACER, so you won't be expected to evaluate radicals such as $\sqrt{5}$.

Now, surely you are asking yourself, "What about something like $27^{\frac{1}{3}}$ or $32^{\frac{2}{5}}$ even?" Well, it follows that if $b^{\frac{1}{2}}$ asks, "What number times itself gives us b?", then $27^{\frac{1}{3}}$ asks, "What number times itself times itself again gives us 27?" Another way of thinking of this is, "What number has 27 as its third power, or what value of x satisfies $x^3 = 27$?"

$$3^3 = 27, \text{ so } 27^{\frac{1}{3}} = 3$$

If $a^{\frac{1}{2}} = \sqrt{a}$, then we must be able to write $a^{\frac{1}{3}}$ in a similar fashion. Roots other than square roots are written like this:

index

$\sqrt[n]{a}$

radicand

The vocabulary, **index** and **radicand**, are not incredibly important, but what is important is that the **index**, or n, tells us the root that we are taking, and the **radicand**, or a, is the number that we are taking the root on.

➡ Example 3

$\sqrt[3]{64} = ?$

Solution

$\sqrt[3]{64}$ is asking us to find the cube root of 64. Put another way, what number has 64 as its third power?

$$4^3 = 4 \times 4 \times 4 = 64, \text{ so } \sqrt[3]{64} = 4$$

Answer: 4

A value with a rational exponent with a numerator of 1 is written like this in radical form:

$$a^{\frac{1}{n}} = \sqrt[n]{a}$$

➡ Example 4

$10,000^{\frac{1}{4}} = ?$

Solution

$$10,000^{\frac{1}{4}} = \sqrt[4]{10,000}$$

$$10^4 = 10 \times 10 \times 10 \times 10 = 10,000$$

Therefore, $\sqrt[4]{10,000} = \sqrt[4]{10 \times 10 \times 10 \times 10} = 10$, and $10,000^{\frac{1}{4}} = 10$.

Answer: 10

➡ Example 5

$8^{\frac{1}{3}} = ?$

Solution

$$8^{\frac{1}{3}} = \sqrt[3]{8}$$

$$2^3 = 2 \times 2 \times 2 = 8$$

Therefore, $\sqrt[3]{8} = \sqrt[3]{2 \times 2 \times 2} = 2$, and $8^{\frac{1}{3}} = 2$.

Answer: 2

As for how to handle an exponent that is a fraction with a numerator other than 1, such as $32^{\frac{2}{5}}$, let's first consider how to represent $a^{\frac{2}{5}}$ in radical form.

$$a^{\frac{2}{5}} = a^{\frac{1}{5} + \frac{1}{5}} = a^{\frac{1}{5}} \bullet a^{\frac{1}{5}} = \left(a^{\frac{1}{5}}\right)^2 = \left(\sqrt[5]{a}\right)^2$$

From this example, we can make a rule.

To Change a Rational Exponent to Radical Form

$$a^{\frac{m}{n}} = \left(\sqrt[n]{a}\right)^m$$

➡ Example 6

Write $5^{\frac{4}{7}}$ in radical form.

Solution

To write $5^{\frac{4}{7}}$ in radical form, the denominator, 7, becomes the index, the base, 5, becomes the radicand, and the numerator, 4, becomes the exponent outside of the parentheses.

$$5^{\frac{4}{7}} = \left(\sqrt[7]{5}\right)^4$$

Answer: $\left(\sqrt[7]{5}\right)^4$

➡ Example 7

Write $32^{\frac{2}{5}}$ in radical form.

Solution

To write $32^{\frac{2}{5}}$ in radical form, the denominator, 5, becomes the index, the base, 32, becomes the radicand, and the numerator, 2, becomes the exponent outside of the parentheses.

$$32^{\frac{2}{5}} = \left(\sqrt[5]{32}\right)^2$$

Answer: $\left(\sqrt[5]{32}\right)^2$

If this question were worded differently, we would still have more work to do. The fifth root of 32 is a quantity that we can determine and one that you should be familiar with. Earlier in this chapter, we listed some powers you should commit to memory. The powers of 2 are key and $2 \times 2 \times 2 \times 2 \times 2 = 32$, or $2^5 = 32$.

ACCUPLACER TIPS

In addition to becoming familiar with lesser powers of lesser numbers, such as $2^4 = 16$, you can also determine if you can simplify a root by factoring. In Chapter 7, we used factoring to do this with square roots. We can also use this method for cube roots and roots with higher indices.

Consider the following:

$$\sqrt{2 \bullet 2} = 2, \text{ which is to say } \sqrt{2^2} = 2$$

$$\sqrt[3]{2 \bullet 2 \bullet 2} = 2, \text{ which is to say } \sqrt[3]{2^3} = 2$$

$$\sqrt[4]{2 \bullet 2 \bullet 2 \bullet 2} = 2, \text{ which is to say } \sqrt[4]{2^4} = 2$$

We can then generalize that $\sqrt[n]{a^n} = a$. Let's look at these two roots $\sqrt[4]{625}$ and $\sqrt[4]{24}$. Which of these numbers results in a whole number when evaluated? The number 625 ends in a 5, so it must be divisible by 5. We can continually break down 625 under the radical sign until we have the prime factors of 5.

$$\sqrt[4]{625} = \sqrt[4]{125 \bullet 5} = \sqrt[4]{25 \bullet 5 \bullet 5} = \sqrt[4]{5 \bullet 5 \bullet 5 \bullet 5} = 5$$

The number 24 ends in an even number, so it must be divisible by 2. We can continually break down 24 under the radical sign until we have the prime factors of 24.

$$\sqrt[4]{24} = \sqrt[4]{2 \bullet 12} = \sqrt[4]{2 \bullet 2 \bullet 6} = \sqrt[4]{2 \bullet 2 \bullet 2 \bullet 3}$$

We were able to get a whole number after evaluating $\sqrt[4]{625}$ because $625 = 5^4$, so $\sqrt[4]{625} = \sqrt[4]{5^4} = 5$. However, we were not able to get a whole number after evaluating $\sqrt[4]{24}$ because the prime factors of 24 are $2 \bullet 2 \bullet 2 \bullet 3$. We did not have the same factor four times so $\sqrt[4]{24}$ does not result in a whole number.

LESSON 19.2A—SKILLS CHECK

> **Directions**: For questions 1–5, write each expression below in radical form $\left(\sqrt[n]{a}\right)^m$.

1. $x^{\frac{1}{4}}$

2. $a^{\frac{2}{5}}$

3. $y^{\frac{4}{3}}$

4. $2^{\frac{1}{3}}$

5. $3^{\frac{7}{3}}$

> **Directions**: For questions 6–10, write each expression below in exponential form $a^{\frac{m}{n}}$.

6. $\left(\sqrt[3]{x}\right)^2$

7. $\left(\sqrt{y}\right)^3$

8. $\left(\sqrt[5]{a}\right)^4$

9. $\left(\sqrt[8]{b}\right)^4$

10. $\left(\sqrt[5]{c}\right)^{10}$

> **Directions**: For questions 11–15, evaluate each expression.

11. $\left(\sqrt[4]{16}\right)^2$

12. $\left(\sqrt[3]{8}\right)^4$

13. $27^{\frac{4}{3}}$

14. $125^{\frac{2}{3}}$

15. $1000^{\frac{2}{3}}$

(Answers are on page 465.)

Lesson 19.2B—Multiplying Powers with Rational Exponents

By the product of powers rule, we can simplify the product of two terms with the same base and different exponents into a single term by keeping the common base and adding the exponents. For example, $3^5 \times 3^2 = 3^{5+2} = 3^7$. The same rule holds true for powers with rational exponents.

To Multiply Powers of the Same Base:

1. Keep the common base.
2. Add the exponents.

➡ Example 1

Simplify $g^{\frac{2}{5}} \times g^{\frac{1}{5}}$.

Solution

To simplify, keep the common base, g, and add the exponents together.

$$g^{\frac{2}{5}} \times g^{\frac{1}{5}} = g^{\frac{2}{5} + \frac{1}{5}} = g^{\frac{3}{5}}$$

Answer: $g^{\frac{3}{5}}$

➡ Example 2

Simplify $\left(\sqrt[5]{b}\right)^4 \left(\sqrt[3]{b}\right)^2$.

Solution

To simplify, we'll first need to change the expression to exponential form.

$$\left(\sqrt[5]{b}\right)^4 \left(\sqrt[3]{b}\right)^2 = b^{\frac{4}{5}} \times b^{\frac{2}{3}}$$

Next, keep the common base, b, and add the exponents together.

$$b^{\frac{4}{5}} \times b^{\frac{2}{3}} = b^{\frac{4}{5} + \frac{2}{3}}$$

Unlike the previous example, we cannot immediately add the exponents because they have different denominators. Change the rational exponents to equivalent fractions with common denominators, and then find their sum.

$$b^{\frac{4}{5} + \frac{2}{3}} = b^{\frac{12}{15} + \frac{10}{15}} = b^{\frac{22}{15}}$$

Answer: $b^{\frac{22}{15}}$

LESSON 19.2B—SKILLS CHECK

Directions: Express each product in exponential form $a^{\frac{m}{n}}$.

1. $x^{\frac{2}{7}} \cdot x^{\frac{3}{7}}$

2. $m^{\frac{1}{4}} \cdot m^{\frac{3}{8}}$

3. $\left(\sqrt[8]{a}\right)^5 \left(\sqrt[4]{a}\right)^3$

4. $\left(\sqrt[3]{16}\right)^{10} \left(\sqrt[5]{16}\right)$

5. $\left(\sqrt[3]{8g}\right)^2 \left(\sqrt[3]{8g}\right)^4$

(Answers are on page 465.)

LESSON 19.2—ACCUPLACER CHALLENGE

> **Directions**: For each of the questions below, choose the best answer from the five choices given.

1. $\left(\sqrt[4]{x}\right)^8 =$

   (A) 2

   (B) $x^{\frac{1}{2}}$

   (C) x^2

   (D) x^{16}

   (E) x^{32}

2. $81^{\frac{3}{4}} =$

   (A) 3

   (B) 9

   (C) 27

   (D) 60.75

   (E) 108

3. Given $f(x) = x^{\frac{1}{3}}$, what does $f(64)$ equal?

   (A) 2

   (B) 4

   (C) 4^3

   (D) 8

   (E) 16

4. $\left(\sqrt[3]{a}\right)^5 \left(\sqrt[6]{a}\right)^4 =$

   (A) $a^{\frac{2}{3}}$

   (B) $a^{\frac{3}{2}}$

   (C) $a^{\frac{7}{3}}$

   (D) $a^{\frac{9}{10}}$

   (E) $a^{\frac{21}{10}}$

5. $2^{\frac{7}{3}} - 2^{\frac{4}{3}} =$

   (A) 0

   (B) 2

   (C) $2^{\frac{1}{3}}$

   (D) $2^{\frac{4}{3}}$

   (E) $2^{\frac{7}{3}}$

(Answers are on page 466.)

> To complete a set of practice questions that reviews all of Chapter 19, go to:
> *barronsbooks.com/TP/accuplacer/math29qw/*

Lesson 19.1A—Skills Check (page 448)

1. $\log_2 32 = 5$

   $2^5 = 32$

   Base = 2
   Exponent = 5
   Power = 32

   $\log_2 32 = 5$

2. $\log_3 81 = 4$

   $3^4 = 81$

   Base = 3
   Exponent = 4
   Power = 81

   $\log_3 81 = 4$

3. $\log_{10} 1{,}000{,}000 = 6$
   or $\log 1{,}000{,}000 = 6$

   $10^6 = 1{,}000{,}000$

   Base = 10
   Exponent = 6
   Power = 1,000,000

   $\log_{10} 1{,}000{,}000 = 6$

4. $\log_3 45 = x$

   $3^x = 45$

   Base = 3
   Exponent = x
   Power = 45

   $\log_3 45 = x$

5. $\log_7 b = a$

   $7^a = b$

   Base = 7
   Exponent = a
   Power = b

   $\log_7 b = a$

6. $2^5 = 32$

   $\log_2 32 = 5$

   Base = 2
   Exponent = 5
   Power = 32

   $2^5 = 32$

7. $10^2 = 100$

   $\log 100 = 2$

   Remember, a logarithm written with no visible base is always base 10, the *common logarithm*.

   Base = 10
   Exponent = 2
   Power = 100

   $10^2 = 100$

8. $3^0 = 1$

   $\log_3 1 = 0$

   Base = 3
   Exponent = 0
   Power = 1

   $3^0 = 1$

9. $b^3 = 1{,}000$

   $\log_b 1{,}000 = 3$

   Base = b
   Exponent = 3
   Power = 1,000

   $b^3 = 1{,}000$

10. $b^4 = a$

    $\log_b a = 4$

    Base = b
    Exponent = 4
    Power = a

    $b^4 = a$

11. $x = 4$

$$\log_2 16 = x$$
$$2 \times 2 \times 2 \times 2 = 16$$
$$2^4 = 16$$
$$x = 4$$

12. $x = 3$

$$\log_4 64 = x$$
$$4 \times 4 \times 4 = 64$$
$$4^3 = 64$$
$$x = 3$$

13. $x = 5$

$$\log_{10} 100{,}000 = x$$
$$10 \times 10 \times 10 \times 10 \times 10 = 100{,}000$$
$$10^5 = 100{,}000$$
$$x = 5$$

14. $b = 6$

$$\log_b 36 = 2$$
$$6^2 = 36$$
$$b = 6$$

15. $a = 243$

$$\log_3 a = 5$$
$$a = 3^5 = 3 \times 3 \times 3 \times 3 \times 3 = 243$$

Lesson 19.1B—Skills Check (page 450)

1. $x = 10$

$$\log_b (a \bullet c) = \log_b a + \log_b c$$
$$\log_{10} (100)(10) = \log_{10} 100 + \log_{10} x$$
$$a = 100$$
$$c = 10$$
$$x = c = 10$$
$$\log_{10} (100)(10) = \log_{10} 100 + \log_{10} 10$$

2. $x = 16$

$$\log_b (a \bullet c) = \log_b a + \log_b c$$
$$\log_2 64 = \log_2 4 + \log_2 x$$
$$ac = 64$$
$$a = 4$$
$$c = x$$
$$4x = 64$$
$$x = 16$$
$$\log_2 64 = \log_2 4 + \log_2 16$$

3. $x = 15$

$$\log_b a + \log_b c = \log_b (a \bullet c)$$
$$\log_2 5 + \log_2 3 = \log_2 x$$
$$a = 5$$
$$c = 3$$
$$x = ac = (5)(3) = 15$$
$$\log_2 5 + \log_2 3 = \log_2 15$$

4. $x = 8$

$$\log_b a + \log_b c = \log_b (a \bullet c)$$
$$\log_a 2 + \log_a 4 = \log_a x$$
$$a = 2$$
$$c = 4$$
$$x = ac = (2)(4) = 8$$
$$\log_a 2 + \log_a 4 = \log_a 8$$

5. $x = 6$

$$\log_b (a \bullet c) = \log_b a + \log_b c$$
$$\log_b 3x = \log_b 3 + \log_b 6$$
$$ac = 3x$$
$$a = 3$$
$$x = c = 6$$
$$\log_b 3(6) = \log_b 3 + \log_b 6$$

Lesson 19.1C—Skills Check (page 451)

1. $x = 100$

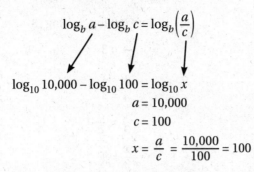

$$\log_b a - \log_b c = \log_b\left(\frac{a}{c}\right)$$

$$\log_{10} 10{,}000 - \log_{10} 100 = \log_{10} x$$
$$a = 10{,}000$$
$$c = 100$$
$$x = \frac{a}{c} = \frac{10{,}000}{100} = 100$$
$$\log_{10} 10{,}000 - \log_{10} 100 = \log_{10} 100$$

2. $x = 4$

$$\log_b a - \log_b c = \log_b\left(\frac{a}{c}\right)$$
$$\log_2 32 - \log_2 8 = \log_2 x$$
$$a = 32$$
$$c = 8$$
$$x = \frac{a}{c} = \frac{32}{8} = 4$$
$$\log_2 32 - \log_2 8 = \log_2 4$$

3. $x = 8$

$$\log_b a - \log_b c = \log_b\left(\frac{a}{c}\right)$$
$$\log_y 16 - \log_y 2 = \log_y x$$
$$a = 16$$
$$c = 2$$
$$x = \frac{a}{c} = \frac{16}{2} = 8$$
$$\log_y 16 - \log_y 2 = \log_y 8$$

4. $x = 3$

$$\log_b\left(\frac{a}{c}\right) = \log_b a - \log_b c$$
$$\log_b 8 = \log_b 24 - \log_b x$$
$$\frac{a}{c} = 8$$
$$a = 24$$
$$c = x$$
$$\frac{24}{x} = \frac{a}{c}$$
$$\frac{24}{x} = 8$$
$$x = 3$$
$$\log_b 8 = \log_b 24 - \log_b 3$$

5. $x = 9$

$$\log_b \left(\frac{a}{c} \right) = \log_b a - \log_b c$$

$$\log_a 3 = \log_a x - \log_a 3$$

$$\frac{a}{c} = 3$$

$$a = x$$

$$c = 3$$

$$\frac{x}{3} = \frac{a}{c}$$

$$\frac{x}{3} = 3$$

$$x = 9$$

$$\log_a 3 = \log_a 9 - \log_a 3$$

Lesson 19.1D—Skills Check (page 453)

1. $x = 3$

$$\log_b a^c = c(\log_b a)$$
$$\log_2 5^3 = x \log_2 5$$
$$a = 5$$
$$c = 3$$
$$x = c = 3$$
$$\log_2 5^3 = 3\log_2 5$$

2. $x = 2$

$$c(\log_b a) = \log_b a^c$$
$$2\log_{10} 10 = \log_{10} 10^x$$
$$c = 2$$
$$a = 10$$
$$a^c = 10^x$$
$$x = c = 2$$
$$2\log_{10} 10 = \log_{10} 10^2$$

3. $x = 4$

$$\log_b a^c = c(\log_b a)$$
$$\log_2 16 = 2\log_2 x$$
$$a^c = 16$$
$$c = 2$$
$$a = x$$
$$x^2 = 16$$
$$x = 4$$
$$\log_2 4^2 = 2\log_2 4$$
$$\log_2 16 = 2\log_2 4$$

4. $x = 125$

$$\log_b a^c = c(\log_b a)$$
$$\log_b x = 3\log_b 5$$
$$a^c = x$$
$$c = 3$$
$$a = 5$$
$$x = 5^3 = 125$$
$$\log_b 5^3 = 3\log_b 5$$
$$\log_b 125 = 3\log_b 5$$

5. $x = 4$

$$\log_b a^c = c(\log_b a)$$
$$\log_b 64 = 3\log_b x$$
$$a^c = 64$$
$$c = 3$$
$$a = x$$
$$x^3 = 64$$
$$x = 4$$
$$\log_b 4^3 = 3\log_b 4$$
$$\log_b 64 = 3\log_b 4$$

Lesson 19.1—ACCUPLACER Challenge (page 453)

1. **(E)**
$$\text{Base} = 2$$
$$\text{Exponent} = 4$$
$$\text{Power} = 2^4 = 16$$
$$\log_2 16 = 4$$

2. **(D)**
$$\text{Base} = 3$$
$$\text{Exponent} = x$$
$$\text{Power} = y$$
$$\log_3 y = x$$

3. **(B)** By the power rule, choice B, $2\log_b 32 = \log_b 32^2$, which is equivalent to $\log_b 1024$. This expression is not equivalent to $\log_b 64$.

4. **(C)** Written exponentially, $\log_b y = x$ is $b^x = y$. Choice C, 0, is undefined for y as y is a power of b, and there is no exponent that we can raise b to in order to have an outcome of 0. $b^0 = 1$. b raised to a positive exponent gives us a positive outcome, and b raised to a negative exponent also gives us a positive outcome.

5. **(B)** By the power rule, $3\log_4 a = \log_4 a^3$. We can then change that expression to $\log_4 a^3 - \log_4 b$. By the quotient rule, $\log_4 a^3 - \log_4 b = \log_4\left(\dfrac{a^3}{b}\right)$.

Lesson 19.2A—Skills Check (page 458)

1. $\sqrt[4]{x}$ $\quad x^{\frac{1}{4}} = \sqrt[4]{x}$

2. $\left(\sqrt[5]{a}\right)^2$ $\quad a^{\frac{2}{5}} = \left(\sqrt[5]{a}\right)^2$

3. $\left(\sqrt[3]{y}\right)^4$ $\quad y^{\frac{4}{3}} = \left(\sqrt[3]{y}\right)^4$

4. $\sqrt[3]{2}$ $\quad 2^{\frac{1}{3}} = \sqrt[3]{2}$

5. $\left(\sqrt[3]{3}\right)^7$ $\quad 3^{\frac{7}{3}} = \left(\sqrt[3]{3}\right)^7$

6. $x^{\frac{2}{3}}$ $\quad \left(\sqrt[3]{x}\right)^2 = x^{\frac{2}{3}}$

7. $y^{\frac{3}{2}}$ $\quad \left(\sqrt{y}\right)^3 = y^{\frac{3}{2}}$

8. $a^{\frac{4}{5}}$ $\quad \left(\sqrt[5]{a}\right)^4 = a^{\frac{4}{5}}$

9. $b^{\frac{1}{2}}$ $\quad \left(\sqrt[8]{b}\right)^4 = b^{\frac{4}{8}} = b^{\frac{1}{2}}$

10. c^2 $\quad \left(\sqrt[5]{c}\right)^{10} = c^{\frac{10}{5}} = c^2$

11. **4** $\quad \left(\sqrt[4]{16}\right)^2 = 2^2 = 4$

12. **16** $\quad \left(\sqrt[3]{8}\right)^4 = 2^4 = 16$

13. **81** $\quad 27^{\frac{4}{3}} = \left(\sqrt[3]{27}\right)^4 = 3^4 = 81$

14. **25** $\quad 125^{\frac{2}{3}} = \left(\sqrt[3]{125}\right)^2 = 5^2 = 25$

15. **100** $\quad 1000^{\frac{2}{3}} = \left(\sqrt[3]{1000}\right)^2 = 10^2 = 100$

Lesson 19.2B—Skills Check (page 459)

1. $x^{\frac{5}{7}}$ $\quad x^{\frac{2}{7}} \bullet x^{\frac{3}{7}} = x^{\frac{2}{7}+\frac{3}{7}} = x^{\frac{5}{7}}$

2. $m^{\frac{5}{8}}$ $\quad m^{\frac{1}{4}} \bullet m^{\frac{3}{8}} = m^{\frac{1}{4}+\frac{3}{8}} = m^{\frac{2}{8}+\frac{3}{8}} = m^{\frac{5}{8}}$

3. $a^{\frac{11}{8}}$ $\left(\sqrt[8]{a}\right)^5\left(\sqrt[4]{a}\right)^3$

First, change the expression to exponential form.

$$\left(\sqrt[8]{a}\right)^5\left(\sqrt[4]{a}\right)^3 = a^{\frac{5}{8}} \times a^{\frac{3}{4}}$$

Keep the common base, a, and add the exponents together.

$$a^{\frac{5}{8}} \times a^{\frac{3}{4}} = a^{\frac{5}{8}+\frac{3}{4}}$$

Change the rational exponents to equivalent fractions with common denominators, and then find their sum.

$$a^{\frac{5}{8}+\frac{3}{4}} = a^{\frac{5}{8}+\frac{6}{8}} = a^{\frac{11}{8}}$$

4. $16^{\frac{53}{15}}$ $\left(\sqrt[3]{16}\right)^{10}\left(\sqrt[5]{16}\right)$

First, change the expression to exponential form.

$$\left(\sqrt[3]{16}\right)^{10}\left(\sqrt[5]{16}\right) = 16^{\frac{10}{3}} \times 16^{\frac{1}{5}}$$

Keep the common base, 16, and add the exponents together.

$$16^{\frac{10}{3}} \times 16^{\frac{1}{5}} = 16^{\frac{10}{3}+\frac{1}{5}}$$

Change the rational exponents to equivalent fractions with common denominators, and then find their sum.

$$16^{\frac{10}{3}+\frac{1}{5}} = 16^{\frac{50}{15}+\frac{3}{15}} = 16^{\frac{53}{15}}$$

5. $64g^2$ $\left(\sqrt[3]{8g}\right)^2\left(\sqrt[3]{8g}\right)^4$

First, change the expression to exponential form.

$$\left(\sqrt[3]{8g}\right)^2\left(\sqrt[3]{8g}\right)^4 = (8g)^{\frac{2}{3}} \times (8g)^{\frac{4}{3}}$$

Keep the common base, $(8g)$, and add the exponents together.

$$(8g)^{\frac{2}{3}} \times (8g)^{\frac{4}{3}} = (8g)^{\frac{2}{3}+\frac{4}{3}}$$

Add the exponents, and simplify.

$$(8g)^{\frac{2}{3}+\frac{4}{3}} = (8g)^{\frac{6}{3}} = (8g)^2 = 64g^2$$

Lesson 19.2—ACCUPLACER Challenge (page 460)

1. **(C)** $\left(\sqrt[4]{x}\right)^8 = x^{\frac{8}{4}} = x^2$

2. **(C)** $81^{\frac{3}{4}} = \left(\sqrt[4]{81}\right)^3 = 3^3 = 27$

3. **(B)** To find $f(64)$, substitute 64 in place of x in the function.

$$f(64) = 64^{\frac{1}{3}} = \sqrt[3]{64} = 4$$

4. **(C)** Change each radical to exponential form.

$$\left(\sqrt[3]{a}\right)^5 \left(\sqrt[6]{a}\right)^4 = a^{\frac{5}{3}} \times a^{\frac{4}{6}}$$

Keep the common base, change the rational exponents to equivalents with common denominators, and add them together.

$$a^{\frac{5}{3}} \times a^{\frac{4}{6}} = a^{\frac{5}{3} + \frac{4}{6}} = a^{\frac{10}{6} + \frac{4}{6}} = a^{\frac{14}{6}} = a^{\frac{7}{3}}$$

5. **(D)** We cannot subtract terms with different exponents. We can, however, factor out their common factor $2^{\frac{4}{3}}$, which allows us to simplify the expression.

$$2^{\frac{7}{3}} - 2^{\frac{4}{3}} =$$
$$2^{\frac{4}{3}} \left(2^{\frac{3}{3}} - 1\right) =$$
$$2^{\frac{4}{3}} (2 - 1) =$$
$$2^{\frac{4}{3}} (1) = 2^{\frac{4}{3}}$$

Trigonometry

20

Another word for a triangle is **trigon**, and **trigonometry** is quite simply the study of triangles. Although the triangle has the least number of sides of any polygon, trigonometry, the study of this humble shape, has directly contributed to innumerable human achievements. We will only be able to scratch upon the surface of this deep discipline, but we will get you where you need to be to excel on ACCUPLACER trigonometry questions.

> In this chapter, we will work to master
>
> - understanding the Pythagorean theorem and the six trigonometric ratios, and
> - interpreting the graphs of the sine and cosine functions.

LESSON 20.1—THE PYTHAGOREAN THEOREM AND THE SIX TRIGONOMETRIC RATIOS

The right triangle is one of the most studied figures in geometry. This is partly because all other polygons, and even shapes with curves, can essentially be divided into right triangles. That means that we can learn a lot about right triangles and then apply that information to almost any other figure we want to study. Therefore, let's get to know right triangles a little better.

Right Triangle Facts:

- One angle measures 90°, and the two other angles are **acute** (<90°) and **complementary** (two angles with a sum of 90°). All three angles add up to 180°.
- The **hypotenuse** is the longest side, and it is always found opposite the 90° angle. The other two sides are called the **legs**.
- The lengths of the three sides of a right triangle always satisfy the Pythagorean theorem, $a^2 + b^2 = c^2$, where c represents the hypotenuse.

Here are two examples of right triangles.

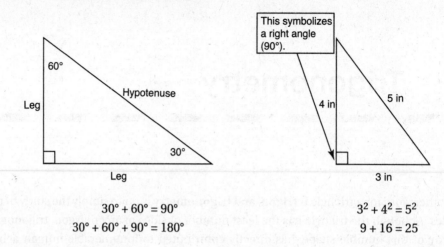

$$30° + 60° = 90°$$
$$30° + 60° + 90° = 180°$$

$$3^2 + 4^2 = 5^2$$
$$9 + 16 = 25$$

Notice that we did not draw one triangle that has angles of 30°-60°-90° and sides that measured lengths of 3 in, 4 in, and 5 in. Such a triangle does not exist. That's because a triangle with 30°-60°-90° angles has sides that are in a very certain relationship, which is not 3:4:5, and a 3:4:5 triangle has angles that are very specific but are not 30°-60°-90°. How angles and sides relate to one another is the basis for trigonometric functions, which is what you need to master for ACCUPLACER trigonometry questions. Before we get there, we need to practice a few other skills.

Lesson 20.1A—Finding the Missing Side of a Right Triangle Using the Pythagorean Theorem

If the lengths of two sides of a right triangle are already known, we can use $a^2 + b^2 = c^2$ to find the length of the third side by substituting the two known lengths into the equation, and then solving for the missing length.

To Find a Side of a Right Triangle Using the Pythagorean Theorem:

1. Substitute the lengths of the two known sides into $a^2 + b^2 = c^2$, making sure that c represents the hypotenuse and a and b represent the lengths of the legs.
2. Evaluate the equation following the order of operations.
3. Solve for the missing value.

➡ **Example 1**

For the right triangle below, find the length of x in feet.

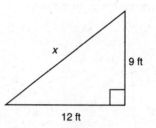

Solution

The side of length x is the hypotenuse, so substitute x in place of c in $a^2 + b^2 = c^2$. Use 9 and 12 in place of a and b.

$$9^2 + 12^2 = x^2$$

Evaluate the equation.

$$81 + 144 = x^2$$
$$225 = x^2$$
$$\sqrt{225} = 15 = x$$

Answer: $x = 15$ ft

Here, evaluating the equation also solved the equation because x was already by itself. To check your work, substitute all three lengths back into $a^2 + b^2 = c^2$.

$$9^2 + 12^2 = 15^2$$
$$81 + 144 = 225$$
$$225 = 225 \text{ ✔}$$

➡ Example 2

For the right triangle below, find the length of side $\overline{AC}$ in meters.

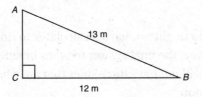

Solution

Side $\overline{AB}$ is the hypotenuse, so substitute 13 in place of c in $a^2 + b^2 = c^2$. Use 12 in place of a. Here we will have to solve for b.

$$12^2 + b^2 = 13^2$$

Evaluate the equation.

$$144 + b^2 = 169$$

Now, we must solve for b, first by subtracting 144 from both sides to get b^2 by itself.

$$b^2 = 169 - 144$$
$$b^2 = 25$$
$$b = \sqrt{25} = 5$$

Answer: $b = 5$ m, which means that $\overline{AC} = 5$ meters.

To check your work, substitute all three lengths back into $a^2 + b^2 = c^2$.

$$12^2 + 5^2 = 13^2$$
$$144 + 25 = 169$$
$$169 = 169 \text{ ✔}$$

➡ Example 3

The two legs of a right triangle measure 3 cm and 5 cm. Find the length of the hypotenuse of the triangle in centimeters.

Solution

Let's start by drawing a diagram. You should always draw a diagram when working with figures if you are not already provided with one.

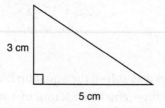

Substitute 3 and 5 for a and b in $a^2 + b^2 = c^2$. Then, evaluate the problem, and solve for c.

$$3^2 + 5^2 = c^2$$
$$9 + 25 = c^2$$
$$34 = c^2$$
$$\sqrt{34} = c$$

Answer: $c = \sqrt{34}$ centimeters

On the ACCUPLACER, you might not have a calculator to find, nor will you need to find, the square root of 34. Therefore, the final answer remains in radical form as $\sqrt{34}$ centimeters. In fact, most right triangles will not have three sides that are whole numbers. Having at least one side as a radical is common.

LESSON 20.1A—SKILLS CHECK

> **Directions**: Find the missing side of the right triangle as directed. Simplify any radicals that result and leave them in radical form.

1. The two legs of a right triangle measure 6 in and 8 in. Find the length of the hypotenuse in inches.

2. One leg of a right triangle measures 15 m, and the hypotenuse measures 17 m. Find the length of the other leg in meters.

3. The two legs of a right triangle measure 2 cm and 4 cm. Find the length of the hypotenuse in centimeters.

4. One leg of a right triangle measures 5 yd, and the hypotenuse measures 7 yd. Find the length of the other leg in yards.

5. One leg of a right triangle measures $\sqrt{3}$ km, and the hypotenuse measures 2 km. Find the length of the other leg in kilometers.

(Answers are on page 491.)

Lesson 20.1B—Using an Angle to Name the Legs of a Right Triangle

Imagine for a second that three friends, Tom, Linette, and Carlos, are sitting in a circle. If we were to say, "the friend to the right," it would be impossible to know which person we are talking about because the friend to the right of Tom would be different from the friend to the right of Linette. In trigonometry, we have the same problem when referring to a leg of a right triangle. Since there are two legs, we use the angles of the triangle to describe the position of the legs. The legs will be either next to an angle (**adjacent**) or across from an angle (**opposite**).

Take a look at the triangle below.

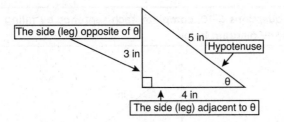

The 90° angle is the only angle in a right triangle that is formed by both legs. Opposite this angle, we always find the hypotenuse. The other two angles are formed by either one of the legs and the hypotenuse.

The symbol θ, read "theta," is commonly used, as it is above, to stand for an angle of undetermined measure. The leg that measures 4 in and the 5 in hypotenuse form the angle θ. Both sides are adjacent (next to) angle θ, but the 5 in side is already referred to as the hypotenuse, so we will call the 4 in side "adjacent θ." The third side, the 3 in side, is not next to θ. In fact, it is on the far side of the triangle from θ. We can refer to this side as "opposite θ."

Will the bottom of the triangle always be the adjacent side? Let's take a look below.

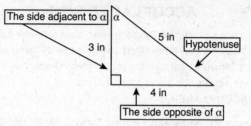

We have a **congruent** (identical) triangle, but this time we used α, read "alpha," the angle formed by the 3 in side and the hypotenuse, to name the sides. The hypotenuse is still the same, but the "adjacent" and "opposite" sides have switched.

LESSON 20.1B—SKILLS CHECK

Directions: For questions 1–5, complete each sentence by filling in the appropriate length found on the diagram below.

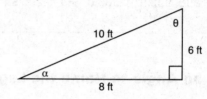

1. The hypotenuse is _____ in length.

2. The side adjacent θ is _____ in length.

3. The side opposite θ is _____ in length.

4. The side adjacent α is _____ in length.

5. The side opposite α is _____ in length.

Directions: For questions 6–10, complete each sentence by filling in the appropriate length found on the diagram below.

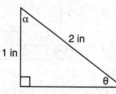

6. The hypotenuse is _____ in length.

7. The side adjacent θ is _____ in length.

8. The side opposite θ is _____ in length.

9. The side adjacent α is _____ in length.

10. The side opposite α is _____ in length.

(Answers are on page 491.)

Lesson 20.1C—The Six Trigonometric Ratios

A ratio is a comparison of two quantities, and we usually write a ratio as a fraction. When we use trigonometric ratios, we compare two sides of a right triangle to one another as a fraction.

Take a look at the triangle below. How many different ways can we compare two sides as fractions?

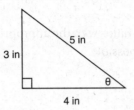

The only possible fractions we can make using the three numbers 3, 4, and 5 are

Distinct Fractions: $\dfrac{3}{5}, \dfrac{4}{5}, \dfrac{3}{4}$

Reciprocals: $\dfrac{5}{3}, \dfrac{5}{4}, \dfrac{4}{3}$

There you have it. We just came up with examples for all six of the trigonometric ratios.

To facilitate communication about trigonometric ratios, names have been given to each of these six possible comparisons. Using the symbol θ to represent one of the acute angles found in a right triangle, the three primary trig ratios are:

$$\text{sine } \theta = \frac{\text{side opposite } \theta}{\text{hypotenuse}}$$

$$\text{cosine } \theta = \frac{\text{side adjacent } \theta}{\text{hypotenuse}}$$

$$\text{tangent } \theta = \frac{\text{side opposite } \theta}{\text{side adjacent } \theta}$$

For convenience, sine θ is often written as sin θ, cosine θ is written as cos θ, and tangent θ is written as tan θ.

Students often use SOH, CAH, TOA as a mnemonic device to help them remember how to set up the sine, cosine, and tangent ratios.

SOH tells us that **S**ine is **O**pposite over **H**ypotenuse, or $S = \dfrac{O}{H}$

CAH tells us that **C**osine is **A**djacent over **H**ypotenuse, or $C = \dfrac{A}{H}$

TOA tells us that **T**angent is **O**pposite over **A**djacent, or $T = \dfrac{O}{A}$

The other three functions are known as the **reciprocal trig functions** since their ratios are the reciprocals of sine, cosine, and tangent.

$$\text{cosecant } \theta = \frac{\text{hypotenuse}}{\text{side opposite } \theta} \ or \ \frac{1}{\sin \theta}$$

$$\text{secant } \theta = \frac{\text{hypotenuse}}{\text{side adjacent } \theta} \ or \ \frac{1}{\cos \theta}$$

$$\text{cotangent } \theta = \frac{\text{side adjacent } \theta}{\text{side opposite } \theta} \ or \ \frac{1}{\tan \theta}$$

For convenience, cosecant θ is often written as csc θ, secant θ is written as sec θ, and cotangent θ is written as cot θ.

To Write Trigonometric Ratios:

1. Identify the sides of the right triangle as the hypotenuse, adjacent, or opposite as they relate to θ.
2. Complete the trigonometric ratios with the appropriate sides and angles.
3. Reduce fractions whenever possible.

➡ Example 1

Given the triangle below, find sin θ, cos θ, tan θ, csc θ, sec θ, and cot θ.

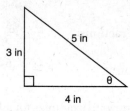

Solution

The first step is labeling the three sides as they relate to θ.

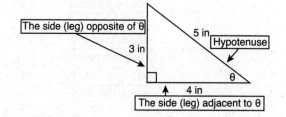

Next, substitute the appropriate values into the trigonometric ratios as they are defined.

Answer:

$\sin \theta = \dfrac{3}{5}$	$\csc \theta = \dfrac{5}{3}$
$\cos \theta = \dfrac{4}{5}$	$\sec \theta = \dfrac{5}{4}$
$\tan \theta = \dfrac{3}{4}$	$\cot \theta = \dfrac{4}{3}$

➡ Example 2

Given the triangle below, find the tangent of α.

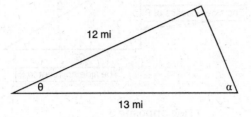

Solution

The first step is labeling the three sides as they relate to α.

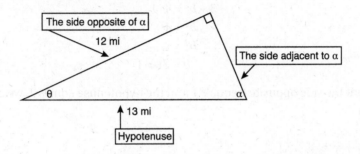

To complete the tangent function, we need tangent $\alpha = \dfrac{\text{side opposite } \alpha}{\text{side adjacent } \alpha}$. We were not given the side adjacent α, but we do know two of the sides of the right triangle, so we can use the Pythagorean theorem to find the missing side of the triangle. Since 12 is a leg, we need to substitute 12 for a, and since 13 is the hypotenuse, we'll need to substitute 13 for c in $a^2 + b^2 = c^2$.

$$12^2 + b^2 = 13^2$$
$$144 + b^2 = 169$$
$$b^2 = 169 - 144$$
$$b^2 = 25$$
$$b = \sqrt{25} = 5$$

Answer: Now we can complete the ratio, tangent $\alpha = \dfrac{12}{5}$.

➡ Example 3

If θ is an acute angle and $\cos \theta = \dfrac{\sqrt{3}}{2}$, then what is sine θ equal to?

Solution

To start this problem, we have to remember and apply the fact that cosine $\theta = \dfrac{\text{side adjacent } \theta}{\text{hypotenuse}}$.

We can then establish that the side adjacent θ is represented by $\sqrt{3}$, and the hypotenuse is represented by 2.

A diagram using this information could look something like this:

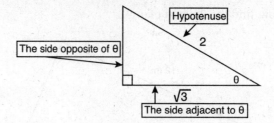

To complete the ratio sine $\theta = \dfrac{\text{side opposite } \theta}{\text{hypotenuse}}$, we need to use the Pythagorean theorem and the two sides we already know to find the length of the side opposite θ.

$$\left(\sqrt{3}\right)^2 + b^2 = 2^2$$
$$3 + b^2 = 4$$
$$b^2 = 4 - 3$$
$$b^2 = 1$$
$$b = \sqrt{1} = 1$$

Knowing that the side opposite θ equals 1 and the hypotenuse equals 2, we can now complete our ratio.

Answer: sine $\theta = \dfrac{1}{2}$

LESSON 20.1C—SKILLS CHECK

Directions: For questions 1–6, complete the trigonometric ratios as directed using the diagram below.

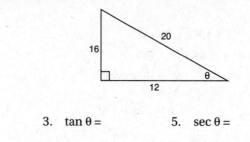

1. $\sin \theta =$ 3. $\tan \theta =$ 5. $\sec \theta =$

2. $\cos \theta =$ 4. $\csc \theta =$ 6. $\cot \theta =$

Directions: For questions 7–14, complete the trigonometric ratios as directed using the diagram below.

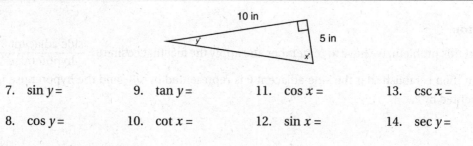

7. $\sin y =$ 9. $\tan y =$ 11. $\cos x =$ 13. $\csc x =$

8. $\cos y =$ 10. $\cot x =$ 12. $\sin x =$ 14. $\sec y =$

(Answers are on page 492.)

LESSON 20.1—ACCUPLACER CHALLENGE

Directions: For each of the questions below, choose the best answer from the five choices given.

1. Which of the following ratios represents sin θ in the accompanying diagram of △ABC?

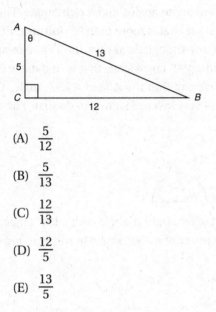

(A) $\dfrac{5}{12}$

(B) $\dfrac{5}{13}$

(C) $\dfrac{12}{13}$

(D) $\dfrac{12}{5}$

(E) $\dfrac{13}{5}$

2. Which of the following ratios represents cos β in the triangle below?

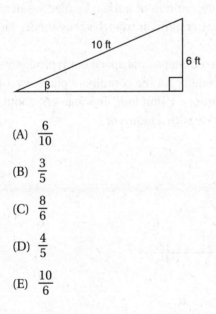

(A) $\dfrac{6}{10}$

(B) $\dfrac{3}{5}$

(C) $\dfrac{8}{6}$

(D) $\dfrac{4}{5}$

(E) $\dfrac{10}{6}$

3. If sin θ = $\dfrac{4}{5}$ and 0 ≤ θ ≤ 90°, then csc θ =

(A) $\dfrac{4}{5}$

(B) $\dfrac{4}{3}$

(C) $\dfrac{3}{5}$

(D) $\dfrac{3}{4}$

(E) $\dfrac{5}{4}$

4. If θ is an acute angle and tan θ = $\dfrac{\sqrt{7}}{3}$, then cot θ =

(A) $\dfrac{3\sqrt{7}}{7}$

(B) $\dfrac{\sqrt{21}}{7}$

(C) $\dfrac{3}{7}$

(D) $\dfrac{4}{3}$

(E) $\dfrac{9}{7}$

5. If θ is an acute angle and cos θ = $\dfrac{15}{17}$, then sin θ =

(A) $\dfrac{15}{17}$

(B) $\dfrac{17}{15}$

(C) $\dfrac{8}{15}$

(D) $\dfrac{8}{17}$

(E) $\dfrac{\sqrt{514}}{17}$

(Answers are on page 493.)

LESSON 20.2—GRAPHS OF THE SINE AND COSINE FUNCTIONS

Until now, we have only dealt with right triangles that have two acute angles and one right angle. For example, the 30°-60°-90° triangle as pictured below.

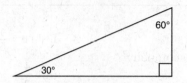

In actuality, a right triangle is always made of two acute angles and a right angle. This is because you cannot have a 90° angle and another angle that is more than 90° (**obtuse**) in the same triangle since the total sum of the angles of any triangle is always 180°. For example, we cannot have a right triangle with a 90° angle and a 150° angle since that would make 240° without even including the third angle. However, we can find the sine of a 150° angle. That may not seem to make much sense, but mathematicians invented a way to do that. The diagram below demonstrates part of the idea.

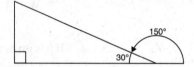

The triangle in the diagram above is not thought of as a right triangle with a 150° angle. It's thought of as a right triangle that is formed as the result of a 150° angle. In this way, we can use trigonometric ratios for obtuse angles.

Take a look at the two congruent triangles below.

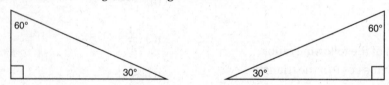

If we include direction and orientation in our description of a triangle, then we need to think of the two triangles above differently. They are congruent triangles only they're facing a different direction.

In mathematics, position and orientation in two-dimensional space are typically measured on the coordinate plane. To consider trigonometry on the coordinate plane, we often make use of a **unit circle**, which is formed by rotating a 1-unit long line segment about the origin in a counterclockwise direction to form a circle with a radius of 1.

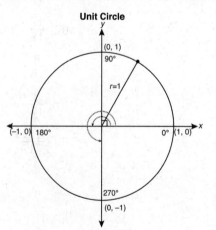

From where the radius touches the circle, a vertical line is drawn to the x-axis to form a right triangle that is perpendicular to the x-axis.

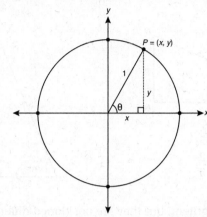

The central angle, θ, the angle we use as our point of reference for finding trigonometric ratios such as sine, cosine, and tangent, is measured from the positive x-axis to the radius.

For example, the unit circle below shows us both a right triangle where $\theta = 30°$ and another right triangle where $\theta = 150°$.

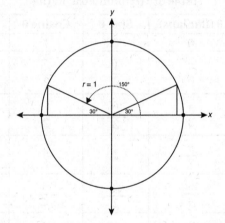

As the 1-unit long radius rotates around the origin, the triangle that is formed changes in shape as shown below.

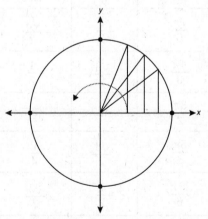

Look at $\triangle AMN$ and $\triangle AJK$ on the next page. With a hypotenuse of 13, we would not find these triangles on the unit circle, but this example can help us explore how $\sin \theta$ and $\cos \theta$ are affected as we rotate a radius around the unit circle. Both triangles have a hypotenuse of

13 and legs that measure 5 and 12. However, θ increases in order to allow for this change from △*AMN* to △*AJK*. How does this increase in the angle θ change sine θ and cosine θ?

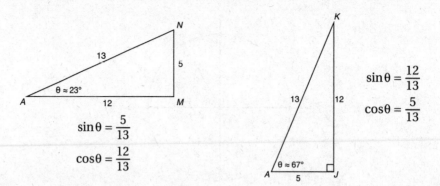

$$\sin\theta = \frac{5}{13}$$

$$\cos\theta = \frac{12}{13}$$

$$\sin\theta = \frac{12}{13}$$

$$\cos\theta = \frac{5}{13}$$

△*AMN* and △*AJK* are congruent, but they are positioned differently. As θ increases, sin θ also increases. However, as θ increases, cos θ decreases. Will this continue indefinitely?

Another way to study the behavior of sin θ and cos θ is to study a trig table such as the one below. The table is a recording of what happens as θ increases.

Table of Trigonometric Ratios

θ (Degrees)	θ (Radians)	Sine θ	Cosine θ	Tangent θ
0°	0	0	1	0
30°	$\frac{\pi}{6}$	$\frac{1}{2}$	$\frac{\sqrt{3}}{2}$	$\frac{\sqrt{3}}{3}$
60°	$\frac{\pi}{3}$	$\frac{\sqrt{3}}{2}$	$\frac{1}{2}$	$\sqrt{3}$
90°	$\frac{\pi}{2}$	1	0	undefined
120°	$\frac{2\pi}{3}$	$\frac{\sqrt{3}}{2}$	$-\frac{1}{2}$	$-\sqrt{3}$
150°	$\frac{5\pi}{6}$	$\frac{1}{2}$	$-\frac{\sqrt{3}}{2}$	$-\frac{\sqrt{3}}{3}$
180°	π	0	−1	0
210°	$\frac{7\pi}{6}$	$-\frac{1}{2}$	$-\frac{\sqrt{3}}{2}$	$\frac{\sqrt{3}}{3}$
240°	$\frac{4\pi}{3}$	$-\frac{\sqrt{3}}{2}$	$-\frac{1}{2}$	$\sqrt{3}$
270°	$\frac{3\pi}{2}$	−1	0	undefined
300°	$\frac{5\pi}{3}$	$-\frac{\sqrt{3}}{2}$	$\frac{1}{2}$	$-\sqrt{3}$
330°	$\frac{11\pi}{6}$	$-\frac{1}{2}$	$\frac{\sqrt{3}}{2}$	$-\frac{\sqrt{3}}{3}$
360°	2π	0	1	0
390°	$\frac{13\pi}{6}$	$\frac{1}{2}$	$\frac{\sqrt{3}}{2}$	$\frac{\sqrt{3}}{3}$

The degrees start at 0° and increase by 30° increments all the way to 360° and beyond. Looking down the sine θ column, we notice that the sine θ values start at 0, go up to 1, back down to 0, down to –1 and back up to 0. Then, the pattern starts all over again. Cosine θ follows a similar pattern except that it starts at 1. We are going to focus on these two trig functions, but the data for the tangent function is also there to see. You do not need to memorize all of these values, but you do need to be familiar with the behavior of the values. You probably also noticed the θ (Radians) column. Don't worry; we'll clear that up too.

All the trig functions are **periodic**. This means that over a certain **period**, or interval, the function travels through a cycle and then starts all over again. Try to visualize the radius of the unit circle rotating around the origin, completing the circle, and then starting all over again.

A table of values is great for organizing data, but a graph is another great way to look at data. Here is the graph of $y = \sin x$, where x stands in place for θ measured in degrees.

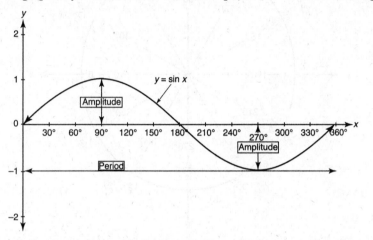

It took from 0° to 360° for $y = \sin x$ to complete a cycle, so the period is 360°.

The distance from the maximum height of the cycle to the median height of the cycle is called the **amplitude**. When the median height of the cycle is 0, we can think of the amplitude as the height. The highest value sin θ reached was 1, and the lowest value sin θ reached was –1. In the middle was 0, so the amplitude was 1.

Now, let's review those radians we saw in the Table of Trigonometric Ratios. Angles can be measured using degrees, but angles can alternatively be measured in terms of radians. A radian, or 1 radian, is the length of a part of the circumference of a circle, an **arc**, that is the same length as the radius of the circle.

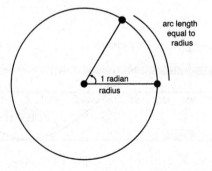

The circumference of a circle can be found using $C = 2\pi r$, where C stands for circumference and r stands for the length of the radius of the circle. The symbol π is approximately 3.14. If we evaluate $C = 2\pi r$ to be $C = 2(3.14)r$, then we find that $C \approx 6.28r$. This means that the circumference of a circle is slightly longer than 6 times the length of the radius. It's not exactly 6, so measuring the circumference this way is messy. Here is where the unit circle is convenient. On this circle, the radius is 1, so $C = 2\pi(1)$, and finally $C = 2\pi$. Therefore, the distance around the outside edge of a unit circle is exactly 2π, halfway around is $\frac{2\pi}{2} = \pi$, one fourth of the way around is $\frac{2\pi}{4} = \frac{\pi}{2}$, and so forth.

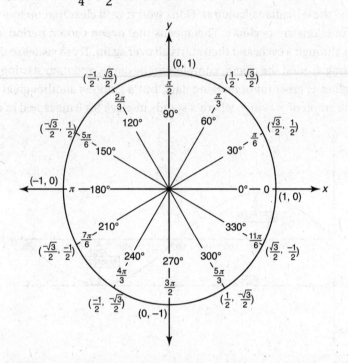

Notice the locations where the radian values correspond with the degree values. For example, $\frac{\pi}{6} = 30°$ and $\frac{\pi}{2} = 90°$. Essentially, when converting from radian measure to degree measure, we can say that $\pi = 180°$. That is to say, $\frac{\pi}{6} = \frac{180°}{6} = 30°$ and $\frac{\pi}{2} = \frac{180°}{2} = 90°$, and so forth.

Now, let's compare the graph of $y = \sin x$ in degrees, that we looked at previously, to the graph of $y = \sin x$ in radians.

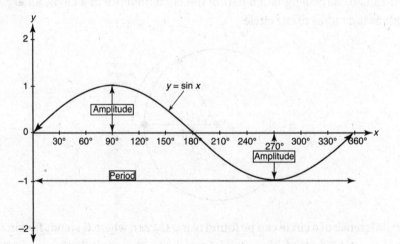

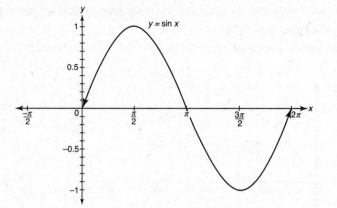

You may notice that the graphs are exactly the same; only the period of the sine wave is measured in different units. The period of $y = \sin x$ in degrees is 360°; the period of $y = \sin x$ in radians is 2π. The amplitude in both graphs is 1. For the purposes of the ACCUPLACER, you should be familiar with thinking about the graphs of both $y = \sin x$ and $y = \cos x$ in degrees and in radians. Let's now get familiar with the graph of $y = \cos x$.

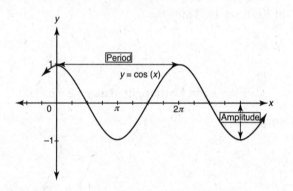

The period for the graph of $y = \cos x$ is also 2π, or 360°, and the amplitude is 1. You may notice that the sine wave and the cosine wave are the same, only the cosine wave is shifted 90°, or $\dfrac{\pi}{2}$, from the sine wave.

Here is what they look like on the same graph.

$y = \sin x$ and $y = \cos x$

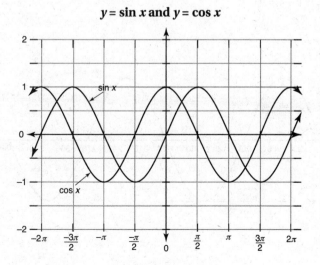

In addition to the standard $y = \sin x$ and $y = \cos x$ graphs, you will need to be familiar with variations of these functions.

For example, here is the graph of $y = 2 \sin x$.

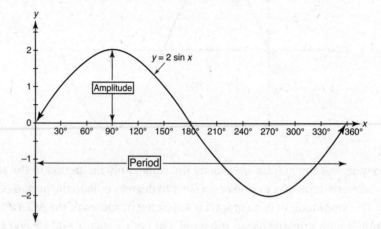

The period of $y = 2 \sin x$ is 360°, or 2π, and the amplitude is 2. If we consider a sine function in the form of $y = A \sin Bx$, then $A = 2$ and $B = 1$.

$$A \qquad B$$

$y = 2 \sin x$ can be thought of as $y = 2 \sin (1)x$

This is different from $y = \sin x$.

$$A \qquad B$$

$y = \sin x$ can be thought of as $y = 1 \sin (1)x$

By changing A from 1 to 2, we increased the amplitude from 1 to 2. As a rule then, for the graphs of $y = A \sin Bx$ and $y = A \cos Bx$:

$$\textbf{Amplitude} = |A|$$

Next, let's look at $y = \sin 2x$.

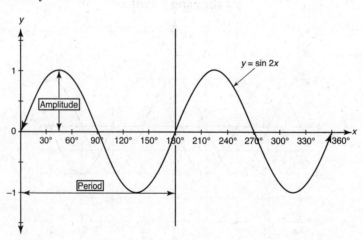

The amplitude of $y = \sin 2x$ is 1; however the period is 180° or π radians. Think of it this way: Whatever x is, we multiply it by 2, and then we take the sine of that new, doubled value. This means that a full period, or wave, is completed twice as fast. In other words, a full period, or wave, is completed in half the interval.

If we consider the sine function $y = \sin 2x$ in the form of $y = A \sin Bx$, then $A = 1$ and $B = 2$. By changing B from 1 to 2, we decreased the period from 360° to 180°, or from 2π to π radians. As a rule then, for the graphs of $y = A \sin Bx$ and $y = A \cos Bx$:

$$\text{Period} = \frac{360°}{B} \quad or \quad \text{Period} = \frac{2\pi}{B} \text{ radians}$$

➡ Example 1

Identify the period and amplitude for the function $y = 2 \cos 3x$ where the period is expressed in radians.

Solution

First, let's identify A and B.

$$y = 2 \cos 3x$$

Given that $A = 2$, the amplitude is $|2|$ or 2. Given that $B = 3$, the period is $\frac{2\pi}{3}$ radians.

We did not need to graph the function in order to answer the question. However, let's look at the graph to see what this means.

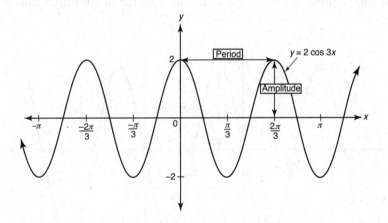

Since A is 2, the amplitude, or height, of the graph reaches up to 2 and down to –2. The graph completes an entire cycle, that is start back to start again, in an interval of $\frac{2\pi}{B} = \frac{2\pi}{3}$. Thinking about the same function in degrees, we could say that the function completes a cycle in an interval of $\frac{360°}{3} = 120°$.

Answer: The amplitude is 2, and the period is $\frac{2\pi}{3}$ radians.

➡ Example 2

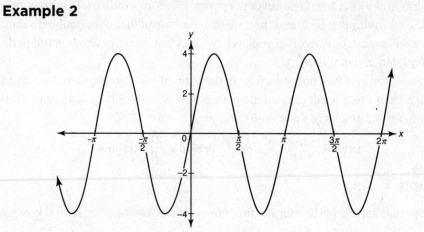

The figure above shows a portion of the graph of

(A) $\sin 4x$

(B) $2 \sin \dfrac{1}{4} x$

(C) $4 \sin \dfrac{1}{2} x$

(D) $4 \sin x$

(E) $4 \sin 2x$

Solution

Let's look at the graph to establish the values of A and B to substitute into the equation $y = A \sin Bx$.

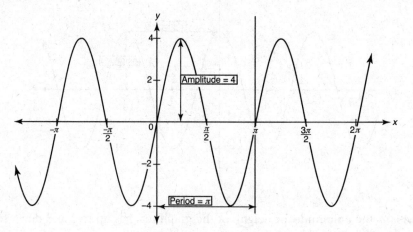

We can see from the graph that the amplitude is 4, so $A = 4$. We can also tell from the graph that the period is π. However, to find B, we have to solve for period $= \dfrac{2\pi}{B}$. Doing so, we find that $B = 2$ since $\dfrac{2\pi}{2} = \pi$. We can then substitute 4 for A and 2 for B into $y = A \sin Bx$ to get $y = 4 \sin 2x$, choice E.

<u>Answer:</u> The amplitude is 4, and the period is π radians, so choice E is the correct answer.

LESSON 20.2—SKILLS CHECK

> **Directions**: Name the period and amplitude of each function as directed.

1. For the graph of the function $y = 2 \cos x$, the period is _____ radians and the amplitude is _____.

2. For the graph of the function $y = 3 \sin 4x$, the period is _____ degrees and the amplitude is _____.

3. For the graph of the function $y = \sin \frac{1}{2} x$, the period is _____ radians and the amplitude is _____.

4. For the graph of the function below, the period is _____ radians and the amplitude is _____.

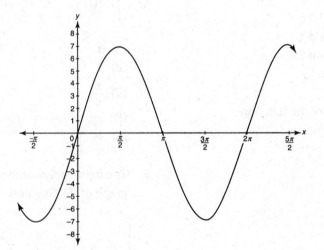

5. For the graph of the function below, the period is _____ degrees and the amplitude is _____.

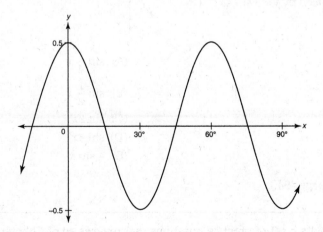

(Answers are on page 494.)

> **Directions**: For each of the questions below, choose the best answer from the five choices given.

1. The figure below shows a portion of the graph of which function?

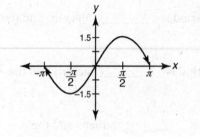

 (A) $y = \sin x$
 (B) $y = \sin 1.5x$
 (C) $y = 1.5 \sin x$
 (D) $y = 1.5 \sin 2x$
 (E) $y = 2 \sin x$

2. The period of the function $y = 3 \sin 5x$ is

 (A) 3
 (B) 5
 (C) 2π
 (D) $\dfrac{2\pi}{5}$
 (E) $\dfrac{2\pi}{3}$

3. What is the amplitude of the curve produced by $y = 3 \sin 4x$?

 (A) 1.5
 (B) 3
 (C) 4
 (D) $\dfrac{\pi}{2}$
 (E) 2π

4. The period of the function found in the graph below is

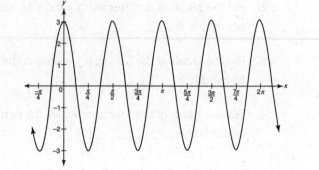

 (A) 3
 (B) 4
 (C) $\dfrac{\pi}{2}$
 (D) 2π
 (E) 4π

5. The figure below shows a portion of the graph of the function

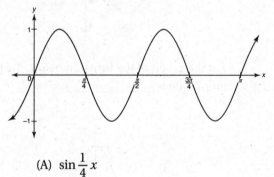

 (A) $\sin \dfrac{1}{4} x$
 (B) $\sin \dfrac{1}{2} x$
 (C) $\sin 2x$
 (D) $\sin 4x$
 (E) $4 \sin x$

(Answers are on page 494.)

> To complete a set of practice questions that reviews all of Chapter 20, go to:
> *barronsbooks.com/TP/accuplacer/math29qw/*

Lesson 20.1A—Skills Check (page 472)

1. **10 inches**

$$6^2 + 8^2 = c^2$$
$$36 + 64 = c^2$$
$$100 = c^2$$
$$\sqrt{100} = c$$
$$10 \text{ inches} = c$$

2. **8 meters**

$$15^2 + b^2 = 17^2$$
$$225 + b^2 = 289$$
$$b^2 = 289 - 225$$
$$b^2 = 64$$
$$b = \sqrt{64}$$
$$b = 8 \text{ meters}$$

3. **$2\sqrt{5}$ centimeters**

$$2^2 + 4^2 = c^2$$
$$4 + 16 = c^2$$
$$20 = c^2$$
$$\sqrt{20} = c$$
$$2\sqrt{5} \text{ centimeters} = c$$

4. **$2\sqrt{6}$ yards**

$$5^2 + b^2 = 7^2$$
$$25 + b^2 = 49$$
$$b^2 = 49 - 25$$
$$b^2 = 24$$
$$b = \sqrt{24}$$
$$b = 2\sqrt{6} \text{ yards}$$

5. **1 kilometer**

$$\left(\sqrt{3}\right)^2 + b^2 = 2^2$$
$$3 + b^2 = 4$$
$$b^2 = 4 - 3$$
$$b^2 = 1$$
$$b = \sqrt{1}$$
$$b = 1 \text{ kilometer}$$

Lesson 20.1B—Skills Check (page 474)

1. **10 ft** The hypotenuse is <u>10 ft</u> in length.

2. **6 ft** The side adjacent θ is <u>6 ft</u> in length.

3. **8 ft** The side opposite θ is <u>8 ft</u> in length.

4. **8 ft** The side adjacent α is <u>8 ft</u> in length.

5. **6 ft** The side opposite α is <u>6 ft</u> in length.

6. **2 in** The hypotenuse is <u>2 in</u> in length.

7. **$\sqrt{3}$ in** To find the length of the side adjacent θ, we need to substitute 1 for a and 2 for c in the Pythagorean theorem.

$$1^2 + b^2 = 2^2$$
$$1 + b^2 = 4$$
$$b^2 = 4 - 1 = 3$$
$$b = \sqrt{3}$$

8. **1 in** The side opposite θ is <u>1 in</u> in length.

9. **1 in** The side adjacent α is <u>1 in</u> in length.

10. $\sqrt{3}$ **in** To find the length of the side opposite α, we need to substitute 1 for a and 2 for c in the Pythagorean theorem.

$$1^2 + b^2 = 2^2$$
$$1 + b^2 = 4$$
$$b^2 = 4 - 1 = 3$$
$$b = \sqrt{3}$$

Lesson 20.1C—Skills Check (page 478)

1. $\dfrac{4}{5}$ $\sin\theta = \dfrac{\text{opposite }\theta}{\text{hypotenuse}} = \dfrac{16}{20} = \dfrac{4}{5}$

2. $\dfrac{3}{5}$ $\cos\theta = \dfrac{\text{adjacent }\theta}{\text{hypotenuse}} = \dfrac{12}{20} = \dfrac{3}{5}$

3. $\dfrac{4}{3}$ $\tan\theta = \dfrac{\text{opposite }\theta}{\text{adjacent }\theta} = \dfrac{16}{12} = \dfrac{4}{3}$

4. $\dfrac{5}{4}$ $\csc\theta = \dfrac{\text{hypotenuse}}{\text{opposite }\theta} = \dfrac{20}{16} = \dfrac{5}{4}$

5. $\dfrac{5}{3}$ $\sec\theta = \dfrac{\text{hypotenuse}}{\text{adjacent }\theta} = \dfrac{20}{12} = \dfrac{5}{3}$

6. $\dfrac{3}{4}$ $\cot\theta = \dfrac{\text{adjacent }\theta}{\text{opposite }\theta} = \dfrac{12}{16} = \dfrac{3}{4}$

7. $\dfrac{\sqrt{5}}{5}$ First, use the Pythagorean theorem to find the hypotenuse, which you will need for several of the other questions.

$$a^2 + b^2 = c^2$$
$$10^2 + 5^2 = c^2$$
$$100 + 25 = c^2$$
$$125 = c^2$$
$$\sqrt{125} = c$$
$$c = \sqrt{125} = 5\sqrt{5}$$

Therefore, the hypotenuse equals $5\sqrt{5}$.

$$\sin y = \frac{\text{opposite } y}{\text{hypotenuse}} = \frac{5}{5\sqrt{5}} = \frac{1}{\sqrt{5}}$$

However, we cannot leave a radical in the denominator, so we must rationalize the denominator.

$$\sin y = \frac{1}{\sqrt{5}} = \frac{1}{\sqrt{5}} \cdot \frac{\sqrt{5}}{\sqrt{5}} = \frac{\sqrt{5}}{5}$$

8. $\dfrac{2\sqrt{5}}{5}$ $\cos y = \dfrac{\text{adjacent } y}{\text{hypotenuse}} = \dfrac{10}{5\sqrt{5}} = \dfrac{2}{\sqrt{5}}$

However, we cannot leave a radical in the denominator, so we must rationalize the denominator.

$$\cos y = \frac{2}{\sqrt{5}} \cdot \frac{\sqrt{5}}{\sqrt{5}} = \frac{2\sqrt{5}}{5}$$

9. $\dfrac{1}{2}$ $\qquad \tan y = \dfrac{\text{opposite } y}{\text{adjacent } y} = \dfrac{5}{10} = \dfrac{1}{2}$

10. $\dfrac{1}{2}$ $\qquad \cot x = \dfrac{\text{adjacent } x}{\text{opposite } x} = \dfrac{5}{10} = \dfrac{1}{2}$

11. $\dfrac{\sqrt{5}}{5}$ $\qquad \cos x = \dfrac{\text{adjacent } x}{\text{hypotenuse}} = \dfrac{5}{5\sqrt{5}}$

However, we cannot leave a radical in the denominator, so we must rationalize the denominator.

$$\cos x = \dfrac{5}{5\sqrt{5}} \bullet \dfrac{\sqrt{5}}{\sqrt{5}} = \dfrac{5\sqrt{5}}{5 \bullet 5} = \dfrac{\sqrt{5}}{5}$$

12. $\dfrac{2\sqrt{5}}{5}$ $\qquad \sin x = \dfrac{\text{opposite } x}{\text{hypotenuse}} = \dfrac{10}{5\sqrt{5}} = \dfrac{2}{\sqrt{5}}$

However, we cannot leave a radical in the denominator, so we must rationalize the denominator.

$$\sin x = \dfrac{2}{\sqrt{5}} \bullet \dfrac{\sqrt{5}}{\sqrt{5}} = \dfrac{2\sqrt{5}}{5}$$

13. $\dfrac{\sqrt{5}}{2}$ $\qquad \csc x = \dfrac{\text{hypotenuse}}{\text{opposite } x} = \dfrac{5\sqrt{5}}{10} = \dfrac{\sqrt{5}}{2}$

14. $\dfrac{\sqrt{5}}{2}$ $\qquad \sec y = \dfrac{\text{hypotenuse}}{\text{adjacent } y} = \dfrac{5\sqrt{5}}{10} = \dfrac{\sqrt{5}}{2}$

Lesson 20.1—ACCUPLACER Challenge (page 479)

1. **(C)** $\sin\theta = \dfrac{\text{opposite } \theta}{\text{hypotenuse}} = \dfrac{12}{13}$

2. **(D)** We are given the side opposite β and the hypotenuse. To complete the ratio $\cos\beta = \dfrac{\text{adjacent } \beta}{\text{hypotenuse}}$, we need to first use the Pythagorean theorem to find the side adjacent β. Using the given sides, we establish that $6^2 + b^2 = 10^2$. From there, we determine that the side adjacent β measures 8. We now complete the ratio, and then reduce.

$$\cos\beta = \dfrac{\text{adjacent } \beta}{\text{hypotenuse}} = \dfrac{8}{10} = \dfrac{4}{5}$$

3. **(E)** If $\sin\theta = \dfrac{\text{opposite } \theta}{\text{hypotenuse}} = \dfrac{4}{5}$, then $\csc\theta = \dfrac{\text{hypotenuse}}{\text{opposite } \theta} = \dfrac{5}{4}$.

4. **(A)** If $\tan\theta = \dfrac{\text{opposite } \theta}{\text{adjacent } \theta} = \dfrac{\sqrt{7}}{3}$, then $\cot\theta = \dfrac{\text{adjacent } \theta}{\text{opposite } \theta} = \dfrac{3}{\sqrt{7}}$. However, it is customary to rationalize the denominator.

$$\dfrac{3}{\sqrt{7}} = \dfrac{3}{\sqrt{7}} \bullet \dfrac{\sqrt{7}}{\sqrt{7}} = \dfrac{3\sqrt{7}}{7}$$

5. **(D)** Given that $\cos\theta = \dfrac{15}{17}$, we can determine that the side adjacent θ is 15 and the hypotenuse is 17. To complete the ratio $\sin\theta = \dfrac{\text{opposite } \theta}{\text{hypotenuse}}$, we need to first use the Pythagorean theorem to find the side opposite θ. Using the given sides, we establish that $15^2 + b^2 = 17^2$. From there, we determine that the side opposite θ measures 8. We can now complete the ratio.

$$\sin\theta = \dfrac{8}{17}$$

Lesson 20.2—Skills Check (page 489)

1. **The period is 2π radians. The amplitude is 2.**

   Referencing $y = A\cos Bx$, for the function $y = 2\cos x$, $A = 2$ and $B = 1$.

   $$\text{Amplitude} = |A| = 2$$

   $$\text{Period} = \frac{2\pi}{B} = \frac{2\pi}{1} = 2\pi$$

2. **The period is 90°. The amplitude is 3.**

   Referencing $y = A\sin Bx$, for the function $y = 3\sin 4x$, $A = 3$ and $B = 4$.

   $$\text{Amplitude} = |A| = 3$$

   $$\text{Period} = \frac{360°}{B} = \frac{360°}{4} = 90°$$

3. **The period is 4π radians. The amplitude is 1.**

   Referencing $y = A\sin Bx$, for the function $y = \sin \frac{1}{2}x$, $A = 1$ and $B = \frac{1}{2}$.

   $$\text{Amplitude} = |A| = 1$$

   $$\text{Period} = \frac{2\pi}{B} = \frac{2\pi}{\frac{1}{2}} = 2\pi \cdot \frac{2}{1} = 4\pi$$

4. **The period is 2π radians. The amplitude is 7.**

   The function goes through one complete cycle in the interval from 0 to 2π and reaches a height of 7. Therefore, the period is 2π, and the amplitude is 7.

5. **The period is 60°. The amplitude is 0.5.**

   The function goes through one complete cycle in the interval from 0 to 60° and reaches a height of 0.5. Therefore, the period is 60°, and the amplitude is 0.5.

Lesson 20.2—ACCUPLACER Challenge (page 490)

1. **(C)** The graph reaches a height of 1.5, making the amplitude, $|A| = 1.5$. The graph completes one full cycle in the interval from $-\pi$ to π, making the period 2π.

   $\frac{2\pi}{B} = 2\pi$, so $B = 1$. Substitute 1.5 for A and 1 for B into $y = A\sin Bx$.

   $$y = 1.5\sin 1x$$
   $$y = 1.5\sin x$$

2. **(D)** In the function $y = 3\sin 5x$, $A = 3$ and $B = 5$.

   $$\text{Period} = \frac{2\pi}{B} = \frac{2\pi}{5}$$

3. **(B)** In the function $y = 3\sin 4x$, $A = 3$ and $B = 4$.

   $$\text{Amplitude} = |A| = 3$$

4. **(C)** The graph of the function completes one full cycle in the interval from 0 to $\frac{\pi}{2}$, so the period is $\frac{\pi}{2}$.

5. **(D)** The graph reaches a height of 1, so the amplitude is 1 and $A = 1$. The graph completes one full cycle in the interval from 0 to $\frac{\pi}{2}$, so the period is $\frac{\pi}{2}$.

   Period $= \frac{2\pi}{B} = \frac{\pi}{2}$, so $B = 4$. Substitute 1 for A and 4 for B into $y = A\sin Bx$.

   $$y = 1\sin 4x$$
   $$y = \sin 4x$$

Other Advanced Topics

21

There are a few topics that you can expect to see on the ACCUPLACER that didn't fit anywhere else in this book, so we will cover them here. If you've made it this far through the book, then you are very well on your way to succeeding on the ACCUPLACER! Just make sure that you work through the problems in this chapter first before moving on to the practice tests.

> In this chapter, we will work to master
>
> - working with complex numbers,
> - solving 2 by 2 matrices,
> - understanding the basic principle of counting, and
> - differentiating between combinations and permutations.

LESSON 21.1—THE COMPLEX NUMBER SYSTEM

Throughout this book, we have referred to **real numbers** countless times. Well, if a number isn't real, then what is it? It's **imaginary** of course! Let's back up, and explore a few things we know to be true to help us begin to consider "imaginary" numbers.

We know:

- The second power of a positive number is always positive (e.g., $5^2 = 5 \times 5 = 25$).
- The second power of a negative number is always positive (e.g., $(-5)^2 = (-5) \times (-5) = 25$).

Therefore, no number times itself can have a negative outcome. In other words, there is no such thing as a second power being negative. This means that there is no real solution to a problem such as $x = \sqrt{-25}$. In fact, there is no real solution to the square root of any negative number since no number times itself can have a negative outcome.

In mathematics, though, we sometimes come across a need to use numbers such as $\sqrt{-1}$. To make computations with such numbers possible, mathematicians use the symbol i to represent $\sqrt{-1}$. In this context, the letter i is not a variable; its value is always the same. The symbol i takes the place of $\sqrt{-1}$. It is not a **real** number; it is an **imaginary**, or **complex**, number!

There is an interesting pattern in the powers of i.

$$i^1 = i$$
$$i^2 = \sqrt{-1} \times \sqrt{-1} = -1$$
$$i^3 = i^2 \times i = (-1) \times i = -i$$
$$i^4 = i^3 \times i = -i \times i = -i^2 = -(-1) = 1$$
$$i^5 = i^4 \times i = 1 \times i = i$$
$$i^6 = i^5 \times i = i \times i = -1$$
$$i^7 = i^6 \times i = (-1) \times i = -i$$
$$i^8 = i^7 \times i = (-i) \times i = -i^2 = -(-1) = 1$$

Notice that when we found the value of i^5, the pattern cycled back to where we started, and we found that $i^5 = i$. Experimenting further, we discover that there are only four outcomes to the powers of i, and the values repeat every fourth power. Let's look at some other examples.

➡ Example 1

Simplify $(-3i)^2$.

Solution

We need to square both the constant -3 and i.

$$(-3i)^2 = (-3)^2(i^2)$$

Keep in mind that $i^2 = \sqrt{-1} \times \sqrt{-1} = -1$.

$$(-3)^2(i^2) = (9)(-1) = -9$$

Answer: -9

To add or subtract complex numbers, we combine real numbers with real numbers, and we combine the imaginary parts with imaginary parts.

➡ Example 2

Simplify $15 + 6i - 11 + 4i$.

Solution

We combine the real parts together, and then we combine the imaginary parts together.

$$15 + 6i - 11 + 4i =$$
$$(15 - 11) + (6i + 4i) = 4 + 10i$$

Answer: The sum is $4 + 10i$.

➡ Example 3

Multiply $(2 + 3i)(3 - 5i)$.

Solution

Multiply as you normally would when multiplying two binomials. The terms become $2(3) + 2(-5i) + 3i(3) + (3i)(-5i)$. We can then combine like terms and simplify. Notice how, in the steps below, $i^2 = -1$.

$$2(3) + 2(-5i) + 3i(3) + (3i)(-5i) =$$
$$2(3) + 2(-5i) + 3i(3) - 15(i^2) =$$
$$6 - 10i + 9i - 15(-1) =$$
$$6 - i + 15 = 21 - i$$

Answer: The product is $21 - i$.

➡ Example 4

Simplify $\sqrt{-50}$.

Solution

When there is a negative number under the radical, we should factor out -1 immediately.

$$\sqrt{-50} = \sqrt{50} \times \sqrt{-1}$$

Next, simplify $\sqrt{50}$ like you normally would.

$$\sqrt{-50} = \sqrt{50} \times \sqrt{-1}$$
$$\sqrt{-50} = \sqrt{25} \times \sqrt{2} \times \sqrt{-1}$$
$$\sqrt{-50} = 5\sqrt{2} \times \sqrt{-1}$$

Now convert $\sqrt{-1}$ into i.

$$\sqrt{-50} = 5\sqrt{2} \times i$$
$$\sqrt{-50} = 5i\sqrt{2}$$

Answer: $5i\sqrt{2}$

LESSON 21.1—SKILLS CHECK

Directions: Simplify each expression.

1. i^8

2. $(-3i)^3$

3. $12i(11i)$

4. $\frac{2}{3}i + \frac{5}{3}i$

5. $(4 + i)^2$

6. $(6 + i)(6 - i)$

7. $\sqrt{-36}$

8. $\sqrt{-200}$

9. $\sqrt{-12} + \sqrt{-75}$

10. $\sqrt{-125} - \sqrt{5}$

(Answers are on page 507.)

LESSON 21.1—ACCUPLACER CHALLENGE

> **Directions**: For each of the questions below, choose the best answer from the five choices given.

1. Which of the following choices is equal to i^{10}?

   (A) i
   (B) $-i$
   (C) 1
   (D) -1
   (E) None of these

2. Which of the following choices is equal to $\sqrt{-72x^2y^2}$?

   (A) $6xy\sqrt{2}$
   (B) $-6xy\sqrt{2}$
   (C) $6i\sqrt{xy}$
   (D) $6xyi\sqrt{2}$
   (E) $12xyi$

3. What is the product of $(2i)^4$ and $(-3i)^2$?

   (A) 144
   (B) -144
   (C) $144i$
   (D) $-144i$
   (E) None of these

4. Which of the following choices is equal to $\dfrac{\sqrt{-48}}{\sqrt{-4}}$?

   (A) $2\sqrt{3}$
   (B) $2i\sqrt{3}$
   (C) $2i$
   (D) 2
   (E) 12

5. Let $y = -3i^x$. For which values of x will y be a negative, real number?

   (A) x = multiples of 2
   (B) x = multiples of 3
   (C) x = multiples of 4
   (D) x = multiples of 5
   (E) x = multiples of 6

(Answers are on page 507.)

LESSON 21.2—CALCULATING DETERMINANTS

A **determinant** is a kind of matrix that plays a major role in advanced algebra courses like calculus and linear algebra. On the ACCUPLACER, you only need to know how to calculate basic 2 × 2 determinants. Thankfully, the process is pretty straightforward.

A 2 × 2 determinant looks like $\begin{vmatrix} a & b \\ c & d \end{vmatrix}$. To calculate the value of the determinant, we multiply the two diagonals. First, multiply the ad diagonal; then multiply the bc diagonal. The last step is to calculate $ad - bc$. Let's sum this up by outlining the process one more time, and then we will look at a few examples.

To Calculate the Value of a 2 × 2 Determinant:

1. Multiply the a and d values.
2. Multiply the b and c values.
3. Find the difference of $ad - bc$.

➡ Example 1

Find the value of the determinant $\begin{vmatrix} 4 & 3 \\ 1 & 7 \end{vmatrix}$.

Solution

In this determinant, $a = 4$, $b = 3$, $c = 1$, and $d = 7$.

$$\begin{vmatrix} a & b \\ c & d \end{vmatrix} = \begin{vmatrix} 4 & 3 \\ 1 & 7 \end{vmatrix}$$

Use the formula $ad - bc$.

$$ad - bc =$$
$$(4)(7) - (3)(1) =$$
$$28 - 3 = 25$$

Answer: The value of the determinant is 25.

➡ Example 2

Calculate the determinant $\begin{vmatrix} 2n & 4n \\ -3 & 7 \end{vmatrix}$.

Solution

The determinant has variables in it, but the process remains the same. We still need to use the formula $ad - bc$.

$$\begin{vmatrix} a & b \\ c & d \end{vmatrix} = \begin{vmatrix} 2n & 4n \\ -3 & 7 \end{vmatrix}$$

Here, $a = 2n$, $b = 4n$, $c = -3$, and $d = 7$.

$$ad - bc =$$
$$2n(7) - (4n)(-3) =$$
$$14n - (-12n) = 26n$$

Answer: The determinant has a value of $26n$.

➡ Example 3

Find the value of x when $\begin{vmatrix} -3 & x \\ 2 & x \end{vmatrix} = 40$.

Solution

To find the value of x, we first need to multiply our diagonals, and then use the formula $ad - bc$. Here, $a = -3$, $b = x$, $c = 2$, and $d = x$.

$$ad - bc =$$
$$(-3)(x) - (x)(2) =$$
$$-3x - 2x = -5x$$

Now we know that $-5x = 40$, so we can solve for x.

Answer: The value of x is -8.

LESSON 21.2—SKILLS CHECK

Directions: Calculate the value of the determinants below.

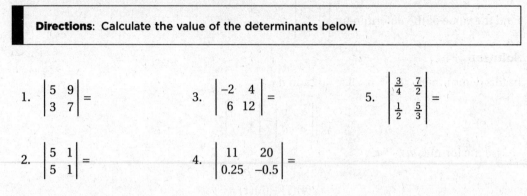

1. $\begin{vmatrix} 5 & 9 \\ 3 & 7 \end{vmatrix} =$

3. $\begin{vmatrix} -2 & 4 \\ 6 & 12 \end{vmatrix} =$

5. $\begin{vmatrix} \frac{3}{4} & \frac{7}{2} \\ \frac{1}{2} & \frac{5}{3} \end{vmatrix} =$

2. $\begin{vmatrix} 5 & 1 \\ 5 & 1 \end{vmatrix} =$

4. $\begin{vmatrix} 11 & 20 \\ 0.25 & -0.5 \end{vmatrix} =$

(Answers are on page 507.)

LESSON 21.2—ACCUPLACER CHALLENGE

Directions: For each of the questions below, choose the best answer from the five choices given.

1. $\begin{vmatrix} 1 & 1 \\ 1 & 1 \end{vmatrix} =$

   (A) −2
   (B) −1
   (C) 0
   (D) 1
   (E) 2

4. $\begin{vmatrix} 9a & 6 \\ a^2 & a \end{vmatrix} =$

   (A) $a^2 + 10a + 6$
   (B) $a^2 - 9a$
   (C) $-3a$
   (D) $-3a^2$
   (E) $3a^2$

2. $\begin{vmatrix} 4 & 6 \\ -2 & 3 \end{vmatrix} =$

   (A) 24
   (B) −24
   (C) −20
   (D) 20
   (E) 0

5. $\begin{vmatrix} 2x & 2 \\ 10 & -3 \end{vmatrix} = \begin{vmatrix} 1 & 4 \\ 7 & -5x \end{vmatrix}$. Then $x =$

   (A) −3
   (B) 1
   (C) 8
   (D) 11
   (E) 12

3. $\begin{vmatrix} 6 & 3 \\ -1 & x \end{vmatrix} = 27$. Then $x =$

   (A) 4
   (B) 5
   (C) 6
   (D) 9
   (E) 12

(Answers are on page 508.)

LESSON 21.3—COUNTING, PERMUTATIONS, AND COMBINATIONS

Although the College-Level Mathematics test fittingly places a high emphasis on algebra, you should expect to see a question or two involving counting, combinations, or permutations. Each of these is a little different, so we will look at a few examples and introduce some key formulas that will help you succeed on the exam. Be sure to spend some time learning them!

Lesson 21.3A—Counting

We'll start by looking at **counting**. If you've made it this far in the book, then you certainly have no trouble counting, as we usually think of it. However, this kind of counting involves determining the total number of possible outcomes given any number of distinct events. Let's look at an example.

➡ Example 1

A taco truck offers four different kinds of meat/fish—chicken, steak, pork, and fish—and three different kinds of salsa—mild, medium, and hot. How many different tacos could you order?

Solution

Let's make a list to illustrate an important principle. We could order:

chicken/mild	steak/mild	pork/mild	fish/mild
chicken/medium	steak/medium	pork/medium	fish/medium
chicken/hot	steak/hot	pork/hot	fish/hot

It's clear from our list that there are 12 possible tacos that you could order. Listing them all isn't the most efficient way to solve a problem like this one. Let's think about it another way.

There are four different ways that you can choose the taco meat/fish, and for each of those there are three different ways that you can choose the salsa, or $4 \times 3 = 12$. This is called the **basic counting principle**.

Answer: There are 12 different tacos that you could order.

To clarify, the basic counting principle says that if there are n ways of doing one thing, and m ways of doing another independent thing, then there are $n \times m$ ways of doing both things together. This can be extended for 3, 4, 5, or any other number of things.

➡ Example 2

In how many different ways can the letters ABCDEF be arranged?

Solution

We already have one arrangement, but there are several others. By the basic counting principle, there are

> 6 ways to choose the first letter,
> 5 remaining ways to choose the second letter,
> 4 remaining ways to choose the third letter,
> 3 remaining ways to choose the fourth letter,
> 2 remaining ways to choose the fifth letter,
> 1 remaining way to choose the sixth letter.

Therefore, there are $6 \times 5 \times 4 \times 3 \times 2 \times 1 = 720$ ways of arranging these six letters. Imagine trying to solve a problem like this one by writing everything in a list!

Answer: There are 720 ways of arranging the letters ABCDEF.

In math, when we multiply a set of numbers like $6 \times 5 \times 4 \times 3 \times 2 \times 1$, in which each number is one less than the one before it, we refer to this as a factorial. Therefore, $6 \times 5 \times 4 \times 3 \times 2 \times 1 = 6!$ Here, the exclamation point denotes the **factorial**.

Now that we know how the basic counting principle works, let's look at some other kinds of problems.

Lesson 21.3B—Permutations and Combinations

A **permutation** is a set of things or events in which order *does* matter, and a **combination** is a set of things or events in which order *does not* matter. Here's a helpful way to remember the difference between the two: At some point in your life, you may have used a combination lock. You had to enter a set of three numbers, in the correct order, to open the lock. These locks are actually called by the wrong name! They should be **permutation** locks because, if you enter the right three numbers in the wrong order, the lock won't open.

Let's look at some examples to see the difference between the two.

➡ Example 1

A math study group has 8 members. The group needs to elect a president, a secretary, and a treasurer. In how many different ways can the group assign these roles?

Solution

This is an example of a permutation because each role is different, and even if three members were chosen from the group to fill these roles, there would still be several different ways that those three members could be president, secretary, and treasurer.

We can use the basic counting principle to help us answer this question. Of the 8 people in the group, there are 8 ways to choose the person who will be president. After this role is filled, there are 7 ways to choose the secretary. Then, after that role is filled, there are still 6 ways to choose the treasurer.

Therefore, by the basic counting principle, there are $8 \times 7 \times 6$ ways that the group could choose these members.

Answer: There are $8 \times 7 \times 6 = 336$ ways that the group can fill the roles of president, secretary, and treasurer.

Permutations are often denoted as $P(n, r)$, or $_nP_r$.

$$_nP_r$$

total number of items/events   **number of items/events being permuted**

Although we can calculate permutations intuitively as we did in the last example, you should learn this formula:

$$_nP_r = \frac{n!}{(n-r)!}$$

In the previous example, we found that there were 336 ways of assigning roles in the study group. The value of n is 8 because there are 8 people in the group. The value of r is 3 because we are "permuting" 3 people from within the group.

$$_nP_r = \frac{n!}{(n-r)!}$$

$$_8P_3 = \frac{8!}{(8-3)!}$$

$$_8P_3 = \frac{8!}{5!} = \frac{8 \times 7 \times 6 \times 5!}{5!} = 8 \times 7 \times 6 = 336$$

Notice how 5! was canceled out. This is an important step in solving problems like this—mostly because it keeps the numbers more manageable.

➡ Example 2

Rachel has six albums by her favorite band. She wants to listen to four of them on Saturday. In how many different orders could she listen to these four albums?

Solution

Here, $n = 6$ and $r = 4$ because she has six albums in total and is permuting four of them. This means that there are 6 ways she could choose the first album, 5 ways she could choose the second, 4 ways she could choose the third, and 3 ways she could choose the fourth.

$$_nP_r = \frac{n!}{(n-r)!}$$

$$_6P_4 = \frac{6!}{(6-4)!}$$

$$_6P_4 = \frac{6!}{2!}$$

$$_6P_4 = \frac{6 \times 5 \times 4 \times 3 \times 2!}{2!} = 6 \times 5 \times 4 \times 3 = 360$$

Answer: There are 360 different orders in which Rachel could listen to the four albums.

Let's look at some examples of combinations. Remember, combinations differ from permutations in that, with combinations, order doesn't matter.

➡ Example 3

A math study group has eight members, and it needs to select a committee of three members to choose which problems the group works on. In how many ways can it choose these three members?

Solution

This problem looks a lot like Example 1, in which the math group elected a president, a secretary, and a treasurer. This time, however, the group needs to choose any three members. Since order doesn't matter, this is a combination instead of a permutation.

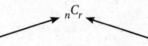

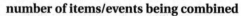

total number of items/events   **number of items/events being combined**

We read this as "n choose r," and it is sometimes also denoted as $C(n, r)$ or simply $\binom{n}{r}$.

The formula for finding combinations is a little different. Notice that, this time, we have to divide by $r!$ to eliminate all the different "orderings" within the group that is chosen.

$$_nC_r = \frac{n!}{r!(n-r)!}$$

To find the number of ways that the group can choose a committee of three members, we use $n = 8$ and $r = 3$.

$$_nC_r = \frac{n!}{r!(n-r)!}$$

$$_8C_3 = \frac{8!}{3!(8-3)!}$$

$$_8C_3 = \frac{8!}{3!(5!)}$$

$$_8C_3 = \frac{8 \times 7 \times 6 \times 5!}{3!(5!)}$$

$$_8C_3 = \frac{8 \times 7 \times 6}{3 \times 2 \times 1}$$

$$_8C_3 = 8 \times 7 = 56$$

Answer: There are 56 ways to choose who is in the committee.

When order/role *did* matter, there were 336 unique possibilities of selecting three people. When their order/role *didn't* matter, there were only 56 unique ways.

➡ Example 4

Bianca has 6 movies on her shelf, and she randomly chooses three to watch over the weekend. In how many different ways can she choose these movies?

Solution

Here, the order in which Bianca watches the movies doesn't matter, so we know this is a combination problem, with $n = 6$ movies in total and $r = 3$ being chosen.

$$_nC_r = \frac{n!}{r!(n-r)!}$$

$$_6C_3 = \frac{6!}{3!(6-3)!}$$

$$_6C_3 = \frac{6 \times 5 \times 4 \times 3!}{3!(3!)}$$

$$_6C_3 = \frac{6 \times 5 \times 4}{3 \times 2 \times 1}$$

$$_6C_3 = 5 \times 4 = 20$$

Answer: There are 20 ways that Bianca could choose three movies from six potential choices.

LESSON 21.3—SKILLS CHECK

Directions: Use the permutation and combination formulas to solve the problems below.

1. $_7P_5$

2. $_{11}P_3$

3. $_5P_2$

4. $_7C_5$

5. $_{11}C_3$

6. $_5C_2$

(Answers are on page 509.)

LESSON 21.3—ACCUPLACER CHALLENGE

Directions: For each of the questions below, choose the best answer from the five choices given.

1. A family of five is having a family photo taken in which each member sits in a different chair. In how many different ways can the family be seated for the photo?

   (A) 10
   (B) 20
   (C) 50
   (D) 120
   (E) 720

2. Nigel is going on tour with his band, and he needs to select three electric guitars to take with him. He has seven electric guitars altogether. In how many different ways can he choose which guitars to take on tour?

   (A) 6
   (B) 21
   (C) 35
   (D) 70
   (E) 210

3. In how many different ways can the numbers 1234 be arranged?

   (A) 6
   (B) 16
   (C) 24
   (D) 64
   (E) 256

4. A basketball team has 8 players. The team needs to choose one player to be the captain and another player to schedule the team's practices. In how many different ways can they do this?

   (A) 28
   (B) 56
   (C) 72
   (D) 112
   (E) 40,320

5. A math teacher wrote 10 problems and plans to work on 4 of them in class. How many different ways can she choose which 4 problems to work on in class?

   (A) 24
   (B) 40
   (C) 210
   (D) 540
   (E) 5,040

6. There are 9 players on a basketball team, 5 of whom will start the game. How many ways can the team choose its 5 starters?

   (A) 45
   (B) 126
   (C) 210
   (D) 512
   (E) 15,120

(Answers are on page 509.)

To complete a set of practice questions that reviews all of Chapter 21,
and to complete a set of practice questions that covers all of Unit 3, go to:
barronsbooks.com/TP/accuplacer/math29qw/

Lesson 21.1—Skills Check (page 497)

1. **1**   We know that $i^4 = 1$, so $i^8 = i^4 \times i^4 = 1$.

2. **27i** $(-3i)^3 = (-3^3)(i^3) = (-27)(-i) = 27i$

3. **−132** $12i(11i) = 132i^2 = 132(-1) = -132$

4. $\dfrac{7}{3}i$ $\dfrac{2}{3}i + \dfrac{5}{3}i = \left(\dfrac{2}{3} + \dfrac{5}{3}\right)i = \dfrac{7}{3}i$

5. **15 + 8i** $(4+i)^2 = (4+i)(4+i) = 16 + 4i + 4i + i^2 = 16 + 8i - 1 = 15 + 8i$

6. **37** $(6+i)(6-i) = 36 - 6i + 6i - i^2 = 36 - (-1) = 37$

7. **6i** $\sqrt{-36} = \sqrt{36} \times \sqrt{-1} = 6\sqrt{-1} = 6i$

8. **10$i\sqrt{2}$** $\sqrt{-200} = \sqrt{100} \times \sqrt{2} \times \sqrt{-1} = 10i\sqrt{2}$

9. **7$i\sqrt{3}$** $\sqrt{-12} + \sqrt{-75} = \left(\sqrt{4} \times \sqrt{3} \times \sqrt{-1}\right) + \left(\sqrt{25} \times \sqrt{3} \times \sqrt{-1}\right) = 2i\sqrt{3} + 5i\sqrt{3} = 7i\sqrt{3}$

10. **(5i − 1)$\sqrt{5}$** $\sqrt{-125} - \sqrt{5} = \left(\sqrt{25} \times \sqrt{5} \times \sqrt{-1}\right) - \sqrt{5} = 5i\sqrt{5} - \sqrt{5} = (5i - 1)\sqrt{5}$

Lesson 21.1—ACCUPLACER Challenge (page 498)

1. **(D)** We know that $i^2 = -1$ and $i^4 = 1$, so $i^{10} = i^4 \times i^4 \times i^2 = 1 \times 1 \times (-1) = -1$.

2. **(D)** $\sqrt{-72x^2y^2} = \sqrt{36} \times \sqrt{2} \times \sqrt{-1} \times \sqrt{x^2} \times \sqrt{y^2} = 6xyi\sqrt{2}$

3. **(B)** $(2i)^4 \times (-3i)^2 = \left(2^4 i^4\right) \times \left((-3)^2 i^2\right) = (16 \times 1) \times (9 \times (-1)) = -144$

4. **(A)** $\dfrac{\sqrt{-48}}{\sqrt{-4}} = \dfrac{\sqrt{-4} \times \sqrt{12}}{\sqrt{-4}} = \sqrt{12} = \sqrt{4} \times \sqrt{3} = 2\sqrt{3}$

5. **(C)** The value of y will be a negative, real number whenever $i^x = 1$. This will result in $y = -3(1)$. We know that $i^x = 1$ whenever x is a multiple of 4.

Lesson 21.2—Skills Check (page 500)

1. **8** $\begin{vmatrix} 5 & 9 \\ 3 & 7 \end{vmatrix} = (5 \times 7) - (9 \times 3) = 35 - 27 = 8$

2. **0** $\begin{vmatrix} 5 & 1 \\ 5 & 1 \end{vmatrix} = (5 \times 1) - (1 \times 5) = 5 - 5 = 0$

3. **−48** $\begin{vmatrix} -2 & 4 \\ 6 & 12 \end{vmatrix} = (-2 \times 12) - (4 \times 6) = -24 - 24 = -48$

4. **−10.5** $\begin{vmatrix} 11 & 20 \\ 0.25 & -0.5 \end{vmatrix} = (11 \times (-0.5)) - (20 \times 0.25) = -5.5 - 5 = -10.5$

5. $-\dfrac{1}{2}$ $\begin{vmatrix} \frac{3}{4} & \frac{7}{2} \\ \frac{1}{2} & \frac{5}{3} \end{vmatrix} = \dfrac{3}{4} \times \dfrac{5}{3} - \dfrac{7}{2} \times \dfrac{1}{2} = \dfrac{5}{4} - \dfrac{7}{4} = -\dfrac{2}{4} = -\dfrac{1}{2}$

Lesson 21.2—ACCUPLACER Challenge (page 500)

1. **(C)** $\begin{vmatrix} 1 & 1 \\ 1 & 1 \end{vmatrix} = (1 \times 1) - (1 \times 1) = 1 - 1 = 0$

2. **(A)** $\begin{vmatrix} 4 & 6 \\ -2 & 3 \end{vmatrix} = (4 \times 3) - (6 \times -2) = 12 - (-12) = 24$

3. **(A)**

$$\begin{vmatrix} 6 & 3 \\ -1 & x \end{vmatrix} = 27$$

$$(6x) - (3 \times -1) = 27$$

$$6x - (-3) = 27$$

$$6x + 3 = 27$$

Now solve the resulting equation.

$$6x + 3 = 27$$
$$\underline{ -3 \quad -3}$$
$$\frac{6x}{6} = \frac{24}{6}$$
$$x = 4$$

4. **(E)** $\begin{vmatrix} 9a & 6 \\ a^2 & a \end{vmatrix} = (9a \times a) - (6 \times a^2) = 9a^2 - 6a^2 = 3a^2$

5. **(C)**

$$\begin{vmatrix} 2x & 2 \\ 10 & -3 \end{vmatrix} = \begin{vmatrix} 1 & 4 \\ 7 & -5x \end{vmatrix}$$

$$(2x \times (-3)) - (2 \times 10) = (1 \times (-5x)) - (4 \times 7)$$

$$-6x - 20 = -5x - 28$$

Now solve the equation for x.

$$-6x - 20 = -5x - 28$$
$$\underline{+6x \qquad\qquad +6x}$$
$$-20 = x - 28$$
$$\underline{+28 \qquad +28}$$
$$8 = x$$

Lesson 21.3—Skills Check (page 505)

1. **2,520**

$$_nP_r = \frac{n!}{(n-r)!}$$

$$_7P_5 = \frac{7!}{(7-5)!}$$

$$_7P_5 = \frac{7!}{2!}$$

$$_7P_5 = \frac{7 \times 6 \times 5 \times 4 \times 3 \times 2!}{2!}$$

$$_7P_5 = 7 \times 6 \times 5 \times 4 \times 3 = 2,520$$

2. **990**

$$_nP_r = \frac{n!}{(n-r)!}$$

$$_{11}P_3 = \frac{11!}{(11-3)!}$$

$$_{11}P_3 = \frac{11!}{8!}$$

$$_{11}P_3 = \frac{11 \times 10 \times 9 \times 8!}{8!}$$

$$_{11}P_3 = 11 \times 10 \times 9 = 990$$

3. **20**

$$_nP_r = \frac{n!}{(n-r)!}$$

$$_5P_2 = \frac{5!}{(5-2)!}$$

$$_5P_2 = \frac{5!}{3!}$$

$$_5P_2 = \frac{5 \times 4 \times 3!}{3!}$$

$$_5P_2 = 5 \times 4 = 20$$

4. **21**

$$_nC_r = \frac{n!}{r!(n-r)!}$$

$$_7C_5 = \frac{7!}{5!(7-5)!}$$

$$_7C_5 = \frac{7 \times 6 \times 5!}{5!(2!)}$$

$$_7C_5 = \frac{7 \times 6}{2 \times 1}$$

$$_7C_5 = 21$$

5. **165**

$$_nC_r = \frac{n!}{r!(n-r)!}$$

$$_{11}C_3 = \frac{11!}{3!(11-3)!}$$

$$_{11}C_3 = \frac{11!}{3!(8!)}$$

$$_{11}C_3 = \frac{11 \times 10 \times 9 \times 8!}{3!(8!)}$$

$$_{11}C_3 = \frac{11 \times 10 \times 9}{3 \times 2 \times 1}$$

$$_{11}C_3 = \frac{990}{6} = 165$$

6. **10**

$$_nC_r = \frac{n!}{r!(n-r)!}$$

$$_5C_2 = \frac{5!}{2!(5-2)!}$$

$$_5C_2 = \frac{5!}{2!(3!)}$$

$$_5C_2 = \frac{5 \times 4 \times 3!}{2!(3!)}$$

$$_5C_2 = \frac{20}{2}$$

$$_5C_2 = 10$$

Lesson 21.3—ACCUPLACER Challenge (page 506)

1. **(D)** Imagine that the chairs are numbered from 1 to 5. There are five ways different people could sit in the first chair. After that, there are four ways to fill the second chair, three ways to fill the third chair, two ways to fill the fourth chair, and only one way to fill the last chair. By the basic principle of counting, the total number of ways to arrange the family is $5 \times 4 \times 3 \times 2 \times 1 = 120$.

2. **(C)** The order in which Nigel chooses his guitars doesn't matter, so this is a combination in which $n = 7$ guitars and $r = 3$ that he chooses.

$$_nC_r = \frac{n!}{r!(n-r)!}$$

$$_7C_3 = \frac{7!}{3!(7-3)!}$$

$$_7C_3 = \frac{7 \times 6 \times 5 \times 4!}{3!(4!)}$$

$$_7C_3 = \frac{7 \times 6 \times 5}{3 \times 2 \times 1}$$

$$_7C_3 = 35$$

3. **(C)** There are 4 numbers that could occupy the first position, 3 that could occupy the second position, 2 that could occupy the third position, and 1 that could occupy the last position. By the basic counting principle, the total number of arrangements is $4 \times 3 \times 2 \times 1 = 24$.

4. **(B)** In this case, the roles of captain and scheduler are distinct. Since there are 8 players on the team, there are 8 ways to choose the captain. After the captain is chosen, there are 7 ways to choose the scheduler. Therefore, there are $8 \times 7 = 56$ ways to fill these roles.

5. **(C)** Here, the order in which the problems are chosen and given to the class is irrelevant. The teacher just needs to choose 4 problems from 10, so this is a combination where $n = 10$ problems and $r = 4$ that are chosen.

$$_nC_r = \frac{n!}{r!(n-r)!}$$

$$_{10}C_4 = \frac{10!}{4!(10-4)!}$$

$$_{10}C_4 = \frac{10!}{4!(6!)}$$

$$_{10}C_4 = \frac{10 \times 9 \times 8 \times 7 \times 6!}{4!(6!)}$$

$$_{10}C_4 = \frac{10 \times 9 \times 8 \times 7}{4 \times 3 \times 2 \times 1}$$

$$_{10}C_4 = \frac{5040}{24}$$

$$_{10}C_4 = 210$$

6. **(B)** Here, the order in which the players are chosen is irrelevant. The team just needs to choose 5 starters from 9 players, so this is a combination where $n = 9$ players and $r = 5$ that are chosen.

$$_nC_r = \frac{n!}{r!(n-r)!}$$

$$_9C_5 = \frac{9!}{5!(9-5)!}$$

$$_9C_5 = \frac{9!}{5!(4!)}$$

$$_9C_5 = \frac{9 \times 8 \times 7 \times 6 \times 5!}{5!(4!)}$$

$$_9C_5 = \frac{9 \times 8 \times 7 \times 6}{4 \times 3 \times 2 \times 1}$$

$$_9C_5 = 126$$

PRACTICE TESTS

ANSWER SHEET
Practice Test 1

ARITHMETIC

1. Ⓐ Ⓑ Ⓒ Ⓓ
2. Ⓐ Ⓑ Ⓒ Ⓓ
3. Ⓐ Ⓑ Ⓒ Ⓓ
4. Ⓐ Ⓑ Ⓒ Ⓓ
5. Ⓐ Ⓑ Ⓒ Ⓓ
6. Ⓐ Ⓑ Ⓒ Ⓓ
7. Ⓐ Ⓑ Ⓒ Ⓓ
8. Ⓐ Ⓑ Ⓒ Ⓓ
9. Ⓐ Ⓑ Ⓒ Ⓓ
10. Ⓐ Ⓑ Ⓒ Ⓓ
11. Ⓐ Ⓑ Ⓒ Ⓓ
12. Ⓐ Ⓑ Ⓒ Ⓓ
13. Ⓐ Ⓑ Ⓒ Ⓓ
14. Ⓐ Ⓑ Ⓒ Ⓓ
15. Ⓐ Ⓑ Ⓒ Ⓓ
16. Ⓐ Ⓑ Ⓒ Ⓓ
17. Ⓐ Ⓑ Ⓒ Ⓓ

ELEMENTARY ALGEBRA

1. Ⓐ Ⓑ Ⓒ Ⓓ
2. Ⓐ Ⓑ Ⓒ Ⓓ
3. Ⓐ Ⓑ Ⓒ Ⓓ
4. Ⓐ Ⓑ Ⓒ Ⓓ
5. Ⓐ Ⓑ Ⓒ Ⓓ
6. Ⓐ Ⓑ Ⓒ Ⓓ
7. Ⓐ Ⓑ Ⓒ Ⓓ
8. Ⓐ Ⓑ Ⓒ Ⓓ
9. Ⓐ Ⓑ Ⓒ Ⓓ
10. Ⓐ Ⓑ Ⓒ Ⓓ
11. Ⓐ Ⓑ Ⓒ Ⓓ
12. Ⓐ Ⓑ Ⓒ Ⓓ

COLLEGE-LEVEL MATHEMATICS

1. Ⓐ Ⓑ Ⓒ Ⓓ Ⓔ
2. Ⓐ Ⓑ Ⓒ Ⓓ Ⓔ
3. Ⓐ Ⓑ Ⓒ Ⓓ Ⓔ
4. Ⓐ Ⓑ Ⓒ Ⓓ Ⓔ
5. Ⓐ Ⓑ Ⓒ Ⓓ Ⓔ
6. Ⓐ Ⓑ Ⓒ Ⓓ Ⓔ
7. Ⓐ Ⓑ Ⓒ Ⓓ Ⓔ
8. Ⓐ Ⓑ Ⓒ Ⓓ Ⓔ
9. Ⓐ Ⓑ Ⓒ Ⓓ Ⓔ
10. Ⓐ Ⓑ Ⓒ Ⓓ Ⓔ
11. Ⓐ Ⓑ Ⓒ Ⓓ Ⓔ
12. Ⓐ Ⓑ Ⓒ Ⓓ Ⓔ
13. Ⓐ Ⓑ Ⓒ Ⓓ Ⓔ
14. Ⓐ Ⓑ Ⓒ Ⓓ Ⓔ
15. Ⓐ Ⓑ Ⓒ Ⓓ Ⓔ
16. Ⓐ Ⓑ Ⓒ Ⓓ Ⓔ
17. Ⓐ Ⓑ Ⓒ Ⓓ Ⓔ
18. Ⓐ Ⓑ Ⓒ Ⓓ Ⓔ
19. Ⓐ Ⓑ Ⓒ Ⓓ Ⓔ
20. Ⓐ Ⓑ Ⓒ Ⓓ Ⓔ

There are three separately scored ACCUPLACER Math tests: Arithmetic, Elementary Algebra, and College-Level Mathematics. Practice Test 1 is broken into the same three tests. You will notice that the question numbers start at 1 for each of the three test sections. You will have completed Practice Test 1 when you have answered all of the test questions for all three sections.

ARITHMETIC

> **Directions:** For each of the questions below, choose the best answer from the four choices given.

1. What is $\dfrac{1257}{99}$ rounded to the nearest integer?

   (A) 12
   (B) 13
   (C) 14
   (D) 15

2. If $\dfrac{5}{4} \div \dfrac{1}{8} = m$, then what whole numbers is m between?

   (A) 7 and 9
   (B) 9 and 11
   (C) 11 and 13
   (D) 13 and 15

3. Roman's bag of candies is comprised of $\dfrac{1}{3}$ yellow candies and $\dfrac{2}{5}$ green candies. The rest of the candies are orange. What fraction of Roman's bag of candies is orange?

   (A) $\dfrac{4}{15}$

   (B) $\dfrac{3}{8}$

   (C) $\dfrac{5}{8}$

   (D) $\dfrac{11}{15}$

4. $7\dfrac{3}{8} + 5\dfrac{3}{4} =$

   (A) $12\dfrac{1}{8}$

   (B) $12\dfrac{1}{2}$

   (C) $12\dfrac{3}{4}$

   (D) $13\dfrac{1}{8}$

5. Which of the following choices places the fractions correctly in order from least to greatest?

   (A) $\dfrac{5}{4}, \dfrac{16}{12}, \dfrac{12}{10}$

   (B) $\dfrac{12}{10}, \dfrac{16}{12}, \dfrac{5}{4}$

   (C) $\dfrac{12}{10}, \dfrac{5}{4}, \dfrac{16}{12}$

   (D) $\dfrac{16}{12}, \dfrac{5}{4}, \dfrac{12}{10}$

6. $4 \div \dfrac{8}{9} =$

   (A) $\dfrac{9}{2}$

   (B) $\dfrac{9}{8}$

   (C) $\dfrac{4}{5}$

   (D) $4\dfrac{1}{9}$

7. Which of the following choices is closest to 21.2 × 19.4?

   (A) 40
   (B) 400
   (C) 450
   (D) 4,000

8. Which of the following statements is true?

   (A) 0.62 < 0.062 < 0.602 < 0.26
   (B) 0.26 < 0.062 < 0.62 < 0.602
   (C) 0.062 < 0.26 < 0.62 < 0.602
   (D) 0.062 < 0.26 < 0.602 < 0.62

9. 4.29 × 3.9 =

   (A) 16.531
   (B) 16.651
   (C) 16.731
   (D) 177.31

10. 10.53 ÷ 6 =

    (A) 1.755
    (B) 0.1755
    (C) 1.6755
    (D) 1.905

11. 15 is 20% of what number?

    (A) 3
    (B) 60
    (C) 75
    (D) 300

12. What is 55% of 60?

    (A) 33
    (B) 32.5
    (C) 22
    (D) 36

13. If 30% of a number, P, is 1.8, what is the value of P?

    (A) 0.54
    (B) 5.4
    (C) 6
    (D) 12.6

14. Which of the following is equivalent to 3.17?

    (A) 0.0317%
    (B) 0.317%
    (C) 31.7%
    (D) 317%

15. $\frac{15}{3} - 2 \times 5 + 16 \times 3 =$

    (A) −53
    (B) 43
    (C) 53
    (D) 93

16. The measures of two angles of a triangle are 52° and 35°. What is the measure of the third angle of the triangle?

    (A) 87°
    (B) 90°
    (C) 93°
    (D) 180°

17. Three of four numbers have a sum of 24. If the average of the four numbers is 9, what is the fourth number?

    (A) 6
    (B) 8
    (C) 10
    (D) 12

ELEMENTARY ALGEBRA

> **Directions:** For each of the questions below, choose the best answer from the four choices given.

1. $\dfrac{-3}{5-(-1)} =$

   (A) $-\dfrac{1}{2}$

   (B) $\dfrac{1}{2}$

   (C) -2

   (D) $-\dfrac{3}{4}$

2. Which of the following lists of numbers is ordered from least to greatest?

   (A) $-\dfrac{2}{3}, -\dfrac{5}{8}, \dfrac{3}{8}, \dfrac{5}{12}$

   (B) $-\dfrac{2}{3}, -\dfrac{5}{8}, \dfrac{5}{12}, \dfrac{3}{8}$

   (C) $-\dfrac{5}{8}, -\dfrac{2}{3}, \dfrac{3}{8}, \dfrac{5}{12}$

   (D) $-\dfrac{5}{8}, -\dfrac{2}{3}, \dfrac{5}{12}, \dfrac{3}{8}$

3. What is the value of the expression $3a^2 - 5ab + 2b^2$ when $a = -1$ and $b = 6$?

   (A) 45

   (B) 57

   (C) 99

   (D) 105

4. If b represents the number of paintbrushes sold for \$12 each, and c represents the number of paint cans sold for \$45 each, which of the following represents the total cost of buying b paintbrushes and c paint cans in dollars?

   (A) $12b + 45c$

   (B) $57bc$

   (C) $45c - 12b$

   (D) $45b + 12c$

5. If $3x - 7 = 21$, then what does x equal?

   (A) 7

   (B) 28

   (C) $9\dfrac{1}{3}$

   (D) 25

6. Tymel and his brother are 4 years apart in age. If the sum of their ages is 42, what is the product of their ages?

   (A) 437

   (B) $1\dfrac{4}{19}$

   (C) 546

   (D) 336

7. The solution set of which of the following inequalities is graphed on the number line below?

   (A) $-4x + 1 > 2$

   (B) $-x + 8 < -4$

   (C) $-4x - 3 \le 13$

   (D) $-2x - 3 > 5$

8. If (x, y) is a solution to the following system of equations, what is the value of y?

   $$2x - y = 20$$
   $$5x + y = 8$$

   (A) -12

   (B) 0

   (C) 4

   (D) 7

9. $(16m^3 - 4m^2 + 5) - (12m^3 - 2m^2 + 5m) =$

(A) $4m^3 - 2m^2$

(B) $4m^3 + 2m^2 - 5m + 5$

(C) $4m^3 - 2m^2 - 5m + 5$

(D) $4m^3 - 2m^2 + 5m + 5$

10. $a^2 - 64x^2 =$

(A) $(a - 8x)^2$

(B) $(a - 8x)(a + 8x)$

(C) $(a - 32x)(a + 32x)$

(D) $8a(a - 8x)$

11. If the quotient $\dfrac{4x^2 + 16x}{2x} \div \dfrac{12x^2 - 8x}{6x}$ is simplified to lowest terms, which of the following is the *numerator* of the resulting expression?

(A) $48x$

(B) $x + 4$

(C) $3x + 12$

(D) $x + 12$

12. For which of the following functions are $x = 2$ and $x = -5$ both solutions?

(A) $x^2 + 3x - 10 = 0$

(B) $x^2 - 3x - 10 = 0$

(C) $x^2 - 7x - 10 = 0$

(D) $x^2 + 7x - 10 = 0$

COLLEGE-LEVEL MATHEMATICS

Directions: For each of the questions below, choose the best answer from the five choices given.

1. Consider the system of equations below.

$$x + y = 10$$
$$x + z = 19$$
$$y + z = 13$$

What is the value of z?

(A) 8

(B) 9

(C) 10

(D) 11

(E) 12

2. $\begin{vmatrix} 7 & -2 \\ -3 & 4 \end{vmatrix} =$

(A) -26

(B) -22

(C) -2

(D) 2

(E) 22

3. $a^3 - b^3 =$

(A) $(a + b)(a^2 - ab + b^2)$

(B) $(a + b)(a^2 - ab - b^2)$

(C) $(a - b)(a^2 + ab + b^2)$

(D) $(a - b)(a^2 + b^2)$

(E) $(a + b)(a^2 - b^2)$

4. What is the solution set to $\sqrt{|x + 2|} = 4$?

(A) $x = -14$ and $x = 18$

(B) $x = -14$ and $x = -18$

(C) $x = 14$ and $x = 18$

(D) $x = 14$ and $x = -18$

(E) $x = 14$ and $x = 22$

5. The area of the square below is $x^2 + 6x + 9$. Which of the expressions below gives the area of the circle inscribed inside the square?

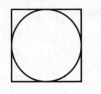

(A) $\pi\left(\dfrac{x+3}{2}\right)^2$

(B) $\dfrac{\pi(x^2 + 6x + 9)}{2}$

(C) $\pi(x^2 + 6x + 9)^2$

(D) $\pi(x + 3)$

(E) $\pi(x + 3)^2$

6. Consider the graph below. Which of the following choices is the equation of the line in slope-intercept form?

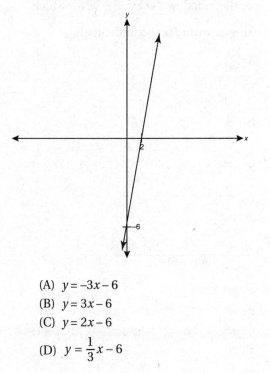

(A) $y = -3x - 6$

(B) $y = 3x - 6$

(C) $y = 2x - 6$

(D) $y = \dfrac{1}{3}x - 6$

(E) $y = -\dfrac{1}{3}x - 6$

7. A math club has seven members, and it must elect one member to be president and another member to be vice president. In how many different ways can the math club fill these two roles?

(A) 5,040
(B) 2,520
(C) 420
(D) 42
(E) 14

8. $6i \times (-11i) =$

(A) 66
(B) –66
(C) –66i
(D) 66i
(E) i

9. If $f(x) = \dfrac{x}{x+1}$, which of the following is true for any two values, a and b, where $a > b > 0$?

(A) $f(a) < f(b)$
(B) $f(a) > f(b)$
(C) $f(a) = f(b)$
(D) $f(a) = f(b + 1)$
(E) $f(a + 1) < f(b)$

10. What are the roots of $x^2 - 50 = 0$?

(A) $x = 50$ and $x = -50$
(B) $x = 5\sqrt{2}$ only
(C) $x = 2\sqrt{5}$ only
(D) $x = 5\sqrt{2}$ and $x = -5\sqrt{2}$
(E) $x = 2\sqrt{5}$ and $x = -2\sqrt{5}$

11. If $x, y, z \neq 0$, then what does $\frac{1}{x} + \frac{1}{y} + \frac{1}{z}$ equal?

(A) $\dfrac{xyz}{x + y + z}$

(B) $\dfrac{3}{xyz}$

(C) $\dfrac{x + y + z}{xyz}$

(D) $\dfrac{3}{x + y + z}$

(E) $\dfrac{xy + xz + yz}{xyz}$

12. $2\sqrt{18} + 6\sqrt{50} =$

(A) 72

(B) $24\sqrt{2}$

(C) $36\sqrt{2}$

(D) $8\sqrt{2}$

(E) $8\sqrt{68}$

13. Carla works two jobs—one as a teacher and another as an editor—for a total of 40 hours each week. She makes $24 per hour as a teacher, and she makes $30 per hour as an editor. If Carla made $1,020 last week, how many hours did she teach?

(A) 32

(B) 30

(C) 28

(D) 26

(E) 24

14. $-3x^2yz^3(7xz - y^2z^2) =$

(A) $-21x^2yz^3 + 3x^2y^2z^2$

(B) $-21x^3yz^4 - 3x^2y^3z^5$

(C) $-21x^3z^4 + 3y^3z^5$

(D) $-21x^3yz^4 + 3x^2y^3z^5$

(E) $-21xyz^8 - 3xyz^{10}$

15. Consider a circle that has a center at the point $(2, -5)$ and a radius of 7. Which of the following lines will intersect the circle at exactly one point?

(A) $y = 7$

(B) $y = -2$

(C) $y = -5$

(D) $y = -7$

(E) $y = -12$

16. $27^{\frac{4}{3}} =$

(A) 81

(B) 49

(C) 36

(D) 18

(E) 9

17. For the function $f(x) = \dfrac{x + 2}{3x - 2}$, for which value of x will $f(x)$ be undefined?

(A) -2

(B) $-\dfrac{3}{2}$

(C) $\dfrac{3}{2}$

(D) $-\dfrac{2}{3}$

(E) $\dfrac{2}{3}$

18. $\left(\dfrac{n^2(n^3 - 2n^2) - n(3n^3 - 6n^2)}{n(n - 5) + 6} \right)^2 =$

(A) n^6

(B) n^4

(C) $n^2 - 1$

(D) n

(E) 1

19. If θ is an acute angle and $\sin \theta = \dfrac{\sqrt{2}}{2}$, then what does $\cos \theta$ equal?

(A) 1

(B) 2

(C) $\dfrac{1}{2}$

(D) $\dfrac{\sqrt{3}}{2}$

(E) $\dfrac{\sqrt{2}}{2}$

20. $3\log_5 m + \log_5 n =$

(A) $\log_5 m^3 n$

(B) $\log_5 3mn$

(C) $\log_5 (mn)^3$

(D) $\log_5 (m+n)^3$

(E) $\log_5 (m^3 + n)$

ANSWER KEY
Practice Test 1

ARITHMETIC	ELEMENTARY ALGEBRA	COLLEGE-LEVEL MATHEMATICS
1. **B**	1. **A**	1. **D**
2. **B**	2. **A**	2. **E**
3. **A**	3. **D**	3. **C**
4. **D**	4. **A**	4. **D**
5. **C**	5. **C**	5. **A**
6. **A**	6. **A**	6. **B**
7. **B**	7. **D**	7. **D**
8. **D**	8. **A**	8. **A**
9. **C**	9. **C**	9. **B**
10. **A**	10. **B**	10. **D**
11. **C**	11. **C**	11. **E**
12. **A**	12. **A**	12. **C**
13. **C**		13. **B**
14. **D**		14. **D**
15. **B**		15. **E**
16. **C**		16. **A**
17. **D**		17. **E**
		18. **A**
		19. **E**
		20. **A**

Arithmetic

1. **(B)** We can first estimate that $\frac{1257}{100} = 12.57$, which would round to 13 because 12.57 is closer to 13 than 12. Dividing by 99 will give us a slightly larger quotient than dividing by 100, so $\frac{1257}{99}$ rounded to the nearest integer would be 13. To be completely sure, we can divide:

$$
\begin{array}{r}
12.6 \\
99\overline{)1257.0} \\
-99\downarrow \\
\overline{267} \\
-198\downarrow \\
\overline{690} \\
-594 \\
\overline{96}
\end{array}
$$

No matter what follows, 12.6 is enough to determine that $\frac{1257}{99}$ rounded to the nearest integer is 13.

2. **(B)** To evaluate $m = \frac{5}{4} \div \frac{1}{8}$, keep the first fraction, change $\div$ to $\times$, and flip the second fraction to its reciprocal.

$$m = \frac{5}{4} \div \frac{1}{8} = \frac{5}{4} \times \frac{8}{1} = \frac{40}{4} = 10$$

3. **(A)** To find the fraction of candies that are either yellow or green, we find $\frac{1}{3} + \frac{2}{5}$. To do so, we change the fractions to equivalent fractions with a common denominator.

$$\frac{1}{3} + \frac{2}{5} = \frac{5}{15} + \frac{6}{15} = \frac{11}{15}$$

Then, to find the fraction of candies that are orange, we subtract from 1, the whole. To do so, change 1 to $\frac{15}{15}$.

$$\frac{15}{15} - \frac{11}{15} = \frac{4}{15}$$

4. **(D)** To add mixed numbers, change the fractions to equivalent fractions with common denominators, and then add the whole numbers together and add the fractions together.

$$
\begin{array}{r}
7\frac{3}{8} \Rightarrow 7\frac{3}{8} \\
+5\frac{3}{4} \Rightarrow +5\frac{6}{8} \\
\hline
12\frac{9}{8} = 13\frac{1}{8}
\end{array}
$$

5. **(C)** To more easily compare the fractions, we change them to equivalent fractions with a common denominator of 60.

$$\frac{5}{4} = \frac{5 \times 15}{4 \times 15} = \frac{75}{60}$$

$$\frac{16}{12} = \frac{16 \times 5}{12 \times 5} = \frac{80}{60}$$

$$\frac{12}{10} = \frac{12 \times 6}{10 \times 6} = \frac{72}{60}$$

With equivalent denominators, we can order the fractions from least to greatest according to their new numerators.

$$\frac{72}{60}, \frac{75}{60}, \frac{80}{60}$$

This correlates with choice C, $\frac{12}{10}, \frac{5}{4}, \frac{16}{12}$.

6. **(A)** To divide by a fraction, keep the first term, change ÷ to ×, and flip the second fraction to its reciprocal.

$$4 \div \frac{8}{9} = 4 \times \frac{9}{8} = \frac{4}{1} \times \frac{9}{8} = \frac{4 \times 9}{1 \times 8} = \frac{36}{8}$$

The result is an improper fraction that does not look like any of the answer choices. We need to reduce the fraction to find an equivalent fraction.

$$\frac{36}{8} = \frac{36 \div 4}{8 \div 4} = \frac{9}{2}$$

7. **(B)**

$$
\begin{array}{r}
21.2 \\
\times\, 19.4 \\
\hline
848 \\
19080 \\
+\, 21200 \\
\hline
411.28
\end{array}
$$

411.28 is closest to 400.

8. **(D)** Using zeros as placeholders, make each number four decimal places long. Then compare the decimals as you normally would compare whole numbers, working left to right. Choice D, 0.062 < 0.26 < 0.602 < 0.62, is correct because 0.0620 < 0.2600 < 0.6020 < 0.6200.

9. **(C)**

$$
\begin{array}{r}
4.29 \\
\times\, 3.9 \\
\hline
3861 \\
+\, 12870 \\
\hline
16.731
\end{array}
$$

10. **(A)**

$$6\overline{)10.530} = 1.755$$

$$\begin{array}{r} 1.755 \\ 6\overline{)10.530} \\ \underline{-6}\downarrow \\ 45 \\ \underline{-42}\downarrow \\ 33 \\ \underline{-30}\downarrow \\ 30 \\ \underline{-30} \\ 0 \end{array}$$

11. **(C)**

$$\frac{\text{part}}{\text{whole}} = \frac{\text{percentage}}{100}$$

$$\frac{15}{x} = \frac{20}{100}$$

$$20x = (15)(100)$$

$$x = \frac{1500}{20} = 75$$

12. **(A)**

$$\frac{\text{part}}{\text{whole}} = \frac{\text{percentage}}{100}$$

$$\frac{x}{60} = \frac{55}{100}$$

$$100x = (55)(60)$$

$$x = \frac{3300}{100} = 33$$

13. **(C)**

$$\frac{\text{part}}{\text{whole}} = \frac{\text{percentage}}{100}$$

$$\frac{1.8}{P} = \frac{30}{100}$$

$$30P = (1.8)(100)$$

$$x = \frac{180}{30} = 6$$

14. **(D)** To change a decimal to a percentage, move the decimal point two places to the right, and add the percent symbol. 3.17 becomes 317%.

15. **(B)** Follow the order of operations.

$$\frac{15}{3} - 2 \times 5 + 16 \times 3 =$$

$$5 - 10 + 48 =$$

$$-5 + 48 = 43$$

16. **(C)** The sum of the measures of the three interior angles of a triangle is 180°. The sum of the two given interior angles is 52° + 35° = 87°. To find the third angle, subtract 87° from 180°.

$$180° - 87° = 93°$$

17. **(D)** The average of four numbers is 9. To find the average of four numbers, we divide their sum by 4. Only $\frac{36}{4} = 9$; therefore, the sum of all four numbers is 36. If we know that the sum of the first three numbers is 24, and the sum of all four numbers is 36, then the fourth number must be 12.

$$\frac{24 + 12}{4} = 9$$
$$\frac{36}{4} = 9$$

Elementary Algebra

1. **(A)** $\dfrac{-3}{5 - (-1)} = \dfrac{-3}{5 + 1} = \dfrac{-3}{6} = -\dfrac{1}{2}$

2. **(A)** Change each fraction to an equivalent fraction with a common denominator, and then put the list in least to greatest order according to their numerators.

$$-\frac{2}{3}, \quad -\frac{5}{8}, \quad \frac{3}{8}, \quad \frac{5}{12}$$
$$\Downarrow \qquad \Downarrow \qquad \Downarrow \qquad \Downarrow$$
$$-\frac{16}{24}, \quad -\frac{15}{24}, \quad \frac{9}{24}, \quad \frac{10}{24}$$

Choice A is in the correct order.

3. **(D)** Substitute –1 for a and 6 for b in the expression $3a^2 - 5ab + 2b^2$, and then evaluate.

$$3(-1)^2 - 5(-1)(6) + 2(6)^2 =$$
$$3(1) - 5(-6) + 2(36) =$$
$$3 + 30 + 72 = 105$$

4. **(A)** b paintbrushes for $12 each would cost $12b$ dollars. c paint cans for $45 each would cost $45c$ dollars. Altogether, the cost of buying b paintbrushes and c paint cans would cost $12b + 45c$.

5. **(C)**

$$3x - 7 = 21$$
$$3x - 7 + 7 = 21 + 7$$
$$3x = 28$$
$$x = \frac{28}{3} = 9\frac{1}{3}$$

6. **(A)** Let x equal Tymel's age. Let $x + 4$ equal Tymel's brother's age. The sum of their ages is 42.

$$x + x + 4 = 42$$
$$2x + 4 = 42$$
$$2x + 4 - 4 = 42 - 4$$
$$2x = 38$$
$$x = 19$$

Tymel is 19 years old, and his brother is $19 + 4 = 23$ years old. The product of their ages is $19 \times 23 = 437$.

7. **(D)** This number line shows all the values less than –4. Since the circle is empty, –4 is not included in the solution set. The solution set shown is $x < -4$. Choice D has the only inequality that is solved by the solution set $x < -4$.

$$-2x - 3 > 5$$
$$-2x - 3 + 3 > 5 + 3$$
$$-2x > 8$$
$$\frac{-2x}{-2} < \frac{8}{-2}$$
$$x < -4$$

Note that when we divided both sides of the inequality by –2, the inequality symbol switches from > to <.

8. **(A)** To solve this system of equations, we need to eliminate one of the variables. By adding both equations together, we eliminate y.

$$(2x - y) + (5x + y) = 20 + 8$$
$$2x + 5x - y + y = 28$$
$$\frac{7x}{7} = \frac{28}{7}$$
$$x = 4$$

To find the value of y, substitute 4 in for x into one, or both, of the original equations.

$$2(4) - y = 20$$
$$8 - y = 20$$
$$8 - 8 - y = 20 - 8$$
$$-y = 12$$
$$y = -12$$

9. **(C)** Remove the parentheses, and distribute the – sign to all the terms in the second polynomial.

$$(16m^3 - 4m^2 + 5) - (12m^3 - 2m^2 + 5m) =$$
$$16m^3 - 4m^2 + 5 - 12m^3 + 2m^2 - 5m =$$
$$16m^3 - 12m^3 - 4m^2 + 2m^2 - 5m + 5 = 4m^3 - 2m^2 - 5m + 5$$

10. **(B)** A difference of two perfect squares always fits the form $x^2 - a^2 = (x + a)(x - a)$.

$$a^2 - 64x^2 = (a - 8x)(a + 8x)$$

11. **(C)** To divide rational expressions, keep the first fraction, change the $\div$ to $\times$, and flip the second fraction to its reciprocal.

$$\frac{4x^2 + 16x}{2x} \div \frac{12x^2 - 8x}{6x}$$

$$\frac{4x^2 + 16x}{2x} \times \frac{6x}{12x^2 - 8x}$$

Factor each expression, and eliminate common factors by dividing them.

$$\frac{4x(x + 4)}{2x} \times \frac{6x}{4x(3x - 2)} =$$

$$\frac{\overset{1}{\cancel{4}}\,\overset{1}{\cancel{x}}(x + 4)}{\underset{1}{\cancel{2}}\,\underset{1}{\cancel{x}}} \times \frac{\overset{3}{\cancel{6}}\,\overset{1}{\cancel{x}}}{\underset{1}{\cancel{4}}\,\underset{1}{\cancel{x}}(3x - 2)} =$$

$$\frac{3(x + 4)}{3x - 2} = \frac{3x + 12}{3x - 2}$$

The resulting expression has a numerator of $3x + 12$.

12. **(A)** After factoring and solving each expression, we find that $x^2 + 3x - 10 = 0$ can be solved both by $x = 2$ and $x = -5$.

$$x^2 + 3x - 10 = 0$$
$$(x + 5)(x - 2) = 0$$

$$x + 5 = 0 \qquad\qquad x - 2 = 0$$
$$x = -5 \qquad\qquad x = 2$$

Alternatively, we can separately substitute 2 for x and -5 for x in each equation to find which equation is solved by these values.

When $x = 2$, $x^2 + 3x - 10 = (2)^2 + 3(2) - 10 = 4 + 6 - 10 = 10 - 10 = 0$

When $x = -5$, $x^2 + 3x - 10 = (-5)^2 + 3(-5) - 10 = 25 - 15 - 10 = 10 - 10 = 0$

College-Level Mathematics

1. **(D)** To find z, solve the first equation in terms of x.

$$\begin{array}{r} x + y = 10 \\ -y \quad -y \\ \hline x = 10 - y \end{array}$$

Now substitute this into the second equation.

$$\begin{array}{l} x + z = 19 \\ (10 - y) + z = 19 \\ 10 + z - y = 19 \\ \underline{-10 \qquad\qquad -10} \\ z - y = 9 \end{array}$$

Finally, use the addition method to add this new equation to the last original equation. This will eliminate y. Then solve.

$$\begin{array}{r} z - y = \ \ 9 \\ + y + z = 13 \\ \hline \dfrac{2z}{2} = \dfrac{22}{2} \\ z = 11 \end{array}$$

2. **(E)** $\begin{vmatrix} 7 & -2 \\ -3 & 4 \end{vmatrix} = (7)(4) - (-2)(-3) = 28 - 6 = 22$

3. **(C)** Solve this problem by working backwards. When multiplied, only choice C will produce $a^3 - b^3$.

$$(a - b)(a^2 + ab + b^2) = a^3 + a^2b + ab^2 - a^2b - ab^2 - b^3$$

Notice that all the middle terms between a^3 and b^3 cancel out.

$$a^3 + a^2b + ab^2 - a^2b - ab^2 - b^3 = a^3 - b^3$$

4. **(D)** Square both sides of the equation.

$$\left(\sqrt{|x + 2|}\right)^2 = 4^2$$
$$|x + 2| = 16$$

Set up two cases for the absolute value equation, and solve each.

$$\begin{array}{ll} \begin{array}{r} x + 2 = 16 \\ -2 \ -2 \\ \hline x = 14 \end{array} & \qquad \begin{array}{r} x + 2 = -16 \\ -2 \quad -2 \\ \hline x = -18 \end{array} \end{array}$$

Verify both solutions by plugging them into the original equation.

$$\sqrt{|x+2|} = 4 \qquad\qquad \sqrt{|x+2|} = 4$$
$$\sqrt{|14+2|} = 4 \qquad\qquad \sqrt{|-18+2|} = 4$$
$$\sqrt{|16|} = 4 \qquad\qquad \sqrt{|-16|} = 4$$
$$\sqrt{16} = 4 \qquad\qquad \sqrt{16} = 4$$
$$4 = 4 \qquad\qquad 4 = 4$$

Both solutions check out.

5. **(A)** If the area of the square is $x^2 + 6x + 9$, then the length of one side of the square can be found by taking the square root $\sqrt{x^2 + 6x + 9} = x + 3$. This means that the diameter of the inscribed circle is also $x + 3$. In order to find the area of the circle, $A = \pi r^2$, we need to know the radius. Since the radius is half of the diameter, $x + 3$, the radius is $r = \dfrac{x+3}{2}$. Plugging this into the formula for the area of a circle, we have

$$A = \pi r^2$$
$$A = \pi\left(\frac{x+3}{2}\right)^2$$

6. **(B)** The slope of the line is the ratio of rise to run. From the y-intercept $b = -6$, the line rises 6 and runs 2. Therefore, the slope of the line, m, is $\dfrac{6}{2} = 3$. The equation of the line is $y = 3x - 6$.

7. **(D)** There are 7 ways that the club can choose its president. After the president is chosen from the club members, there are 6 ways that it can choose its vice president. Therefore, there are $7 \times 6 = 42$ ways to choose the president and vice president. This is also a permutation, where $n = 7$ and $r = 2$.

$$_nP_r = \frac{n!}{(n-r)!}$$
$$_7P_2 = \frac{7!}{(7-2)!}$$
$$_7P_2 = \frac{7!}{5!}$$
$$_7P_2 = \frac{7 \times 6 \times 5 \times 4 \times 3 \times 2 \times 1}{5 \times 4 \times 3 \times 2 \times 1}$$
$$_7P_2 = 7 \times 6 = 42$$

8. **(A)** Multiply the coefficients of i, and then multiply $i \times i$.

$$6i \times (-11i) = -66i^2$$

Remember that $i^2 = -1$. Therefore,

$$6i \times (-11i) = -66i^2 = -66(-1) = 66$$

9. **(B)** The domain of this problem is all positive numbers. Use an example case where $a = 2$ and $b = 1$.

$$f(x) = \frac{x}{x+1} \qquad\qquad f(x) = \frac{x}{x+1}$$

$$f(2) = \frac{2}{2+1} \qquad\qquad f(1) = \frac{1}{1+1}$$

$$f(2) = \frac{2}{3} \qquad\qquad f(1) = \frac{1}{2}$$

We see that $f(a) > f(b)$. This will be true for all other positive numbers a and b. In fact, as x gets closer to infinity, the value of $f(x)$ gets infinitely closer and closer to 1.

10. **(D)** We know that this equation is a parabola that has a y-intercept of -50 and opens upward. It will have two real zeros or roots. To find these roots, isolate the x^2.

$$
\begin{array}{r}
x^2 - 50 = 0 \\
\underline{+50 \quad +50} \\
x^2 = 50
\end{array}
$$

Next, find the square root of both sides. Remember that the square root of a whole number can be either positive or negative.

$$\sqrt{x^2} = \sqrt{50}$$

$$x = \pm\sqrt{50}$$

$$x = \pm 5\sqrt{2}$$

The roots are $x = 5\sqrt{2}$ and $x = -5\sqrt{2}$.

11. **(E)** In order to add these rational expressions, we need them all to have the same denominator. The common denominator will be the product of the three denominators, or xyz.

$$\frac{1}{x} + \frac{1}{y} + \frac{1}{z} = \left[\left(\frac{yz}{yz}\right)\left(\frac{1}{x}\right)\right] + \left[\left(\frac{xz}{xz}\right)\left(\frac{1}{y}\right)\right] + \left[\left(\frac{xy}{xy}\right)\left(\frac{1}{z}\right)\right] = \frac{yz}{xyz} + \frac{xz}{xyz} + \frac{xy}{xyz} = \frac{xy + xz + yz}{xyz}$$

12. **(C)**

$$2\sqrt{18} + 6\sqrt{50} = \left(2\sqrt{9} \times \sqrt{2}\right) + \left(6\sqrt{25} \times \sqrt{2}\right) = \left(2 \times 3 \times \sqrt{2}\right) + \left(6 \times 5 \times \sqrt{2}\right) = 6\sqrt{2} + 30\sqrt{2} = 36\sqrt{2}$$

13. **(B)** Let x be the number of hours that Carla worked as a teacher last week, and let y be the number of hours that she worked as an editor last week. Therefore,

$$x + y = 40 \text{ hours}$$

We also know that she makes \$24 per hour as a teacher and \$30 per hour as an editor. Therefore, to calculate her total weekly salary, we can use the equation:

$$24x + 30y = 1{,}020 \text{ dollars}$$

Set up a system of equations.

$$x + y = 40$$

$$24x + 30y = 1020$$

Solve using substitution. Solve the first equation for y, and substitute the result into the second equation.

$$x + y = 40$$
$$\underline{-x \qquad -x}$$
$$y = 40 - x$$

$$24x + 30y = 1020$$
$$24x + 30(40 - x) = 1020$$
$$24x + 1200 - 30x = 1020$$
$$-6x + 1200 = 1020$$
$$\underline{-1200 \; -1200}$$
$$\frac{-6x}{-6} = \frac{-180}{-6}$$
$$x = 30$$

Carla worked 30 hours last week as a teacher.

14. **(D)**

$$-3x^2yz^3(7xz - y^2z^2) = \left[(-3)(7)x^{2+1}yz^{3+1}\right] - \left[-3x^2y^{1+2}z^{3+2}\right] = -21x^3yz^4 + 3x^2y^3z^5$$

15. **(E)** The line $y = -12$ is horizontal and will intersect the circle at the point $(2, -12)$. All the other horizontal lines in the answer choices will intersect the circle twice or not at all (choice A).

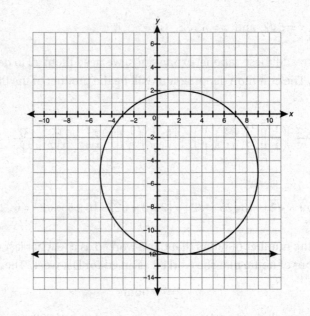

16. **(A)** $27^{\frac{4}{3}} = \left(\sqrt[3]{27}\right)^4 = 3^4 = 3 \times 3 \times 3 \times 3 = 81$

17. **(E)** The function $f(x) = \dfrac{x+2}{3x-2}$ will be undefined when the denominator is equal to 0. Division by zero is undefined. Set the expression in the denominator equal to zero, and then solve.

$$3x - 2 = 0$$
$$\underline{+2 \quad +2}$$
$$\frac{3x}{3} = \frac{2}{3}$$
$$x = \frac{2}{3}$$

The function will be undefined when $x = \dfrac{2}{3}$.

18. **(A)**

$$\left(\frac{n^2(n^3 - 2n^2) - n(3n^3 - 6n^2)}{n(n-5) + 6} \right)^2 =$$

$$\left(\frac{(n^5 - 2n^4) - (3n^4 - 6n^3)}{n^2 - 5n + 6} \right)^2 =$$

$$\left(\frac{n^5 - 5n^4 + 6n^3}{n^2 - 5n + 6} \right)^2 =$$

$$\left(\frac{n^3(n^2 - 5n + 6)}{n^2 - 5n + 6} \right)^2 =$$

$$\left(n^3 \right)^2 = n^6$$

19. **(E)** If $\sin\theta = \dfrac{\sqrt{2}}{2}$, then the side opposite θ equals $\sqrt{2}$, and the hypotenuse equals 2. To find the side adjacent θ, use the Pythagorean theorem.

$$\left(\sqrt{2} \right)^2 + b^2 = 2^2$$
$$2 + b^2 = 4$$
$$b^2 = 2$$
$$b = \sqrt{2}$$

$$\cos\theta = \frac{\text{side adjacent } \theta}{\text{hypotenuse}} = \frac{\sqrt{2}}{2}$$

20. **(A)** Using the power rule of logarithms, $c(\log_b a) = \log_b a^c$, $3\log_5 m$ becomes $\log_5 m^3$. Write the expression as $\log_5 m^3 + \log_5 n$. Then, by the product rule of logarithms, $\log_b a + \log_b c = \log_b (a \times c)$, we can change the expression $\log_5 m^3 + \log_5 n$ to $\log_5 m^3 n$.

ANSWER SHEET
Practice Test 2

ARITHMETIC

1. Ⓐ Ⓑ Ⓒ Ⓓ
2. Ⓐ Ⓑ Ⓒ Ⓓ
3. Ⓐ Ⓑ Ⓒ Ⓓ
4. Ⓐ Ⓑ Ⓒ Ⓓ
5. Ⓐ Ⓑ Ⓒ Ⓓ
6. Ⓐ Ⓑ Ⓒ Ⓓ
7. Ⓐ Ⓑ Ⓒ Ⓓ
8. Ⓐ Ⓑ Ⓒ Ⓓ
9. Ⓐ Ⓑ Ⓒ Ⓓ
10. Ⓐ Ⓑ Ⓒ Ⓓ
11. Ⓐ Ⓑ Ⓒ Ⓓ
12. Ⓐ Ⓑ Ⓒ Ⓓ
13. Ⓐ Ⓑ Ⓒ Ⓓ
14. Ⓐ Ⓑ Ⓒ Ⓓ
15. Ⓐ Ⓑ Ⓒ Ⓓ
16. Ⓐ Ⓑ Ⓒ Ⓓ
17. Ⓐ Ⓑ Ⓒ Ⓓ

ELEMENTARY ALGEBRA

1. Ⓐ Ⓑ Ⓒ Ⓓ
2. Ⓐ Ⓑ Ⓒ Ⓓ
3. Ⓐ Ⓑ Ⓒ Ⓓ
4. Ⓐ Ⓑ Ⓒ Ⓓ
5. Ⓐ Ⓑ Ⓒ Ⓓ
6. Ⓐ Ⓑ Ⓒ Ⓓ
7. Ⓐ Ⓑ Ⓒ Ⓓ
8. Ⓐ Ⓑ Ⓒ Ⓓ
9. Ⓐ Ⓑ Ⓒ Ⓓ
10. Ⓐ Ⓑ Ⓒ Ⓓ
11. Ⓐ Ⓑ Ⓒ Ⓓ
12. Ⓐ Ⓑ Ⓒ Ⓓ

COLLEGE-LEVEL MATHEMATICS

1. Ⓐ Ⓑ Ⓒ Ⓓ Ⓔ
2. Ⓐ Ⓑ Ⓒ Ⓓ Ⓔ
3. Ⓐ Ⓑ Ⓒ Ⓓ Ⓔ
4. Ⓐ Ⓑ Ⓒ Ⓓ Ⓔ
5. Ⓐ Ⓑ Ⓒ Ⓓ Ⓔ
6. Ⓐ Ⓑ Ⓒ Ⓓ Ⓔ
7. Ⓐ Ⓑ Ⓒ Ⓓ Ⓔ
8. Ⓐ Ⓑ Ⓒ Ⓓ Ⓔ
9. Ⓐ Ⓑ Ⓒ Ⓓ Ⓔ
10. Ⓐ Ⓑ Ⓒ Ⓓ Ⓔ
11. Ⓐ Ⓑ Ⓒ Ⓓ Ⓔ
12. Ⓐ Ⓑ Ⓒ Ⓓ Ⓔ
13. Ⓐ Ⓑ Ⓒ Ⓓ Ⓔ
14. Ⓐ Ⓑ Ⓒ Ⓓ Ⓔ
15. Ⓐ Ⓑ Ⓒ Ⓓ Ⓔ
16. Ⓐ Ⓑ Ⓒ Ⓓ Ⓔ
17. Ⓐ Ⓑ Ⓒ Ⓓ Ⓔ
18. Ⓐ Ⓑ Ⓒ Ⓓ Ⓔ
19. Ⓐ Ⓑ Ⓒ Ⓓ Ⓔ
20. Ⓐ Ⓑ Ⓒ Ⓓ Ⓔ

There are three separately scored ACCUPLACER Math tests: Arithmetic, Elementary Algebra, and College-Level Mathematics. Practice Test 2 is broken into the same three tests. You will notice that the question numbers start at 1 for each of the three test sections. You will have completed Practice Test 2 when you have answered all of the test questions for all three sections.

ARITHMETIC

> **Directions:** For each of the questions below, choose the best answer from the four choices given.

1. What is $\dfrac{2045}{101}$ rounded to the nearest integer?

   (A) 20
   (B) 21
   (C) 22
   (D) 23

2. Miho used 12 ounces of soy sauce to make a dish. This was $\dfrac{2}{3}$ of the full bottle. How many ounces of soy sauce were in the bottle when it was full?

   (A) 8
   (B) 18
   (C) 24
   (D) 36

3. The third place finisher in a local race was awarded $\dfrac{1}{6}$ of the prize money, and the second place finisher was awarded $\dfrac{1}{4}$ of the prize money. The first place finisher was awarded the remaining prize money. What fraction of the total prize money was the first place finisher awarded?

   (A) $\dfrac{4}{5}$
   (B) $\dfrac{9}{10}$
   (C) $\dfrac{5}{12}$
   (D) $\dfrac{7}{12}$

4. Of the following choices, which is closest to $\dfrac{15}{7} \div 2$?

   (A) 0
   (B) $\dfrac{1}{2}$
   (C) 1
   (D) $1\dfrac{1}{2}$

5. Of the registered voters in Sullivan County, 45% are registered Democrats and 40% are registered Republicans. The remaining registered voters are considered Independent. What fraction of the registered voters in Sullivan County are considered Independent?

   (A) $\dfrac{3}{20}$
   (B) $\dfrac{17}{20}$
   (C) $\dfrac{1}{15}$
   (D) $\dfrac{1}{6}$

6. $5\dfrac{1}{7} \div \dfrac{12}{7} =$

   (A) 2
   (B) 3
   (C) $5\dfrac{1}{12}$
   (D) $6\dfrac{6}{7}$

7. From a full 10-ounce bottle, a chemist used 3.24 ounces of hydrochloric acid. How many ounces of the fluid are left in the bottle?

(A) 3.14
(B) 6.76
(C) 7.24
(D) 7.86

8. $2.5 - 1.14 =$

(A) 0.64
(B) 1.11
(C) 1.36
(D) 1.44

9. $8.04 \times 6.5 =$

(A) 52.26
(B) 52.31
(C) 52.86
(D) 52.91

10. $40.05 \times 10^{-2} =$

(A) 4,005
(B) −400.5
(C) 0.4005
(D) 4.005

11. What is 22% of 44?

(A) 0.50
(B) 9.68
(C) 17.6
(D) 200

12. $\frac{1}{10}$ is what percent of 40?

(A) 0.25%
(B) 4%
(C) 25%
(D) 400%

13. Out of 50 customers who called their credit card company, only 60% completed a customer satisfaction survey. If all of the next 50 customers complete the survey, what percent of all of the callers will have completed the survey?

(A) 110%
(B) 90%
(C) 85%
(D) 80%

14. Which of the following is the greatest?

(A) 40% of 60
(B) 60% of 40
(C) 30% of 70
(D) 50% of 50

15. $\sqrt{200}$ is between which two whole numbers?

(A) 14 and 15
(B) 15 and 16
(C) 16 and 17
(D) 17 and 18

16. Scott broke a rectangular glass window pane that measured $2\frac{1}{4}$ feet wide and 4 feet high. How many square feet of glass is needed to replace the window pane?

(A) $8\frac{1}{2}$
(B) 9
(C) $9\frac{1}{2}$
(D) 10

17. In order to stay on the football team, Greg needs to maintain an 80 grade point average (GPA) in science class. On his first three tests, his scores were 71, 86, and 77. What is the lowest score he can receive on his fourth test to achieve at least an 80 GPA?

(A) 78
(B) 80
(C) 82
(D) 86

ELEMENTARY ALGEBRA

> **Directions:** For each of the questions below, choose the best answer from the four choices given.

1. $\dfrac{-12-(-3)}{-15} =$

   (A) $\dfrac{3-12}{15}$

   (B) $\dfrac{-3-12}{-15}$

   (C) $\dfrac{12-3}{15}$

   (D) $\dfrac{-3-12}{15}$

2. $|-6+2| =$

   (A) $|6| \times |2|$
   (B) $|6| + |2|$
   (C) $|-6| + |2|$
   (D) $|-6| - |-2|$

3. If $x = 3$, what is the value of $(1-x)^5$?

   (A) 243
   (B) −243
   (C) 32
   (D) −32

4. In the figure below, a rectangle that measures a units long and b units wide is inscribed within a larger rectangle that measures A units long and B units wide. Which of the following choices represents the area of the shaded region in square units?

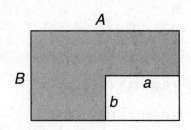

   (A) $Aa - Bb$
   (B) $AB - ab$
   (C) $(A-a)(B-b)$
   (D) $2A + 2B - (2a + 2b)$

5. If $4n - 3(n-4) = 17$, then what does n equal?

   (A) 5
   (B) 7
   (C) 21
   (D) 29

6. Which of the following expressions is equivalent to $-\dfrac{1}{2}x + 6 \le 16$?

   (A) $x \ge -20$
   (B) $x \le -20$
   (C) $x \ge -5$
   (D) $x \le -5$

7. How many solutions (x, y) are there to the following system of equations?

   $$3x + y = 4$$
   $$6x + 2y = 8$$

   (A) 0
   (B) 1
   (C) 2
   (D) More than 2

8. $(2a - 5b)^2 =$

   (A) $4a^2 + 25b^2$
   (B) $2a^2 - 5b^2$
   (C) $4a^2 - 10ab + 25b^2$
   (D) $4a^2 - 20ab + 25b^2$

9. Which of the following choices is a factor of both $x^2 - 3x - 4$ and $x^2 - 9x + 20$?

(A) $x + 4$
(B) $x - 5$
(C) $x - 4$
(D) $x + 1$

10. If $x \neq -3$ or 3, then what does $\dfrac{x^2 + 8x + 15}{x^2 - 9}$ equal?

(A) $\dfrac{x - 5}{x + 3}$

(B) $\dfrac{x + 5}{x - 3}$

(C) $\dfrac{8x + 15}{-9}$

(D) $\dfrac{23}{9}$

11. $\dfrac{\frac{a}{2b}}{\frac{3b}{4a}} =$

(A) $\dfrac{2a^2}{3b^2}$

(B) $\dfrac{6a}{b}$

(C) $\dfrac{6a^2}{b^2}$

(D) $\dfrac{27a}{8b}$

12. A rectangular rug has an area of 64 square feet. If the width of the rug is 12 feet less than the length, what is the perimeter of the rug in feet?

(A) 32
(B) 40
(C) 44
(D) 68

COLLEGE-LEVEL MATHEMATICS

Directions: For each of the questions below, choose the best answer from the five choices given.

1. Consider a circle centered at the point $(0, -3)$ with a radius of 4. How many points of intersection will it have with the parabola $y = x^2 + 1$?

(A) 0
(B) 1
(C) 2
(D) 3
(E) More than 3

2. $4^{\frac{5}{2}} - 4^{\frac{1}{2}} =$

(A) 62
(B) 30
(C) 16
(D) 8
(E) 4

3. Siobhan has 7 books on her shelf, and she randomly selects 4 to take with her on vacation. In how many different ways can she choose the books that she takes with her?

(A) 14
(B) 28
(C) 35
(D) 70
(E) 210

4. $(2x^2 + 8x) - (x + 4) =$

(A) $(2x - 4)(x + 1)$
(B) $(2x + 4)(x - 1)$
(C) $(2x + 1)(x + 4)$
(D) $(2x - 1)(x + 4)$
(E) $(2x - 1)(x - 4)$

5. $i^{11} =$

(A) -1

(B) 1

(C) $-i$

(D) i

(E) $11i$

6. At a school play, the cost of a student ticket is \$6, and the price of a ticket sold to the general public is \$12. In total, 300 tickets were sold. The total revenue from ticket sales was \$2,550. How many tickets were sold to the general public?

(A) 80

(B) 95

(C) 110

(D) 125

(E) 140

7. Let $f(n) = 16n^2 - 3$ and $g(n) = \dfrac{1}{n}$.

What does $f \circ g(4)$ equal?

(A) -2

(B) -1

(C) 0

(D) 1

(E) 2

8. What is the solution set to the equation $\sqrt{x+1} + 8 = 4$?

(A) $x = 15$

(B) $x = -15$

(C) $x = 15$ and $x = -15$

(D) $x = 15$ and $x = -17$

(E) No real solutions exist.

9. What are the roots of $3x^2 + 1 = 8x$?

(A) $\dfrac{4 \pm \sqrt{13}}{3}$

(B) $8 \pm 2\sqrt{13}$

(C) $\dfrac{4 \pm 2\sqrt{13}}{3}$

(D) $\dfrac{8 \pm 2\sqrt{13}}{3}$

(E) $\dfrac{4 \pm 4\sqrt{13}}{3}$

10. For $n > 0$, $\dfrac{n}{n+4} + \dfrac{2n+2}{n^2+5n+4} =$

(A) $\dfrac{3n+2}{n^2+6n+8}$

(B) $\dfrac{3n+2}{n+4}$

(C) $\dfrac{3n+2}{n+1}$

(D) $\dfrac{n^2+3n+3}{n^2+5n+4}$

(E) $\dfrac{n+2}{n+4}$

11. Find the volume of a cube that has sides with a length of $2 + \sqrt{3}$.

(A) $7 + 4\sqrt{3}$

(B) $8 + 3\sqrt{3}$

(C) $11 + 12\sqrt{3}$

(D) $26 + 15\sqrt{3}$

(E) $72\sqrt{3}$

12. The trinomial $x^3 + 4x^2 - 8$ has only one positive real zero, r. Which of the following is true of this zero?

(A) $0 < r < 1$

(B) $1 < r < 2$

(C) $2 < r < 3$

(D) $3 < r < 4$

(E) $4 < r < 5$

13. If $h(x) = \frac{2}{3}x - 4$, then what does $h^{-1}(x)$ equal?

(A) $h^{-1}(x) = \frac{4(x-3)}{2}$

(B) $h^{-1}(x) = -\frac{2}{3}x + 4$

(C) $h^{-1}(x) = \frac{2}{4(x-3)}$

(D) $h^{-1}(x) = \frac{3(x-2)}{4}$

(E) $h^{-1}(x) = \frac{3(x+4)}{2}$

14. Consider the system of equations below.

$$2a + 2b + 2c = 4$$
$$a + c = 7$$
$$a + b = -1$$

What is the value of a?

(A) -5

(B) 2

(C) 3

(D) 4

(E) 5

15. In the form $y = mx + b$, which of the following is the equation of the line below?

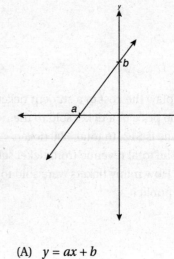

(A) $y = ax + b$

(B) $y = bx + b$

(C) $y = \frac{b}{a}x + b$

(D) $y = \frac{a}{b}x + b$

(E) $y = \frac{b-a}{a-b}x + b$

16. If $6x^2 + x - 2 = 0$, then what does $\left(x + \frac{1}{12}\right)^2$ equal?

(A) $\frac{25}{144}$

(B) $\frac{49}{144}$

(C) $\frac{81}{144}$

(D) $\frac{25}{36}$

(E) $\frac{49}{36}$

PRACTICE TEST 2

17. $\left(x^{\frac{2}{5}}\right)\left(x^{\frac{3}{2}}\right) =$

(A) $\sqrt[5]{x^3}$

(B) $\sqrt[3]{x^5}$

(C) $\sqrt[10]{x^{19}}$

(D) $\sqrt[7]{x^5}$

(E) $\sqrt[15]{x^4}$

18. If $x \geq 0$, $f(x) = \sqrt{x}$, and $g(x) = x^5$, then what does $f \circ g(a^4)$ equal?

(A) $a^4\sqrt{a}$

(B) $4a\sqrt{a}$

(C) $a^9\sqrt{a}$

(D) a^{10}

(E) a^5

19. The figure below shows a portion of which graph?

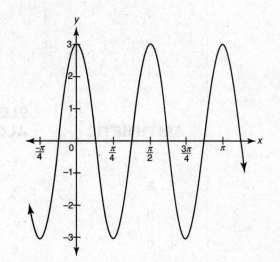

(A) $y = 3 \cos 4x$

(B) $y = 3 \cos \frac{\pi}{2} x$

(C) $y = 3 \cos \frac{\pi}{4} x$

(D) $y = 4 \cos 3 x$

(E) $y = \frac{\pi}{2} \cos x$

20. If $\log_{10} x = 0$, then what does x equal?

(A) 0

(B) $\frac{1}{10}$

(C) 1

(D) 10

(E) 100

ANSWER KEY
Practice Test 2

ARITHMETIC

1. **A**
2. **B**
3. **D**
4. **C**
5. **A**
6. **B**
7. **B**
8. **C**
9. **A**
10. **C**
11. **B**
12. **A**
13. **D**
14. **D**
15. **A**
16. **B**
17. **D**

ELEMENTARY ALGEBRA

1. **C**
2. **D**
3. **D**
4. **B**
5. **A**
6. **A**
7. **D**
8. **D**
9. **C**
10. **B**
11. **A**
12. **B**

COLLEGE-LEVEL MATHEMATICS

1. **B**
2. **B**
3. **C**
4. **D**
5. **C**
6. **D**
7. **A**
8. **E**
9. **A**
10. **E**
11. **D**
12. **B**
13. **E**
14. **D**
15. **C**
16. **B**
17. **C**
18. **D**
19. **A**
20. **C**

Arithmetic

1. **(A)** To find the quotient of $\frac{2045}{101}$, we divide 101 into 2,045.

$$
\begin{array}{r}
20.2 \\
101\overline{)2045.0} \\
\underline{-202}\downarrow \\
25 \\
\underline{-0}\downarrow \\
250 \\
\underline{-202} \\
48
\end{array}
$$

With 20.2, we have enough information to determine that $\frac{2045}{101}$ rounded to the nearest integer is 20.

2. **(B)** The 12 ounces that Miho used was $\frac{2}{3}$ of the full bottle, so we need to determine how much the full bottle held before the 12 ounces were used.

$$\frac{2}{3} = \frac{12}{x}$$

$$2x = (3)(12)$$

$$x = \frac{36}{2}$$

$$x = 18$$

3. **(D)** To figure out the total amount of money that was awarded to the third and second place finishers, we need to add $\frac{1}{6} + \frac{1}{4}$. To do so, we need to change the fractions to equivalent fractions with common denominators.

$$
\begin{aligned}
\frac{1}{6} &= \frac{2}{12} \\
+\frac{1}{4} &= \frac{3}{12} \\
\hline
&\frac{5}{12}
\end{aligned}
$$

The first place finisher was awarded the remaining prize money, so we need to subtract $\frac{5}{12}$ of the prize money from the whole prize money by changing 1 whole to $\frac{12}{12}$, and then subtracting.

$$\frac{12}{12} - \frac{5}{12} = \frac{7}{12}$$

4. **(C)** To divide a fraction by a whole number, first place the whole number over 1. Then, change the division problem to a multiplication problem by keeping the first fraction, changing the ÷ to ×, and flipping the second fraction to its reciprocal.

$$\frac{15}{7} \div 2 = \frac{15}{7} \div \frac{2}{1} = \frac{15}{7} \times \frac{1}{2} = \frac{15}{14}$$

$\frac{15}{14}$ is $\frac{1}{14}$ away from 1 and $\frac{6}{14}$ away from $1\frac{1}{2}$, so $\frac{15}{7} \div 2$ is closest to 1.

5. **(A)** The total percentage of registered voters who are either registered as Democrats or as Republicans is 45% + 40% = 85%. The total percentage of registered voters who are considered Independent is 100% – 85% = 15%. To express 15% as a fraction, we change 15% to $\frac{15}{100}$ which equals $\frac{3}{20}$.

6. **(B)** To divide a mixed number by a fraction, first change the mixed number to an improper fraction. Then, change the division problem to a multiplication problem by keeping the first fraction, changing the ÷ to ×, and flipping the second fraction to its reciprocal.

$$5\frac{1}{7} \div \frac{12}{7} = \frac{36}{7} \times \frac{7}{12} = \frac{36 \times 7}{7 \times 12}$$

At this point, we could multiply the numerators and multiply the denominators, but we can also reduce by common factors found in both the numerator and the denominator.

$$\frac{36 \times 7}{7 \times 12} = \frac{\overset{3}{\cancel{36}} \times \overset{1}{\cancel{7}}}{\underset{1}{\cancel{7}} \times \underset{1}{\cancel{12}}} = \frac{3}{1} = 3$$

7. **(B)** To subtract decimals, be sure to line up the decimal points.

$$\begin{array}{r} 1\,0.0\,0 \\ -\ 3.2\,4 \\ \hline 6.7\,6 \end{array}$$

8. **(C)** To subtract decimals, be sure to line up the decimal points.

$$\begin{array}{r} 2.\overset{4\ 1}{\cancel{3}}0 \\ -1.1\,4 \\ \hline 1.3\,6 \end{array}$$

9. **(A)**

$$\begin{array}{r} 8.04 \\ \times\ 6.5 \\ \hline 4020 \\ +\ 48240 \\ \hline 52.260 \end{array}$$

10. **(C)** To multiply by 10^{-2}, move the decimal point two places to the left.

$$40.05 \times 10^{-2} = 0.4005$$

11. **(B)**

$$\frac{x}{44} = \frac{22}{100}$$

$$100x = (44)(22)$$

$$x = \frac{968}{100}$$

$$x = 9.68$$

12. **(A)**

$$\frac{x}{100} = \frac{\frac{1}{10}}{40}$$

$$40x = (100)\left(\frac{1}{10}\right)$$

$$40x = 10$$

$$x = \frac{10}{40}$$

$$x = 0.25$$

$\frac{1}{10}$ is 0.25% of 40.

13. **(D)** First we need to find 60% of 50 to know how many of the original 50 customers completed the survey.

$$60\% \text{ of } 50 = 0.60 \times 50 = 30 \text{ customers}$$

If all of the next 50 customers complete their survey, then $30 + 50 = 80$ customers will have completed the survey in total out of 100 customers. 80 out of 100 customers is 80%.

14. **(D)** We can evaluate each answer choice by changing the percentage to a decimal, and then multiplying.

$$40\% \text{ of } 60 = 0.40 \times 60 = 24$$
$$60\% \text{ of } 40 = 0.60 \times 40 = 24$$
$$30\% \text{ of } 70 = 0.30 \times 70 = 21$$
$$50\% \text{ of } 50 = 0.50 \times 50 = 25$$

Therefore, choice D is the greatest.

15. **(A)** We can determine the following.

$$14^2 = 196, \text{ so } \sqrt{196} = 14$$

$$15^2 = 225, \text{ so } \sqrt{225} = 15$$

$\sqrt{196} < \sqrt{200} < \sqrt{225}$; therefore $\sqrt{200}$ is between 14 and 15.

16. **(B)** To find the measure of the window pane in square feet, we need to find the area by multiplying the width of the pane by the height of the pane.

$$2\frac{1}{4} \times 4 = \frac{9}{4} \times \frac{4}{1} = \frac{36}{4} = 9$$

17. **(D)** To receive at least an 80 GPA over four tests, the following must be true:

$$\frac{\text{The sum of the test scores}}{4} \geq 80$$

Only $\frac{320}{4} = 80$, so the sum of the test scores must be at least 320 points altogether. Knowing that he has $71 + 86 + 77 = 234$ points so far, we can subtract $320 - 234 = 86$ to find the lowest score that Greg can receive on the fourth test to have at least an 80 GPA.

Elementary Algebra

1. **(C)** Evaluate the given expression.

$$\frac{-12 - (-3)}{-15} = \frac{-12 + 3}{-15} = \frac{-9}{-15} = \frac{3}{5}$$

The only expression presented that is also equal to $\frac{3}{5}$ is $\frac{12 - 3}{15}$.

$$\frac{12 - 3}{15} = \frac{9}{15} = \frac{3}{5}$$

2. **(D)** Evaluate the given expression.

$$|-6 + 2| = |-4| = 4$$

The only expression presented that is also equal to 4 is $|-6| - |-2|$.

$$|-6| - |-2| = 6 - 2 = 4$$

3. **(D)** Substitute 3 for x into the expression $(1 - x)^5$.

$$(1 - 3)^5 = (-2)^5 = -32$$

4. **(B)** To find the area of the shaded region, subtract the area of the smaller rectangle from the area of the larger rectangle. The area of the shaded region equals $AB - ab$.

5. **(A)** Distribute the 3 outside the parentheses to both terms inside the parentheses, and then solve for n.

$$4n - 3(n - 4) = 17$$
$$4n - 3n + 12 = 17$$
$$n + 12 = 17$$
$$n + 12 - 12 = 17 - 12$$
$$n = 5$$

6. **(A)**

$$-\frac{1}{2}x + 6 \le 16$$

$$-\frac{1}{2}x + 6 - 6 \le 16 - 6$$

$$-\frac{1}{2}x \le 10$$

$$\left(-\frac{2}{1}\right)\left(-\frac{1}{2}x\right) \ge (10)\left(-\frac{2}{1}\right)$$

$$x \ge -20$$

Note that, when we multiplied both sides of the inequality by $-\frac{2}{1}$, the inequality symbol switched from $\le$ to $\ge$.

7. **(D)** By multiplying $3x + y = 4$ by 2, we find that both equations are equivalent.

$$2(3x + y) = 2(4)$$
$$6x + 2y = 8$$

Since both equations are equivalent, every solution to $3x + y = 4$ will also solve $6x + 2y = 8$. There are an infinite number of solutions to this system of equations. Therefore, the correct answer is choice D.

8. **(D)**

$$(2a - 5b)^2 =$$
$$(2a - 5b)(2a - 5b) =$$
$$2a(2a) + 2a(-5b) - 5b(2a) - 5b(-5b) =$$
$$4a^2 - 10ab - 10ab + 25b^2 = 4a^2 - 20ab + 25b^2$$

9. **(C)**

$$x^2 - 3x - 4 = (x + 1)(x - 4)$$
$$x^2 - 9x + 20 = (x - 5)(x - 4)$$

Both expressions have $(x - 4)$ as a factor.

10. **(B)** Factor the numerator and the denominator, and then eliminate common factors by division.

$$\frac{x^2 + 8x + 15}{x^2 - 9}$$

$$\frac{(x + 3)(x + 5)}{(x + 3)(x - 3)}$$

$$\frac{\cancel{(x + 3)}^{1}(x + 5)}{\cancel{(x + 3)}_{1}(x - 3)}$$

$$\frac{x + 5}{x - 3}$$

11. **(A)** Write the problem as a numerator divided by a denominator problem.

$$\frac{\frac{a}{2b}}{\frac{3b}{4a}} = \frac{a}{2b} \div \frac{3b}{4a}$$

Then, change the problem to a multiplication problem by keeping the first fraction, changing the $\div$ to $\times$, and flipping the second fraction to its reciprocal.

$$\frac{a}{2b} \div \frac{3b}{4a} = \frac{a}{2b} \times \frac{4a}{3b}$$

Finally, evaluate and reduce.

$$\frac{a}{\underset{1}{\cancel{2}b}} \times \frac{\overset{2}{\cancel{4}a}}{3b} = \frac{2a^2}{3b^2}$$

12. **(B)**

Let x = the length of the rug.
Let $x - 12$ = the width of the rug.
Write an equation for the area of the rug.

(length of the rug) $\times$ (width of the rug) = area of the rug

$$x(x - 12) = 64$$
$$x^2 - 12x = 64$$
$$x^2 - 12x - 64 = 64 - 64$$
$$(x - 16)(x + 4) = 0$$

$$x - 16 = 0 \qquad\qquad x + 4 = 0$$
$$x = 16 \qquad\qquad\quad x = -4$$

The length of the rug cannot be negative, so the length of the rug must be 16 feet. The width of the rug is then $16 - 12 = 4$ feet. The perimeter of a rectangle can be obtained by finding the sum of the lengths of the four sides of the rectangle.

$$16 + 16 + 4 + 4 = 40 \text{ feet}$$

College-Level Mathematics

1. **(B)** The parabola $y = x^2 + 1$ has the y-axis as its axis of symmetry, and it intercepts the y-axis at the point $(0, 1)$. It is a parabola that opens upward, so the point $(0, 1)$ is its minimum. The circle has its center at $(0, -3)$. Since it has a radius of 4, we know that it will intercept the y-axis four units above the center: $(0, 1)$. This is the same point at which the parabola intercepts the y-axis. This is, therefore, the only point the two graphs have in common.

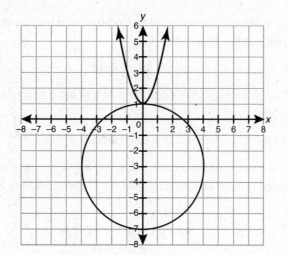

2. **(B)** $4^{\frac{5}{2}} - 4^{\frac{1}{2}} = \left(\sqrt[2]{4}\right)^5 - \left(\sqrt[2]{4}\right)^1 = 2^5 - 2^1 = 32 - 2 = 30$

3. **(C)** Siobhan is choosing a combination of four books. In other words, she is choosing 4 books from 7. The order in which she selects the books is irrelevant, so we use the following formula: $_nC_r = \dfrac{n!}{r!(n-r)!}$

$$_7C_4 = \frac{7!}{4!(7-4)!}$$
$$_7C_4 = \frac{7 \times 6 \times 5 \times \cancel{4!}}{\cancel{4!}(3!)}$$
$$_7C_4 = \frac{7 \times 6 \times 5}{3 \times 2 \times 1}$$
$$_7C_4 = \frac{210}{6} = 35$$

4. **(D)** Factor $2x$ out of the terms inside the first set of parentheses.

$$(2x^2 + 8x) - (x + 4) = 2x(x + 4) - (x + 4)$$

Now we can factor $(x + 4)$ out of $2x(x + 4)$ and out of $1(x + 4)$. Therefore:

$$2x(x + 4) - 1(x + 4) = (x + 4)(2x - 1)$$

5. **(C)** We know that $i^2 = -1$ and $i^4 = 1$. Therefore, we can rewrite i^{11} as $i^4 \times i^4 \times i^3$. Since $i^4 = 1$, we have

$$i^{11} = 1 \times i^3$$
$$i^{11} = i^3$$
$$i^{11} = i^2 \times i$$
$$i^{11} = -1 \times i$$
$$i^{11} = -i$$

6. **(D)** Let x be the number of tickets sold to students, and let y be the number of tickets sold to the general public. Therefore,

$$x + y = 300 \text{ tickets}$$

We also know that a student ticket costs \$6, and a ticket sold to the general public costs \$12. Therefore,

$$6x + 12y = 2,550 \text{ dollars}$$

Set up a system of equations, and solve using substitution. Since we want to know the value of y, we should solve the first equation for x, and substitute it into the second equation.

$$x + y = 300$$
$$\underline{-y \quad -y}$$
$$x = 300 - y$$

$$6x + 12y = 2550$$
$$6(300 - y) + 12y = 2550$$
$$1800 - 6y + 12y = 2550$$
$$1800 + 6y = 2550$$
$$\underline{-1800 \qquad\quad -1800}$$
$$\frac{6y}{6} = \frac{750}{6}$$
$$y = 125$$

There were 125 tickets sold to the general public.

7. **(A)** First, evaluate the function $g(4)$, and use it as the input for the function f.

$$g(n) = \frac{1}{n}$$
$$g(4) = \frac{1}{4}$$

Therefore,

$$f \circ g(4) = f\left(\frac{1}{4}\right)$$

$$f(n) = 16n^2 - 3$$

$$f\left(\frac{1}{4}\right) = 16\left(\frac{1}{4}\right)^2 - 3$$

$$f\left(\frac{1}{4}\right) = 16\left(\frac{1}{16}\right) - 3$$

$$f\left(\frac{1}{4}\right) = 1 - 3$$

$$f\left(\frac{1}{4}\right) = -2$$

Therefore, $f \circ g(4) = -2$.

8. **(E)** Isolate the radical part of the equation by subtracting 8 from both sides.

$$\sqrt{x+1} + 8 = 4$$
$$\underline{\phantom{\sqrt{x+1}}\ -8\ -8}$$
$$\sqrt{x+1} = -4$$

From here, we see that there can be no real solutions. No matter what value of x we try, a square root can never produce a negative number as its result.

9. **(A)** First, set the equation equal to zero.

$$3x^2 + 1 = 8x$$
$$\underline{\ -8x\ -8x\ }$$
$$3x^2 - 8x + 1 = 0$$

The equation is not factorable, so we have to use the quadratic formula. In this equation, $a = 3$, $b = -8$, and $c = 1$.

$$x = \frac{-b \pm \sqrt{b^2 - 4ac}}{2a}$$

$$x = \frac{-(-8) \pm \sqrt{(-8)^2 - 4(3)(1)}}{2(3)}$$

$$x = \frac{8 \pm \sqrt{64 - 12}}{6}$$

$$x = \frac{8 \pm \sqrt{52}}{6}$$

$$x = \frac{8 \pm 2\sqrt{13}}{6}$$

$$x = \frac{4 \pm \sqrt{13}}{3}$$

10. **(E)** The second term in the rational expression can be factored.

$$\frac{2n+2}{n^2+5n+4} = \frac{2(\cancel{n+1})}{(n+4)(\cancel{n+1})} = \frac{2}{n+4}$$

Add.

$$\frac{n}{n+4} + \frac{2}{n+4} = \frac{n+2}{n+4}$$

11. **(D)** To find the volume a cube, use $V = s^3$, where s is the length of the cube's side.

$$V = s^3$$
$$V = \left(2+\sqrt{3}\right)^3$$
$$V = \left(2+\sqrt{3}\right)\left(2+\sqrt{3}\right)\left(2+\sqrt{3}\right)$$
$$V = \left(4 + 2\sqrt{3} + 2\sqrt{3} + 3\right)\left(2+\sqrt{3}\right)$$
$$V = \left(7 + 4\sqrt{3}\right)\left(2+\sqrt{3}\right)$$
$$V = 14 + 7\sqrt{3} + 8\sqrt{3} + 4(3)$$
$$V = 14 + 15\sqrt{3} + 12$$
$$V = 26 + 15\sqrt{3}$$

12. **(B)** It's most efficient to work backwards from the answer choices. In the correct answer, choice B, the real zero is between 1 and 2. Notice that, when we use 1 for x, the result is less than zero, but when we use 2, the result is greater than zero. This means that, somewhere between $x = 1$ and $x = 2$, the curve crosses the x-axis.

$$x^3 + 4x^2 - 8 =$$ $$x^3 + 4x^2 - 8 =$$
$$1^3 + 4(1^2) - 8 =$$ $$2^3 + 4(2^2) - 8 =$$
$$1 + 4 - 8 = -3$$ $$8 + 16 - 8 =$$
$$24 - 8 = 16$$

13. **(E)** To find the inverse function, replace $h(x)$ with y. Then swap the variables.

$$h(x) = \frac{2}{3}x - 4$$
$$y = \frac{2}{3}x - 4$$
$$x = \frac{2}{3}y - 4$$

Solve for y.

$$x = \frac{2}{3}y - 4$$
$$\underline{+4 \qquad\qquad +4}$$
$$x + 4 = \frac{2}{3}y$$
$$\left(\frac{3}{2}\right)(x+4) = \left(\frac{2}{3}y\right)\left(\frac{3}{2}\right)$$
$$\frac{3(x+4)}{2} = y$$

14. **(D)** Multiply both sides of the second equation by –2, and then use the addition method to add the first and second equations together. This will eliminate a and c.

$$-2(a+c) = -2(7)$$
$$-2a - 2c = -14$$

$$
\begin{array}{r}
2a + 2b + 2c = 4 \\
\underline{+\,-2a -2c = -14} \\
\dfrac{2b}{2} = \dfrac{-10}{2} \\
b = -5
\end{array}
$$

Now, plug $b = -5$ into the third equation to find a.

$$
\begin{array}{r}
a + b = -1 \\
a + (-5) = -1 \\
a - 5 = -1 \\
\underline{+5 +5} \\
a = 4
\end{array}
$$

15. **(C)** We can see from the illustration that the y-intercept is b. If we start at point a, we can find the slope by measuring the ratio of rise to run. Notice that we would have to rise b units and move a units to the right. This means that the slope is $\dfrac{b}{a}$. Therefore, the equation of this line is $y = \dfrac{b}{a}x + b$.

16. **(B)** Find the solutions for x by factoring or using the quadratic formula.

$$6x^2 + x - 2 = 0$$
$$(2x - 1)(3x + 2) = 0$$

Setting both binomials equal to 0 and solving for x, we find that $x = \dfrac{1}{2}$ and $x = -\dfrac{2}{3}$.

When we substitute either of these values for x into the expression $\left(x + \dfrac{1}{12}\right)^2$, we get the same result.

$$\left(x + \frac{1}{12}\right)^2 = \left(\frac{1}{2} + \frac{1}{12}\right)^2 = \left(\frac{7}{12}\right)^2 = \frac{49}{144}$$

$$\left(x + \frac{1}{12}\right)^2 = \left(-\frac{2}{3} + \frac{1}{12}\right)^2 = \left(-\frac{7}{12}\right)^2 = \frac{49}{144}$$

17. **(C)** $\left(x^{\frac{2}{5}}\right)\left(x^{\frac{3}{2}}\right) = x^{\frac{2}{5} + \frac{3}{2}} = x^{\frac{4}{10} + \frac{15}{10}} = x^{\frac{19}{10}}$. The numerator of this rational exponent denotes the power to which x is raised, and the denominator represents the 10th root of x^{19}. Therefore, we write the expression as $\sqrt[10]{x^{19}}$.

18. **(D)** First, evaluate $g(a^4)$.

$$g(x) = x^5$$
$$g(a^4) = (a^4)^5$$
$$g(a^4) = a^{20}$$

Use a^{20} as the input for the function f.

$$f \circ g(a^4) = f(a^{20})$$
$$f(a^{20}) = \sqrt{a^{20}} = a^{10}$$

Therefore, $f \circ g(a^4) = a^{10}$.

19. **(A)** This graph shows a portion of a periodic function with an amplitude of 3 and a period of $\frac{\pi}{2}$. A cosine function of this type fits the form $y = A \cos Bx$, where A is the amplitude and the period $= \frac{2\pi}{B}$. In this case, $A = 3$. To determine B, we set the period $\frac{\pi}{2}$ equal to $\frac{2\pi}{B}$, and solve for B.

$$B\pi = 2(2\pi)$$
$$B\pi = 4\pi$$
$$B = 4$$

$$y = A \cos Bx$$
$$y = 3 \cos 4x$$

20. **(C)** Identify the base, power, and exponent that are found in $\log_{10} x = 0$ by following the form $\log_{\text{base}} (\text{power}) = \text{exponent}$.

$$\text{Base} = 10$$
$$\text{Power} = x$$
$$\text{Exponent} = 0$$

Change the expression from logarithmic to exponential form.

$$x = 10^0$$
$$x = 1$$

ANSWER SHEET
Practice Test 3

ARITHMETIC

1. Ⓐ Ⓑ Ⓒ Ⓓ
2. Ⓐ Ⓑ Ⓒ Ⓓ
3. Ⓐ Ⓑ Ⓒ Ⓓ
4. Ⓐ Ⓑ Ⓒ Ⓓ
5. Ⓐ Ⓑ Ⓒ Ⓓ
6. Ⓐ Ⓑ Ⓒ Ⓓ
7. Ⓐ Ⓑ Ⓒ Ⓓ
8. Ⓐ Ⓑ Ⓒ Ⓓ
9. Ⓐ Ⓑ Ⓒ Ⓓ
10. Ⓐ Ⓑ Ⓒ Ⓓ
11. Ⓐ Ⓑ Ⓒ Ⓓ
12. Ⓐ Ⓑ Ⓒ Ⓓ
13. Ⓐ Ⓑ Ⓒ Ⓓ
14. Ⓐ Ⓑ Ⓒ Ⓓ
15. Ⓐ Ⓑ Ⓒ Ⓓ
16. Ⓐ Ⓑ Ⓒ Ⓓ
17. Ⓐ Ⓑ Ⓒ Ⓓ

ELEMENTARY ALGEBRA

1. Ⓐ Ⓑ Ⓒ Ⓓ
2. Ⓐ Ⓑ Ⓒ Ⓓ
3. Ⓐ Ⓑ Ⓒ Ⓓ
4. Ⓐ Ⓑ Ⓒ Ⓓ
5. Ⓐ Ⓑ Ⓒ Ⓓ
6. Ⓐ Ⓑ Ⓒ Ⓓ
7. Ⓐ Ⓑ Ⓒ Ⓓ
8. Ⓐ Ⓑ Ⓒ Ⓓ
9. Ⓐ Ⓑ Ⓒ Ⓓ
10. Ⓐ Ⓑ Ⓒ Ⓓ
11. Ⓐ Ⓑ Ⓒ Ⓓ
12. Ⓐ Ⓑ Ⓒ Ⓓ

COLLEGE-LEVEL MATHEMATICS

1. Ⓐ Ⓑ Ⓒ Ⓓ Ⓔ
2. Ⓐ Ⓑ Ⓒ Ⓓ Ⓔ
3. Ⓐ Ⓑ Ⓒ Ⓓ Ⓔ
4. Ⓐ Ⓑ Ⓒ Ⓓ Ⓔ
5. Ⓐ Ⓑ Ⓒ Ⓓ Ⓔ
6. Ⓐ Ⓑ Ⓒ Ⓓ Ⓔ
7. Ⓐ Ⓑ Ⓒ Ⓓ Ⓔ
8. Ⓐ Ⓑ Ⓒ Ⓓ Ⓔ
9. Ⓐ Ⓑ Ⓒ Ⓓ Ⓔ
10. Ⓐ Ⓑ Ⓒ Ⓓ Ⓔ
11. Ⓐ Ⓑ Ⓒ Ⓓ Ⓔ
12. Ⓐ Ⓑ Ⓒ Ⓓ Ⓔ
13. Ⓐ Ⓑ Ⓒ Ⓓ Ⓔ
14. Ⓐ Ⓑ Ⓒ Ⓓ Ⓔ
15. Ⓐ Ⓑ Ⓒ Ⓓ Ⓔ
16. Ⓐ Ⓑ Ⓒ Ⓓ Ⓔ
17. Ⓐ Ⓑ Ⓒ Ⓓ Ⓔ
18. Ⓐ Ⓑ Ⓒ Ⓓ Ⓔ
19. Ⓐ Ⓑ Ⓒ Ⓓ Ⓔ
20. Ⓐ Ⓑ Ⓒ Ⓓ Ⓔ

There are three separately scored ACCUPLACER Math tests: Arithmetic, Elementary Algebra, and College-Level Mathematics. Practice Test 3 is broken into the same three tests. You will notice that the question numbers start at 1 for each of the three test sections. You will have completed Practice Test 3 when you have answered all of the test questions for all three sections.

ARITHMETIC

Directions: For each of the questions below, choose the best answer from the four choices given.

1. $\dfrac{3}{25} =$

   (A) 0.12
   (B) 0.325
   (C) 0.4
   (D) 3.25

2. A single serving of Red's breakfast cereal measures $\dfrac{2}{3}$ of a cup. How many servings can you get from 6 cups of this cereal?

   (A) 2
   (B) 4
   (C) 7
   (D) 9

3. $\dfrac{5}{8} - \dfrac{1}{3} =$

   (A) $\dfrac{4}{5}$

   (B) $\dfrac{4}{8}$

   (C) $\dfrac{7}{24}$

   (D) $\dfrac{14}{24}$

4. $\left(\dfrac{3}{5} - \dfrac{1}{4}\right) \times \left(\dfrac{5}{6} + \dfrac{7}{10}\right)$ is between which two numbers?

   (A) $\dfrac{1}{2}$ and 1

   (B) 1 and $\dfrac{3}{2}$

   (C) $\dfrac{3}{2}$ and 2

   (D) 2 and $\dfrac{5}{2}$

5. $4\dfrac{2}{3} \times 3\dfrac{4}{5} =$

   (A) $\dfrac{57}{60}$

   (B) $12\dfrac{8}{15}$

   (C) $16\dfrac{3}{5}$

   (D) $17\dfrac{11}{15}$

6. Which of the following is the least?

   (A) 0.302
   (B) 0.203
   (C) 0.23
   (D) 0.032

7. Points *A*, *B*, and *C* all lie on one number line. Point *B* lies exactly halfway between Points *A* and *C*. Which of the following is the coordinate of Point *A* if Point *B* is located at 0.2 and Point *C* is located at 0.25?

    (A) 0.05
    (B) 0.15
    (C) 0.27
    (D) 0.45

8. $3.01 + 0.051 + 0.251 =$

    (A) 0.603
    (B) 3.312
    (C) 3.112
    (D) 17.731

9. $2.4 \div 100 =$

    (A) 0.024
    (B) 0.24
    (C) 24
    (D) 240

10. All of the following are ways to write 5% of *M* EXCEPT

    (A) $0.05M$
    (B) $\frac{1}{20}M$
    (C) $\frac{5M}{100}$
    (D) $0.50M$

11. What is 16% of 80?

    (A) 12.8
    (B) 20
    (C) 64
    (D) 5

12. 90 is what percent of 40?

    (A) 150%
    (B) 200%
    (C) 225%
    (D) 250%

13. A tennis player played 40 matches and lost 15% of them. How many of the matches did the tennis player *win*?

    (A) 60
    (B) 34
    (C) 25
    (D) 6

14. Which of the following is equivalent to 8%?

    (A) $\frac{1}{8}$
    (B) $\frac{1}{12.5}$
    (C) $\frac{2}{5}$
    (D) $\frac{4}{5}$

15. $\sqrt{48} - \sqrt{12} =$

    (A) 2
    (B) 6
    (C) $2\sqrt{3}$
    (D) $12\sqrt{3}$

16. Manny wants to put 2-foot by 2-foot carpet tiles on the floor of his rectangular, 20 feet by 14 feet living room floor. What is the least number of carpet tiles he can use to completely cover his living room floor?

    (A) 7
    (B) 18
    (C) 70
    (D) 280

17. Amy weighs 20 pounds more than her younger sister, and the average of their two weights is 55 pounds. How much does Amy's younger sister weigh?

    (A) 35 pounds
    (B) 40 pounds
    (C) 45 pounds
    (D) 65 pounds

ELEMENTARY ALGEBRA

> **Directions:** For each of the questions below, choose the best answer from the four choices given.

1. Which of the following lists of numbers is ordered from least to greatest?

    (A) $-\dfrac{4}{7}, -\dfrac{2}{3}, \dfrac{1}{3}, \dfrac{5}{7}$

    (B) $-\dfrac{4}{7}, -\dfrac{2}{3}, \dfrac{5}{7}, \dfrac{1}{3}$

    (C) $-\dfrac{2}{3}, -\dfrac{4}{7}, \dfrac{5}{7}, \dfrac{1}{3}$

    (D) $-\dfrac{2}{3}, -\dfrac{4}{7}, \dfrac{1}{3}, \dfrac{5}{7}$

2. $-|8| - |-3| =$

    (A) 11
    (B) −11
    (C) 5
    (D) −5

3. If $y = -2$ and $x = 16$, then what does $|y - x|$ equal?

    (A) 14
    (B) −14
    (C) 18
    (D) −18

4. Rando's Gym charges a monthly membership fee of \$50 plus \$10 for each aerobics class a member takes. Which of the following expresses, in dollars, the cost of belonging to Rando's Gym for one month and taking a aerobics classes?

    (A) $50a + 10$
    (B) $60a$
    (C) $10a + 50$
    (D) $10(a + 50)$

5. If $5b - 2 = 5$, then what does b equal?

    (A) $\dfrac{5}{7}$

    (B) $\dfrac{7}{5}$

    (C) −1

    (D) 7

6. The solution set of which of the following inequalities is graphed on the number line below?

    (A) $4x - 3 \geq 13$
    (B) $3x - 2 < 10$
    (C) $4x + 1 \geq 2$
    (D) $2x + 6 \leq 4$

7. How many solutions (x, y) are there to the system of equations below?

$$2x + y = 3$$
$$-2x - y = 5$$

    (A) 0
    (B) 1
    (C) 2
    (D) More than 2

8. $-4x^3(2x^5 - 5x^3 + 2x - 4) =$

    (A) $-8x^8 - 20x^6 - 8x^4 - 16x^3$
    (B) $-8x^8 - 5x^3 + 2x - 4$
    (C) $-8x^{15} + 20x^9 - 8x^3 - 16x^3$
    (D) $-8x^8 + 20x^6 - 8x^4 + 16x^3$

9. Which of the following is a factor of both $x^2 - x - 6$ and $x^2 + 7x + 10$?

(A) $x + 2$

(B) $x - 2$

(C) $x - 3$

(D) $x + 5$

10. If $m \neq 0$, then what does $\dfrac{n}{2m} + \dfrac{2}{m} - \dfrac{2n}{3m}$ equal?

(A) $\dfrac{2-n}{m}$

(B) $\dfrac{11n}{6m}$

(C) $\dfrac{12-n}{6m}$

(D) $\dfrac{2-n}{6m}$

11. Four identical squares are placed side to side to form a larger square as shown in the figure below. If the individual area covered by each smaller square is 36 square feet, what is the perimeter of the larger square that is formed?

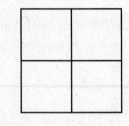

(A) 24 feet

(B) 48 feet

(C) 72 feet

(D) 144 feet

12. For which of the following functions are $x = 9$ and $x = -9$ both solutions?

(A) $x^2 + 81 = 0$

(B) $x^2 - 81 = 0$

(C) $x^2 + 18x - 81 = 0$

(D) $x^2 - 9x + 18 = 0$

COLLEGE-LEVEL MATHEMATICS

Directions: For each of the questions below, choose the best answer from the five choices given.

1. The perimeter of a square is expressed as $12x - 28$. Which expression represents the area of the same square in terms of x?

(A) $3x - 7$

(B) $48x - 112$

(C) $144x^2 - 672x + 784$

(D) $9x^2 - 42x + 49$

(E) $9x^2 - 49$

2. $(a + b)(3a^2 + 2ab + 4b^2) =$

(A) $3a^3b + 2a^2b^2 + 4ab^3$

(B) $3a^3 + 11a^2b + 4b^3$

(C) $3a^3 + 11ab^2 + 4b^3$

(D) $3a^3 + 5a^2b + 6ab^2 + 4b^3$

(E) $3a^3 + 2a^2b + 4ab^2 + 4b^3$

3. If $g(x) = \dfrac{4}{3}x - 10$ and $h(x) = \dfrac{1}{6}x + 11$, then what does $g(a) + h(a)$ equal?

(A) $3a + 1$

(B) $\dfrac{3a - 2}{2}$

(C) $\dfrac{3a + 2}{2}$

(D) $a + 24$

(E) $a - 24$

4. For which values of c will the quadratic equation $0 = x^2 + c$ have no real solutions?

(A) $c < 0$

(B) $c \leq 0$

(C) $c > 0$

(D) $c \geq 0$

(E) $c = 0$

5. What is the solution set to $\sqrt{7x - 5} = x + 1$?

(A) 2 and 3

(B) 2 only

(C) 3 only

(D) –2 only

(E) –2 and 3

6. $\begin{vmatrix} x & -8 \\ y & 3 \end{vmatrix} =$

(A) $24xy$

(B) $3x + 8y$

(C) $3x - 8y$

(D) $xy - 24$

(E) $24 - xy$

7. If $x \neq 0$, then what does

$$\dfrac{x^3 + 7x^2 + 12x}{(x^4 + 10x^3) - (3x^3 - 12x^2)} \quad \text{equal?}$$

(A) 1

(B) x

(C) $\dfrac{1}{x}$

(D) $\dfrac{1}{x^2}$

(E) $x + 3$

8. $\dfrac{\sqrt{3} + 3}{\sqrt{3}} + \dfrac{3\sqrt{3} + 3}{\sqrt{3}} =$

(A) $2(2 + \sqrt{3})$

(B) $2(2\sqrt{3} + 3)$

(C) $6\sqrt{3}$

(D) $2(1 + \sqrt{3})$

(E) $2(3\sqrt{3} + 2)$

9. Six friends each drive to a restaurant with six parking spaces. In how many different ways can they park their cars?

(A) 6

(B) 36

(C) 120

(D) 240

(E) 720

10. How many real zeros does $x^3 - 2x = 0$ have?

(A) 0

(B) 1

(C) 2

(D) 3

(E) 4

11. On which interval is the graph of $2x^2 + 13x + 15$ negative?

 (A) $-5 < x < -\dfrac{3}{2}$

 (B) $-5 < x < \dfrac{3}{2}$

 (C) $-\dfrac{3}{2} < x < 5$

 (D) $\dfrac{3}{2} < x < 5$

 (E) $x < -5$ and $x > -\dfrac{3}{2}$

12. $2^{\frac{9}{2}} - 2^{\frac{7}{2}} =$

 (A) 2

 (B) $2^{\frac{1}{2}}$

 (C) $2^{\frac{3}{2}}$

 (D) $2^{\frac{5}{2}}$

 (E) $2^{\frac{7}{2}}$

13. Find the equation of a line that is parallel to the one that passes through the points $(-2, -4)$ and $(4, 5)$.

 (A) $y = \dfrac{2}{3}x + 3$

 (B) $y = -\dfrac{2}{3}x + 4$

 (C) $y = -3x + 5$

 (D) $y = \dfrac{3}{2}x + 6$

 (E) $y = -\dfrac{3}{2}x + 7$

14. $\dfrac{y^2 + y}{y^3} \cdot \dfrac{y^2}{y^2 + 2y + 1} =$

 (A) 1

 (B) $\dfrac{1}{y}$

 (C) $\dfrac{1}{y+1}$

 (D) $\dfrac{1}{y^2}$

 (E) $\dfrac{y+1}{y}$

15. $\sqrt{-3} \times \sqrt{-6} \times \sqrt{-8} =$

 (A) 12

 (B) -12

 (C) $-12i$

 (D) $12i$

 (E) $144i$

16. What is the solution set to $\dfrac{|x-2|}{2} = x + 1$?

 (A) 0 only

 (B) -4 only

 (C) 0 and -4

 (D) 4 only

 (E) 0 and 4

17. Brian has 5 ties, and he has scheduled 2 job interviews at different companies. If Brian wants to wear a different tie to each interview, how many ways can he choose which tie to wear?

 (A) 10

 (B) 20

 (C) 24

 (D) 60

 (E) 120

18. $\dfrac{b-a}{1-\dfrac{a}{b}} =$

(A) 1

(B) a

(C) b

(D) $\dfrac{1}{a-b}$

(E) $a-b$

19. If $\tan \theta = \dfrac{3}{5}$ and $0 \leq \theta \leq \dfrac{\pi}{2}$, then what does

cot θ equal?

(A) $\dfrac{3}{5}$

(B) $\dfrac{5}{3}$

(C) $\dfrac{3}{4}$

(D) $\dfrac{5}{4}$

(E) $\dfrac{4}{3}$

20. If $\log_2 a = 6$, then what does $\sqrt{a}$ equal?

(A) 3

(B) 8

(C) 12

(D) 36

(E) 64

ANSWER KEY
Practice Test 3

ARITHMETIC	ELEMENTARY ALGEBRA	COLLEGE-LEVEL MATHEMATICS
1. **A**	1. **D**	1. **D**
2. **D**	2. **B**	2. **D**
3. **C**	3. **C**	3. **C**
4. **A**	4. **C**	4. **C**
5. **D**	5. **B**	5. **A**
6. **D**	6. **A**	6. **B**
7. **B**	7. **A**	7. **C**
8. **B**	8. **D**	8. **A**
9. **A**	9. **A**	9. **E**
10. **D**	10. **C**	10. **D**
11. **A**	11. **B**	11. **A**
12. **C**	12. **B**	12. **E**
13. **B**		13. **D**
14. **B**		14. **C**
15. **C**		15. **D**
16. **C**		16. **A**
17. **C**		17. **B**
		18. **C**
		19. **B**
		20. **B**

Arithmetic

1. **(A)** To change $\frac{3}{25}$ to a decimal, divide 25 into 3.

$$\begin{array}{r} 0.12 \\ 25\overline{)3.00} \\ \underline{-2\ 5\downarrow} \\ 50 \\ \underline{-\ 50} \\ 0 \end{array}$$

$$\frac{3}{25} = 0.12$$

2. **(D)** Essentially, we need to know how many $\frac{2}{3}$ cups are in 6 cups. Divide 6 by $\frac{2}{3}$.

$$6 \div \frac{2}{3} = \frac{6}{1} \div \frac{2}{3} = \frac{6}{1} \times \frac{3}{2} = \frac{18}{2} = 9$$

3. **(C)** To subtract the two fractions, change them to equivalent fractions with common denominators. Then, subtract the numerators, and keep the common denominators.

$$\begin{array}{r} \frac{5}{8} \Rightarrow \frac{15}{24} \\ -\frac{1}{3} \Rightarrow -\frac{8}{24} \\ \hline \frac{7}{24} \end{array}$$

4. **(A)** Within the parentheses, change the fractions to equivalent fractions with common denominators in order to simplify.

$$\left(\frac{3}{5} - \frac{1}{4}\right) \times \left(\frac{5}{6} + \frac{7}{10}\right) =$$

$$\left(\frac{12}{20} - \frac{5}{20}\right) \times \left(\frac{25}{30} + \frac{21}{30}\right) = \frac{7}{20} \times \frac{46}{30}$$

Then multiply the fractions together.

$$\frac{7}{20} \times \frac{46}{30} = \frac{7 \times 46}{20 \times 30} = \frac{322}{600}$$

The result, $\frac{322}{600}$, is slightly more than $\frac{1}{2}$, which would be $\frac{300}{600}$. Therefore,

$\left(\frac{3}{5} - \frac{1}{4}\right) \times \left(\frac{5}{6} + \frac{7}{10}\right)$ is between $\frac{1}{2}$ and 1.

5. **(D)** To multiply $4\frac{2}{3} \times 3\frac{4}{5}$, change the mixed numbers to improper fractions.

$$\frac{14}{3} \times \frac{19}{5} = \frac{266}{15} = 17\frac{11}{15}$$

6. **(D)** Use zeros as placeholders to compare the numbers using the same number of places.

$$0.302$$
$$0.203$$
$$0.230$$
$$0.032$$

Choice D has a 0 in the tenths place. The other choices have non-zero digits in the tenths place. Therefore, choice D is the least.

7. **(B)** Draw a number line with Points A, B, and C lying in a row, with Point B exactly in the middle of Points A and C. Then label Point B as 0.2 and Point C as 0.25.

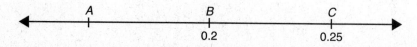

Point B is equidistant from Points A and C. Therefore, Point B is $0.25 - 0.2 = 0.05$ to the left of Point C. As such, Point A is 0.05 to the left of Point B. In other words, Point A is 0.05 less than Point B.

$$0.2 - 0.05 = 0.15$$

8. **(B)** To add the fractions, line up the decimal points, and add vertically using zeros as placeholders where needed.

$$\overset{1}{3}.010$$
$$0.051$$
$$+ \, 0.251$$
$$3.312$$

9. **(A)** To divide a number by 100, move the decimal point two places to the left.

$$2.4 \div 100 = 0.024$$

10. **(D)** To find 5% of M, change 5% to a decimal, and multiply it by M.

$$5\% \text{ of } M = 0.05 \times M = 0.05M$$

Alternatively, we can change 5% to a fraction, and multiply it by M.

$$5\% \text{ of } M = \frac{5}{100}M = \frac{5M}{100}$$

This fraction can be reduced.

$$\frac{5}{100}M = \frac{1}{20}M$$

All of the above are ways to write 5% of M. Choice D, however, is the same as 50% of M since 50% is written as 0.50 as a decimal. Therefore, $0.50M$ is not equivalent to 5% of M.

11. **(A)** To find 16% of 80, change 16% to a decimal, and multiply it by 80.

$$0.16(80) = 12.8$$

12. **(C)**

$$\frac{\text{part}}{\text{whole}} = \frac{\text{percentage}}{100}$$

$$\frac{90}{40} = \frac{x}{100}$$

$$40x = (90)(100)$$

$$x = \frac{9000}{40} = \frac{900}{4} = 225$$

90 is 225% of 40.

13. **(B)** If the tennis player lost 15% of her matches, then she won 100% – 15% = 85% of her matches.

$$85\% \text{ of } 40 = 0.85 \times 40 = 34$$

14. **(B)** $8\% = \frac{8}{100} = \frac{2}{25}$. Choices A, C, and D are not equivalent to $\frac{8}{100}$ or $\frac{2}{25}$.

By dividing the numerator and the denominator of $\frac{2}{25}$ by 2, we find that

$\frac{2 \div 2}{25 \div 2} = \frac{1}{12.5}$. Therefore, $8\% = \frac{1}{12.5}$.

15. **(C)**

$$\sqrt{48} - \sqrt{12} =$$

$$\left(\sqrt{16} \bullet \sqrt{3}\right) - \left(\sqrt{4} \bullet \sqrt{3}\right) =$$

$$4\sqrt{3} - 2\sqrt{3} = 2\sqrt{3}$$

16. **(C)** The area of the living room floor is 20 × 14 = 280 ft². The area of floor that one carpet tile covers is 2 × 2 = 4 ft². To find the least number of carpet tiles that are needed to cover the floor, we divide 280 ÷ 4 = 70 carpet tiles.

17. **(C)** The average of their two weights, 55 pounds, lies exactly halfway between the two sister's weights. Knowing that their weights have to be 20 pounds apart, each sister's weight lies 10 pounds on either side of 55 pounds. Amy weighs 55 + 10 = 65 pounds. Amy's younger sister weighs 55 – 10 = 45 pounds.

Elementary Algebra

1. **(D)** Change each fraction to an equivalent fraction with a common denominator. Then, put the list in least to greatest order according to their numerators.

$$-\frac{2}{3}, \ -\frac{4}{7}, \ \frac{1}{3}, \ \frac{5}{7}$$

$$\Downarrow \quad\quad \Downarrow \quad \Downarrow \quad \Downarrow$$

$$-\frac{14}{21}, -\frac{12}{21}, \frac{7}{21}, \frac{15}{21}$$

Choice D is in the correct order.

2. **(B)**

$$-|8| - |-3| =$$
$$-8 - 3 = -11$$

3. **(C)** Substitute –2 for y and 16 for x into the expression, and then evaluate.

$$|y - x| =$$
$$|-2 - 16| =$$
$$|-18| = 18$$

4. **(C)** The cost of belonging to Rando's Gym for one month plus the cost of aerobics classes at $10 for each class is shown in the table below:

Number of Classes	Total Cost
1	$10 + $50
2	$10 + $10 + $50
3	$10 + $10 + $10 + $50

The number of $10 fees is the same as the number of aerobics classes that a member takes. To take a aerobics classes would cost $10 times a plus $50, or $10a + 50$ dollars.

5. **(B)** Solve for b.

$$5b - 2 = 5$$
$$5b - 2 + 2 = 5 + 2$$
$$5b = 7$$
$$\frac{5b}{5} = \frac{7}{5}$$
$$b = \frac{7}{5}$$

6. **(A)** This number line shows the solution set that includes the number 4 and all of the values greater than 4. That is $x \geq 4$. Choice A has the only inequality that is solved by the solution set $x \geq 4$.

$$4x - 3 \geq 13$$
$$4x - 3 + 3 \geq 13 + 3$$
$$4x \geq 16$$
$$\frac{4x}{4} \geq \frac{16}{4}$$
$$x \geq 4$$

7. **(A)** If we add both equations together, we get a false mathematical statement.

$$(2x + y) + (-2x - y) = 3 + 5$$
$$2x + y - 2x - y = 8$$
$$2x - 2x + y - y = 8$$
$$0 \neq 8$$

Since we cannot solve for x or y this way, there are no solutions to this system of equations.

8. **(D)** Distribute the $-4x^3$ found outside the parentheses to all the terms found inside the parentheses.

$$-4x^3(2x^5 - 5x^3 + 2x - 4)$$
$$-4x^3(2x^5) - 4x^3(-5x^3) - 4x^3(2x) - 4x^3(-4)$$
$$-8x^8 + 20x^6 - 8x^4 + 16x^3$$

9. **(A)**

$$x^2 - x - 6 = (x - 3)(x + 2)$$
$$x^2 + 7x + 10 = (x + 5)(x + 2)$$

Both expressions have $x + 2$ as a factor.

10. **(C)** To add rational expressions, change each fraction to an equivalent fraction with a common denominator.

$$\frac{n}{2m} + \frac{2}{m} - \frac{2n}{3m} =$$
$$\frac{3n}{6m} + \frac{12}{6m} - \frac{4n}{6m} =$$
$$\frac{12 + 3n - 4n}{6m} = \frac{12 - n}{6m}$$

11. **(B)** The smaller squares each have an area of 36 square feet. To find the length of one of the sides of the smaller squares, we find $\sqrt{36} = 6$ feet. The larger square then has sides that are 12 feet long because the sides of the larger square are equal to twice the length of the sides of the smaller squares. With sides that are 12 feet long, the perimeter is found by finding the sum of all four sides, or $12 + 12 + 12 + 12 = 4(12) = 48$ feet.

12. **(B)** A difference of two squares will result in two solutions that are opposites. The only choice that has a difference of two squares is $x^2 - 81 = 0$. To solve for x, first factor.

$$x^2 - 81 = 0$$
$$(x + 9)(x - 9) = 0$$

$$x + 9 = 0 \qquad x - 9 = 0$$
$$x = -9 \qquad x = 9$$

Alternatively, separately substitute -9 for x and 9 for x into each equation.

When $x = 9$, $x^2 - 81 = 9^2 - 81 = 81 - 81 = 0$
When $x = -9$, $x^2 - 81 = (-9)^2 - 81 = 81 - 81 = 0$

College-Level Mathematics

1. **(D)** If the square's perimeter is $12x - 28$, then the length of each side is $\frac{12x - 28}{4} = 3x - 7$. To find the area of the square, square the length of the side.

$$A = (3x - 7)^2$$
$$A = (3x - 7)(3x - 7)$$
$$A = 9x^2 - 21x - 21x + 49$$
$$A = 9x^2 - 42x + 49$$

2. **(D)**

$$(a+b)(3a^2 + 2ab + 4b^2) =$$

$$3a^3 + 2a^2b + 4ab^2 + 3a^2b + 2ab^2 + 4b^3 = 3a^3 + 5a^2b + 6ab^2 + 4b^3$$

3. **(C)** Substitute a into both functions.

$$g(x) = \frac{4}{3}x - 10 \qquad h(x) = \frac{1}{6}x + 11$$

$$g(a) = \frac{4}{3}a - 10 \qquad h(a) = \frac{1}{6}a + 11$$

Then add.

$$g(a) + h(a) = \left(\frac{4}{3}a - 10\right) + \left(\frac{1}{6}a + 11\right)$$

$$g(a) + h(a) = \frac{4}{3}a + \frac{1}{6}a - 10 + 11$$

$$g(a) + h(a) = \frac{3}{2}a + 1 = \frac{3a + 2}{2}$$

4. **(C)** $0 = x^2 + c$ will have no real solutions as long as $c > 0$. Consider the graph of $y = x^2 + c$, where c is any positive number. The parabola will not intersect the x-axis at any point and will have no real zeros. In the equation $0 = x^2 + c$, x^2 can only equal a positive number or 0. There is no positive value of c that could be added to a positive number or 0 to produce a final outcome of zero.

5. **(A)** First, square both sides of the equation.

$$\left(\sqrt{7x - 5}\right)^2 = (x + 1)^2$$

$$7x - 5 = x^2 + 2x + 1$$

Set the equation equal to zero, and solve by factoring.

$$7x - 5 = x^2 + 2x + 1$$

$$\underline{+5 \qquad\qquad\quad +5}$$

$$7x = x^2 + 2x + 6$$

$$\underline{-7x \qquad\quad -7x}$$

$$0 = x^2 - 5x + 6$$

$$0 = (x - 3)(x - 2)$$

Setting each binomial equal to zero and solving for x, we have $x = 3$ and $x = 2$. Check both solutions by substituting them into the original equation.

$$\sqrt{7x - 5} = x + 1 \qquad\qquad \sqrt{7x - 5} = x + 1$$

$$\sqrt{7(3) - 5} = 3 + 1 \qquad\qquad \sqrt{7(2) - 5} = 2 + 1$$

$$\sqrt{21 - 5} = 4 \qquad\qquad \sqrt{14 - 5} = 3$$

$$\sqrt{16} = 4 \qquad\qquad\qquad \sqrt{9} = 3$$

$$4 = 4 \qquad\qquad\qquad\quad 3 = 3$$

Both solutions check out.

6. **(B)**

$$\begin{vmatrix} x & -8 \\ y & 3 \end{vmatrix} = (x)(3) - (-8)(y) = 3x - (-8y) = 3x + 8y$$

7. **(C)** Combine the terms in the denominator, and factor where possible.

$$\frac{x^3 + 7x^2 + 12x}{(x^4 + 10x^3) - (3x^3 - 12x^2)} = \frac{x^3 + 7x^2 + 12x}{x^4 + 7x^3 + 12x^2} = \frac{x(x^2 + 7x + 12)}{x^2(x^2 + 7x + 12)} = \frac{x}{x^2} = \frac{1}{x}$$

8. **(A)**

$$\frac{\sqrt{3} + 3}{\sqrt{3}} + \frac{3\sqrt{3} + 3}{\sqrt{3}} = \frac{6 + 4\sqrt{3}}{\sqrt{3}}$$

Multiply the numerator and the denominator by $\sqrt{3}$.

$$\left(\frac{6 + 4\sqrt{3}}{\sqrt{3}}\right)\frac{\sqrt{3}}{\sqrt{3}} = \frac{6\sqrt{3} + 12}{3} = 2\sqrt{3} + 4 = 2\left(\sqrt{3} + 2\right)$$

9. **(E)** The first friend who pulls up will have his or her choice of 6 spaces. The next friend will have his or her choice of 5 spaces. The next friend will have his or her choice of 4 spaces, and so on. Therefore, the number of ways that the friends can park is calculated by:

$$6! = 6 \times 5 \times 4 \times 3 \times 2 \times 1 = 720$$

10. **(D)** We can find the zeros by factoring.

$$x^3 - 2x = 0$$
$$x(x^2 - 2) = 0$$
$$x\left(x + \sqrt{2}\right)\left(x - \sqrt{2}\right) = 0$$

Setting each of these factors equal to 0, we find that the real zeros of the equation are at $x = \{-\sqrt{2}, 0, \sqrt{2}\}$.

11. **(A)** Find the zeros of the polynomial by first setting it equal to zero, and then factoring or using the quadratic formula.

$$0 = 2x^2 + 13x + 15$$
$$0 = (2x + 3)(x + 5)$$

Setting each binomial equal to 0 and solving for x, we find that the zeros occur at $x = -\frac{3}{2}$ and $x = -5$. This quadratic equation is a parabola that opens upward. At any point between the real zeros, the value of the polynomial will be negative. Therefore, the value is negative when $-5 < x < -\frac{3}{2}$.

12. **(E)** We can factor $2^{\frac{7}{2}}$ out of each term because $2^{\frac{9}{2}} = 2^{\frac{7}{2}} \times 2^{\frac{2}{2}}$. Therefore,

$$2^{\frac{9}{2}} - 2^{\frac{7}{2}} = 2^{\frac{7}{2}}\left(2^{\frac{2}{2}} - 1\right) = 2^{\frac{7}{2}}(2 - 1) = 2^{\frac{7}{2}}(1) = 2^{\frac{7}{2}}$$

13. **(D)** The slope of the line that passes through the points (–2, –4) and (4, 5) can be found using the following formula:

$$m = \frac{y_2 - y_1}{x_2 - x_1} = \frac{5 - (-4)}{4 - (-2)} = \frac{9}{6} = \frac{3}{2}$$

Since parallel lines have the same slope, a line parallel to this one must also have a slope of $\frac{3}{2}$.

14. **(C)** Simplify by multiplying and factoring.

$$\frac{y^2 + y}{y^3} \bullet \frac{y^2}{y^2 + 2y + 1} = \frac{\cancel{(y^3)}\cancel{(y+1)}}{\cancel{(y^3)}\cancel{(y+1)}(y+1)} = \frac{1}{y+1}$$

15. **(D)**

$$\sqrt{-3} \times \sqrt{-6} \times \sqrt{-8} = \sqrt{-144} = 12\sqrt{-1} = 12i$$

16. **(A)** Isolate the absolute value part of the equation.

$$2\left(\frac{|x-2|}{2}\right) = 2(x+1)$$
$$|x-2| = 2x+2$$

Next, set up two cases, remembering to change the sign of both terms on the right side of the equation in the second case.

$$
\begin{array}{ll}
x - 2 = 2x + 2 & \quad x - 2 = -2x - 2 \\
\underline{+2 \qquad +2} & \quad \underline{+2 \qquad \quad +2} \\
x = 2x + 4 & \quad x = -2x \\
\underline{-2x \quad -2x} & \quad x = 0 \\
-x = 4 & \\
x = -4 &
\end{array}
$$

The possible solutions are $x = -4$ and $x = 0$. Test both solutions by substituting them back into the original equation.

$$
\begin{array}{ll}
\dfrac{|x-2|}{2} = x+1 & \qquad \dfrac{|x-2|}{2} = x+1 \\[2mm]
\dfrac{|-4-2|}{2} = -4+1 & \qquad \dfrac{|0-2|}{2} = 0+1 \\[2mm]
\dfrac{|-6|}{2} = -3 & \qquad \dfrac{|-2|}{2} = 1 \\[2mm]
\dfrac{6}{2} = -3 & \qquad \dfrac{2}{2} = 1 \\[2mm]
3 \neq -3 & \qquad 1 = 1
\end{array}
$$

Only the solution $x = 0$ checks out.

17. **(B)** For the first interview, Brian has a choice of 5 ties. For the second interview, he can choose from any of the remaining 4 ties. Therefore, the total number of ways he can choose which tie to wear is $5 \times 4 = 20$. In this case, the order in which he chooses his ties matters, so we are looking for possible orders of 2 ties from 5, or $_5P_2$.

$$_nP_r = \frac{n!}{(n-r)!} = \frac{5!}{(5-2)!} = \frac{5 \times 4 \times \cancel{3!}}{\cancel{3!}} = 5 \times 4 = 20$$

18. **(C)** First multiply the entire expression by $\dfrac{b}{b}$ in order to get the fraction $\dfrac{a}{b}$ out of the denominator, thus making the expression easier to handle.

$$\frac{b}{b} \cdot \frac{b-a}{1-\dfrac{a}{b}} = \frac{b(b-a)}{b\left(1-\dfrac{a}{b}\right)} = \frac{b(b-a)}{b-a}$$

We now have the expression $(b-a)$ common to both the numerator and the denominator, so we can divide them by one another, leaving us with only b.

$$\frac{b\cancel{(b-a)}}{\cancel{b-a}} = b$$

19. **(B)** If $\tan\theta = \dfrac{3}{5}$, then the side opposite $\theta = 3$ and the side adjacent $\theta = 5$.

$$\cot\theta = \frac{\text{side adjacent } \theta}{\text{side opposite } \theta} = \frac{5}{3}$$

20. **(B)** Identify the base, power, and exponent that are found in $\log_2 a = 6$ by following the form $\log_{\text{base}} (\text{power}) = \text{exponent}$.

$$\text{Base} = 2$$
$$\text{Power} = a$$
$$\text{Exponent} = 6$$

Change the expression from logarithmic to exponential form.

$$a = 2^6$$
$$a = 64$$

Then evaluate: $\sqrt{a} = \sqrt{64} = 8$

Summary of Formulas

To excel on the ACCUPLACER Math tests, you should commit the following formulas to memory.

Formulas for Perimeter, Area, and Circumference

Perimeter of a rectangle = (2 × length) + (2 × width)

Area of a rectangle = length × width

Area of a square = $side^2$

Area of a triangle = $\frac{1}{2}$ × base × height

Area of a trapezoid = $\frac{1}{2}$ × (base 1 + base 2) × height

Circumference of a circle = π × diameter

Area of a circle = π × $radius^2$

The Pythagorean Theorem

$a^2 + b^2 = c^2$, where a and b are the lengths of the legs of a right triangle and c is the length of the hypotenuse

Slope-Intercept Form of a Line

$y = mx + b$, where m is the slope and b is the y-intercept

Special Factoring Cases

$a^2 + 2ab + b^2 = (a + b)^2$

$a^2 - 2ab + b^2 = (a - b)^2$

$a^2 - b^2 = (a - b)(a + b)$

The Quadratic Formula

For $ax^2 + bx + c = 0$,

$$x = \frac{-b \pm \sqrt{b^2 - 4ac}}{2a}$$

Trigonometric Identities

$$\text{sine } \theta = \frac{\text{side opposite } \theta}{\text{hypotenuse}}$$

$$\text{cosine } \theta = \frac{\text{side adjacent } \theta}{\text{hypotenuse}}$$

$$\text{tangent } \theta = \frac{\text{side opposite } \theta}{\text{side adjacent } \theta}$$

Permutations and Combinations

$$_nP_r = \frac{n!}{(n - r)!}$$

$$_nC_r = \frac{n!}{r!(n - r)!}$$

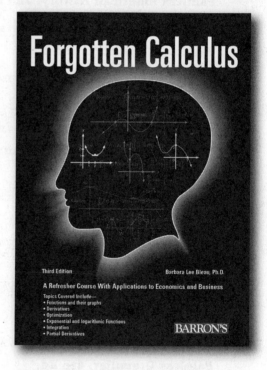